Note to the Reviewer

In assembling this book, I became aware of several technical issues that need some special attention.

The base document for this book was MIL-HDBK-1026/4A, which is to replace DM-26.4 (Fixed Moorings) and DM-26.5 (Fleet Moorings.) In reviewing these documents and DM-26.6, some important changes took place.

The first is that there are several changes in the basic design methodology, both with the basic loads and design conditions (the most substantial differences) and some in the capacity of the load bearing elements (such as the anchor capacities.) We have worked under the assumption that any changes here are for the better (with one exception, which we will mention.) If you do not concur with this, let us know.

The second is that the earlier volumes were primarily for instruction in hand calculations while the latter assumes the use of computer programs for much of the design work. The main result of this is that some of the examples tend to be perfunctory in calculation detail. If we need to expand this, let us know.

One particular issue that we are not entirely at ease about is the static current yaw moment. The newer volume attempts to eliminate chart and table lookup and substitute formulae whenever possible. This is admirable, but with the static current yaw moment, the newer volume uses a linear regression correlation that ignores the fact that the static current yaw moment at $\theta = 0°$ and $\theta = 180°$ is zero, as is clearly shown in the earlier charts. There are a number of mathematical ways to resolve this issue (and we do intend to resolve it one way or another) but your ideas on this would be greatly appreciated.

Finally, we have endeavored to produce a volume that is suitable for both design and maintenance specialists. If our "two-track" approach needs improvement, let us know.

Thanks.

Pilebuck

Preface

Anyone who goes to sea is very familiar with the mooring of ships, both the "fleet mooring" (anchorages, etc.) and the "fixed mooring" (tying up at the dock or pier.) Proper mooring equipment and procedures are an integral part of successful seamanship. Failure in either regard can lead to significant damage to the ship or the vessel finding herself on the bottom.

Traditionally mooring was based on the experience of the officers and mates that actually manned the ships. In recent years engineering principles have been applied, which has moved what once was a pure art to something of a science, a process accelerated by the use of computer programs.

This book discusses both the design of mooring systems for ships and their subsequent maintenance. The following should be kept in mind when using this book:

- A mooring is defined as a compliant structure that restrains a vessel against the action of wind, wave, and current forces.

- The emphasis is on moorings composed of tension members (chain, line, wire rope, etc.) and compression members (fenders, camels, etc.) used to secure vessels (surface ships, submarines, floating drydocks, yard craft, etc.). The term mooring in this book includes anchoring of ships.

- The primary emphasis is the mooring of floating structures or 'vessels,' such as ships, yard craft, submarines, and floating drydocks in harbors. This book does not address systems where the environmental forcing on the mooring members themselves is important, as may be the case for towed underwater bodies, ship-to-ship at-sea mooring, and towing of one vessel by another.

- This book contains background information and procedural guidelines concerning the maintenance of fleet moorings and spare fleet mooring material. This includes mooring installation and recovery procedures, the refurbishing and overhaul of mooring material ashore and afloat, inspection criteria and guidelines, inventory storage criteria, and the utilization of cathodic protection systems to effectively reduce the corrosion rate of mooring material.

- The materials and procedures detailed herein have been prepared to assist the user in establishing and sustaining an effective fleet mooring maintenance program. They have been developed from the best technical sources available and represent many years of practical and successful field experience.

As is customary with Pile Buck books, this volume is a compilation from other sources, which are detailed in the bibliography.

We commend this work to our readers.

Don C. Warrington, P.E.
Editor

TABLE OF CONTENTS

Chapter 1.	**Introduction**	**12**
Chapter 2.	**Mooring Systems**	**13**
2.1.	Introduction	13
2.2.	Purpose Of Mooring	13
2.3.	Types Of Mooring Systems	13
2.3.1.	Fixed Mooring Systems	13
2.3.2.	Fleet Moorings	16
Chapter 3.	**Mooring Equipment**	**28**
3.1.	Introduction	28
3.2.	Key Mooring Components	28
3.2.1.	Tension Members	28
3.2.2.	Compression Members	28
3.3.	Anchors	28
3.3.1.	Navy Standard Stockless	29
3.3.2.	NAVFAC Stato	31
3.3.3.	NAVMOOR Anchor	32
3.3.4.	Stake Piles	33
3.3.5.	"Pearl Harbor" And Other Deadweight Anchors	34
3.3.6.	Driven Plate Anchor	34
3.3.7.	Anchor Pull Test Requirements	38
3.3.7.1.	Setting Drag Anchors	39
3.3.7.2.	Pull Testing	42
3.4.	Chain And Fittings	43
3.4.1.	Types of Chain	43
3.4.1.1.	Cast Chain	45
3.4.1.2.	Di-Loc Chain	45
3.4.1.3.	Stud Link Chain	45
3.4.2.	Grades of Chain	46
3.4.3.	Chain Assembly Accessories	48
3.4.3.1.	Chain Joining Link	49
3.4.3.2.	Anchor Joining Link	52
3.4.3.3.	Shackles	53
3.4.3.4.	Common Stud Link Chain	56
3.4.3.5.	Enlarged Link	58
3.4.3.6.	End Link	59
3.4.3.7.	C-Link	60
3.4.3.8.	Pear Link	60
3.4.3.9.	Swivel	61
3.4.3.10.	Swivel Shackle (Chain)	63
3.4.3.11.	Modified Swivel Shackle (Riser)	65
3.4.3.12.	Ground Ring	67
3.4.3.13.	Spider Plate	67
3.4.3.14.	Chain Equalizer	68
3.5.	Buoys	69
3.5.1.	Steel Buoys	71
3.5.1.1.	Drum Buoy	71
3.5.1.2.	Peg Top Buoy	71
3.5.1.3.	Non-Riser-Type Buoy	71
3.5.2.	Foam Buoys	71

3.6.	Sinkers	74
3.7.	Mooring Lines	74
3.7.1.	Synthetic Fiber Ropes	75
3.7.2.	Wire Ropes	83
3.8.	Fenders	84
3.9.	Pier Fittings	86
3.10.	Vessel Mooring Equipment	90
3.10.1.	Introduction	90
3.10.2.	Types Of Mooring Equipment	90
3.10.3.	Equipment Specification	91
3.10.4.	Fixed Bitts	92
3.10.5.	Recessed Shell Bitts	94
3.10.6.	Exterior Shell Bitts	95
3.10.7.	Chocks	96
3.10.8.	Allowable Hull Pressures	97
3.11.	Sources Of Information	97
Chapter 4.	**Basic Design Procedure**	**100**
4.1.	Design Approach	100
4.2.	General Design Criteria	101
4.2.1.	Mooring Service Types	102
4.2.2.	Facility Design Criteria For Mooring Service Types	103
4.2.3.	Ship Hardware Design Criteria For Mooring Service Types	104
4.2.4.	Strength	105
4.2.5.	Serviceability	105
4.2.6.	General Mooring Integrity	105
4.2.7.	Quasi-Static Safety Factors	106
4.2.8.	Allowable Ship Motions	106
4.3.	Design Methods	108
4.3.1.	Quasi-Static Design	108
4.3.2.	Dynamic Mooring Analysis	109
4.4.	Risk	110
4.5.	Coordinate Systems	112
4.5.1.	Ship Design/Construction Coordinates	112
4.5.2.	Ship Hydrostatics/Hydrodynamics Coordinates	113
4.5.3.	Local Mooring Coordinate System	113
4.5.4.	Global Coordinate System	115
4.6.	Vessel Design Considerations	115
4.7.	Facility Design Considerations	116
4.8.	Environmental Forcing Design Considerations	116
4.8.1.	Winds	117
4.8.2.	Wind Gust Fronts	119
4.8.3.	Storms	123
4.8.4.	Currents	125
4.8.5.	Water Levels	125
4.8.6.	Waves	126
4.8.7.	Water Depths	126
4.8.8.	Environmental Design Information	126
4.9.	Operational Considerations	127
4.10.	Inspection	128
4.11.	Maintenance	129
4.12.	General Mooring Guidelines	130

Chapter 5.	**Static Environmental Forces And Moments On Vessels**	**132**
5.1.	Scope	132
5.2.	Engineering Properties Of Water And Air	132
5.3.	Principal Coordinate Directions	133
5.4.	Static Wind Forces/Moments	133
5.4.1.	Static Transverse Wind Force	135
	Static Longitudinal Wind Force	140
5.4.3.	Static Wind Yaw Moment	142
5.5.	Static Current Forces/Moments	144
5.5.1.	Static Transverse Current Force	144
	Static Longitudinal Current Force	149
5.5.3.	Static Current Yaw Moment	151
5.6.	Wind And Current Forces And Moments On Multiple Ships	152
Chapter 6.	**Mooring System Design Procedures**	**184**
6.1.	Anchor Systems	184
6.1.1.	General Anchor Design Procedure	184
6.1.2.	Drag-Embedment Anchor Specification	189
6.1.3.	Driven-Plate Anchor Design	193
6.2.	Catenary Behavior	194
6.2.1.	Catenary Equations and Hand Solution	194
6.2.2.	Catenary Analysis—Computer Solution	201
6.3.	Example Problems	203
Chapter 7.	**Overview Of Fleet Mooring Maintenance**	**228**
7.1.	Introduction	228
7.2.	Inspections	228
Chapter 8.	**Mooring Installation And Recovery**	**229**
8.1.	Preparation	229
8.1.1.	General.	229
8.1.2.	Planning And Preparation.	229
8.1.2.1.	Preparation	229
8.1.2.2.	Environmental Factors	229
8.1.3.	Tools And Equipment Required	229
8.1.4.	Pre-Installation Layout	230
8.1.5.	Pre-Installation Positioning	230
8.1.5.1.	Marker Buoys	230
8.1.5.2.	Other Positioning Equipment	231
8.1.6.	Pre-Installation Inspection	231
8.1.7.	Field Changes Of Design	231
8.1.8.	As-Built Drawings	232
8.2.	Installation Instructions	232
8.2.1.	General	232
8.2.2.	Riser-Type Mooring System	232
8.2.2.1.	Preinstallation Assembly	232
8.2.3.	Installation Procedures	233
8.2.3.1.	Pull Testing of Anchors	237
8.2.3.2.	Installation Barge	237
8.3.	Recovery Instructions	237
8.3.1.	Riser-Type Mooring System	237
Chapter 9.	**Inspections**	**243**
9.1.	General	243
9.1.1.	Overall Requirements	243

9.1.2.	Inspection Classifications And Types	243
9.1.3.	Purpose	243
9.1.4.	Personnel	243
9.2.	Inspection Procedures	243
9.2.1.	General	243
9.2.2.	Annual Surface Inspections	243
9.2.2.1.	Buoy Inspection	244
9.2.2.2.	Buoy Location	245
9.2.2.3.	Documentation	245
9.2.3.	Underwater Inspections	246
9.2.4.	Life Inspections	246
9.2.5.	Mooring Damage/Failure Inspections	247
9.2.5.1.	Inspection Procedures. Inspection will include the following:	247
9.2.5.2.	Documentation	247
Chapter 10.	**In-Service Maintenance And Repair**	**248**
10.1.	General	248
10.1.1.	Scope.	248
10.1.2.	Equipment	248
10.2.	Procedures	248
10.2.1.	Buoy Replacement (Riser-Type)	248
10.2.2.	Riser Replacement	248
10.2.3.	Buoy Replacement (Non-Riser-Type)	249
10.2.4.	Minor Repairs	249
10.2.4.1.	Welding/Cutting	249
10.2.4.2.	Shackles/Joining Links	249
10.2.5.	Buoy Coatings	250
Chapter 11.	**Ashore Inspection And Refurbishment Of Buoys**	**251**
11.1.	General	251
11.1.1.	Scope	251
11.1.2.	Preparation For Ashore Inspections	251
11.2.	Inspection Procedures	251
11.2.1.	Preliminary Inspection	251
11.2.2.	Detailed Inspection	252
11.2.2.1.	Ultrasonic Testing	253
11.2.2.2.	Pitting Inspection	253
11.2.2.3.	Welds	253
11.2.2.4.	Air Test	253
11.2.2.5.	Documentation	253
11.3.	Buoy Repairs And Modifications	253
11.3.1.	General	253
11.3.2.	Procedures	253
11.3.2.1.	Manhole Cover Replacement	254
11.3.2.2.	Test Plugs and Hull Apertures	254
11.3.2.3.	Fenders and Chafing Strips	254
11.3.2.4.	Welding	255
11.3.2.5.	Buoy air-Pressure Test	256
11.4.	Protective Coatings	256
11.4.1.	Preparation For Application	256
11.4.2.	Foam Filled Elastomer Covered Buoys	256
11.4.3.	Fiberglass Polyester Resin (Fpr) Coating Repairs	256
11.4.4.	Paint Coatings	257

11.4.5.	Quality Of Work	258
Chapter 12.	**Ashore Inspection And Refurbishment Of Chain And Accessories**	**259**
12.1.	General	259
12.1.1.	Scope	259
12.2.	Inspection And Refurbishment	259
12.2.1.	Preliminary Inspection	259
12.2.2.	Detailed Inspection	259
12.2.3.	Protective Coatings	262
Chapter 13.	**Ashore Inspection And Refurbishment Of Anchors**	**263**
13.1.	General	263
13.1.1.	Scope	263
13.2.	Inspection And Refurbishment	263
13.2.1.	General	263
13.2.2.	Preliminary Inspection	263
13.2.3.	Detailed Inspection	264
Chapter 14.	**Cathodic Protection Systems**	**265**
14.1.	General	265
14.1.1.	Scope	265
14.1.2.	Application	265
14.1.3.	Effectiveness	265
14.2.	Cathodic Protection For Buoys	265
14.2.1.	Environmental Considerations	265
14.2.2.	Anodes	265
14.2.3.	Installation Of Anodes	266
14.3.	Cathodic Protection For Mooring Chain	267
14.3.1.	Anodes	267
14.3.1.1.	Chain Stud Anode	268
14.3.1.2.	Link Anode	269
14.3.1.3.	Clump Anode	269
14.3.2.	Installation	269
14.3.2.1.	Chain Stud Anode	269
14.3.2.2.	Link Anode	270
14.3.2.3.	Clump Anode	271
14.3.2.4.	Wire Rope	271
14.3.3.	Anode Replacement	271
Chapter 15.	**Storage Of Mooring Materials**	**272**
15.1.	General	272
15.2.	General Requirements	272
15.2.1.	Storage Area Requirements	272
15.3.	Storage Procedures	272
15.3.1.	Buoys	272
15.3.2.	Chain And Chain Accessories	273
15.3.2.1.	Tiered Chain	273
15.3.2.2.	Palletized Chain	274
15.3.2.3.	Crated Chain	274
15.3.2.4.	Bundled Chain	275
15.3.2.5.	Accessories	275
15.3.3.	Anchors	275
15.3.4.	Cathodic Protection Materials	276
15.3.5.	Marking And Identification	276
15.3.5.1.	General	276

15.3.5.2.	Color Coding	276
15.3.5.3.	Identification	276
15.3.5.4.	Buoys	278
15.3.5.5.	Anchors	278
15.3.6.	Pre-Issue Inspection	278

Chapter 16. Bibliography and References — **280**

16.1.	Publications Directly Used In The Preparation Of This Book	280
16.2.	General Publications	280
16.3.	Specifications And Standards	281
16.4.	Authored Publications	283
16.5.	Glossary	284

Table of Examples

Example 5-1 Transverse Wind Force Drag Coefficient — 138
Example 5-2 Longitudinal Wind Drag Coefficient — 142
Example 5-3 Current Force — 147
Example 5-4 Longitudinal Current Force — 150
Example 6-1 Catenary Analysis and Anchor Design—Hand Solution — 200
Example 6-2 Catenary Analysis—Computer Solution — 201
Example 6-3 Single Point Mooring - Basic Approach. — 203
Example 6-4 *Fixed Mooring - Basic Approach.* — 209
Example 6-5 *Spread Mooring - Basic Approach.* — 218

Chapter 1. Introduction

A mooring, in general terms, is defined as a compliant structure that restrains a vessel against the action of wind, wave, and current forces. For the purposes of this book, the emphasis is on moorings composed of tension members (chain, line, wire rope, etc.) and compression members (fenders, camels, etc.) used to secure vessels (surface ships, submarines, floating drydocks, yard craft, etc.). The term mooring in this book includes anchoring of ships.

Over the design life of a mooring facility, many organizations are involved with the various aspects of a facility. Personnel involved range from policy makers, who set the initial mission requirements for vessels and facilities, to deck personnel (Figure 1-1) securing lines. In addition, all these groups must maintain open communications to ensure safe and effective moorings.

Figure 1-1 Deck Personnel Securing Lines

Safe use of moorings is of particular importance for the end users (the ship's personnel and facility operators). They must understand the safe limits of a mooring to properly respond to significant events, such as a sudden storm, and to be able to meet mission requirements.

It is equally important for all organizations and personnel to understand moorings. For example, if the customer setting the overall mission requirement states "We need a ship class and associated facilities to meet mission X, and specification Y will be used to obtain these assets" and there is a mismatch between X and Y, the ship and facility operators can be faced with a lifetime of problems, mishaps, and/or serious accidents.

Chapter 2. Mooring Systems

2.1. Introduction

Mooring systems can be categorized into two types of moorings:

a) Fixed Moorings - Fixed moorings are defined as systems that include tension and compression members. Typical fixed mooring systems include moorings at piers and wharves.

b) Fleet Moorings - Fleet moorings are defined as systems that include primarily tension members. Mooring loads are transferred into the earth via anchors. Examples of fleet moorings include fleet mooring buoys and ship's anchor systems.

The more common types of moorings are discussed in this section.

2.2. Purpose Of Mooring

The purpose of a mooring is to safely hold a ship in a certain position to accomplish a specific mission. A key need is to safely hold the vessel to protect the ship, life, the public interest, and to preserve the capabilities of the vessel and surrounding facilities. Ship moorings are provided for:

a) Loading/Unloading - Loading and unloading items such as stores, cargo, fuel, personnel, ammunition, etc.

b) Ship Storage - Storing the ship in a mooring reduces fuel consumption and personnel costs. Ships in an inactive or reserve status are stored at moorings.

c) Maintenance/Repairs - Making a variety of repairs or conducting maintenance on the ship is often performed with a ship moored.

d) Mission - Moorings are used to support special mission requirements, such as surveillance, tracking, training, etc.

Most moorings are provided in harbors to reduce exposure to waves, reduce ship motions, and reduce dynamic mooring loads. Mooring in harbors also allows improved access to various services and other forms of transportation.

2.3. Types Of Mooring Systems

There are two basic types of mooring systems.

2.3.1. Fixed Mooring Systems

Examples of typical fixed moorings are given in Table 2-1 and illustrated in Figure 2-1 through Figure 2-5.

Table 2-1 Examples of Fixed Moorings

a. Single Vessel Secured at Multiple Points

MOORING TYPE	FIGURE NUMBER	DESCRIPTION
Pier/Wharf	Figure 2-1, Figure 2-2	Multiple tension lines are used to secure a vessel next to a pier/wharf. Compliant fenders, fender piles and/or camels keep the vessel offset from the structure. A T-pier may be used to keep the ship parallel to the current, where the current speed is high.
Spud Mooring	Figure 2-3	Multiple vertical structural steel beams are used to secure the vessel, such as a floating drydock. This type of mooring is especially effective for construction barges temporarily working in shallow water. Spud moorings can be especially susceptible to dynamic processes, such as harbor seiches and earthquakes.

b. Multiple Vessel Moorings

MOORING TYPE	FIGURE NUMBER	DESCRIPTION
Opposite Sides of a Pier	Figure 2-4	Vessels can be placed adjacent to one another on opposite sides of a pier to provide some blockage of the environmental forces/moments on the downstream vessel.
Multiple Vessels Next to One Another	Figure 2-5	Vessels can be placed adjacent to one another to provide significant blockage of the environmental forces/ moments on the downstream vessel(s).

Figure 2-1 Single Ship, Offset From a Pier With Camels

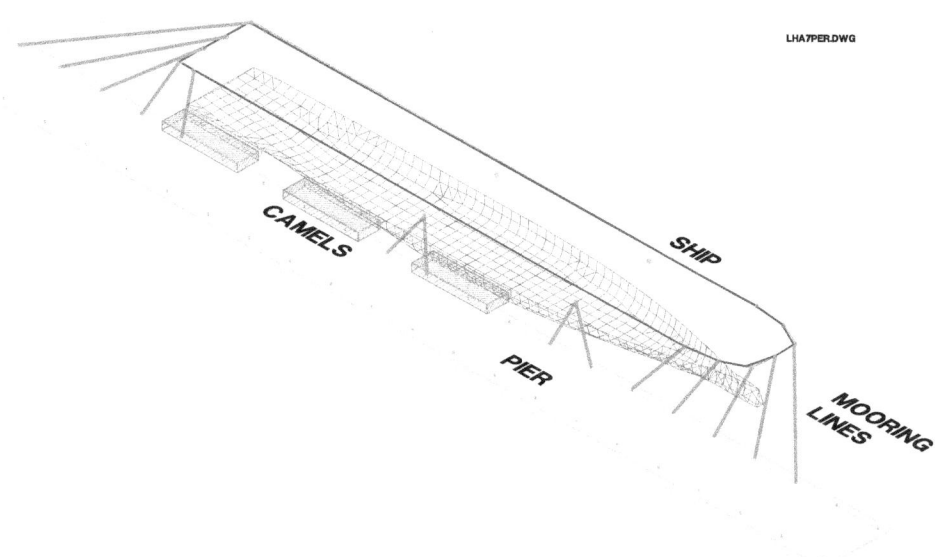

Figure 2-2 Ship at a T-Pier (plan view)

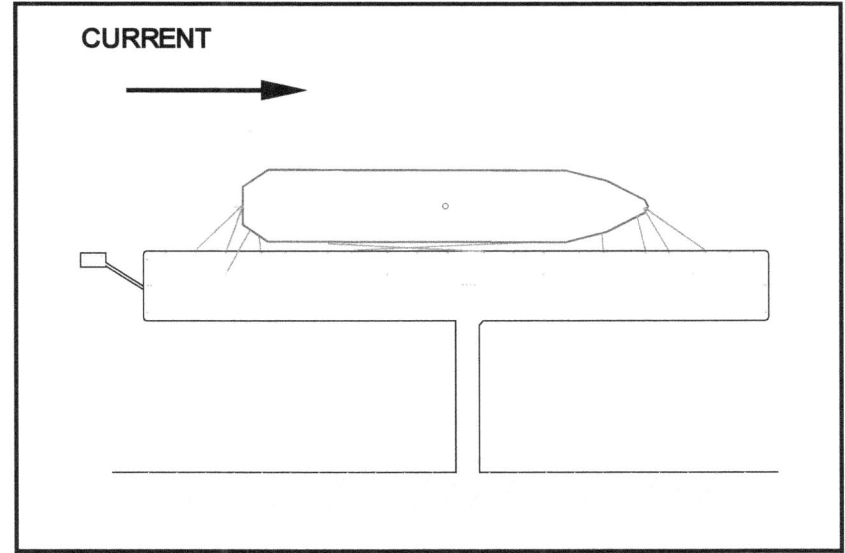

Figure 2-3 Floating Drydock Spud Moored
(spuds are secured to a pier, which is not shown, and the floating drydock rides up and down on the spuds; profile view is shown)

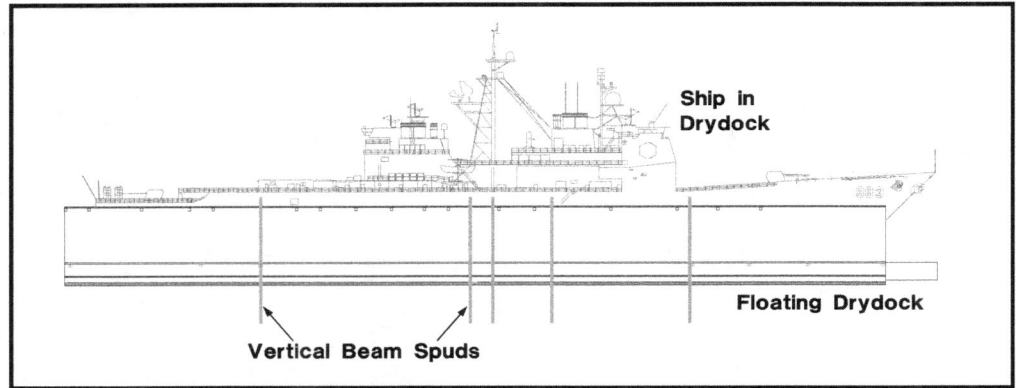

Figure 2-4 Ships on Both Sides of a Pier (plan view)

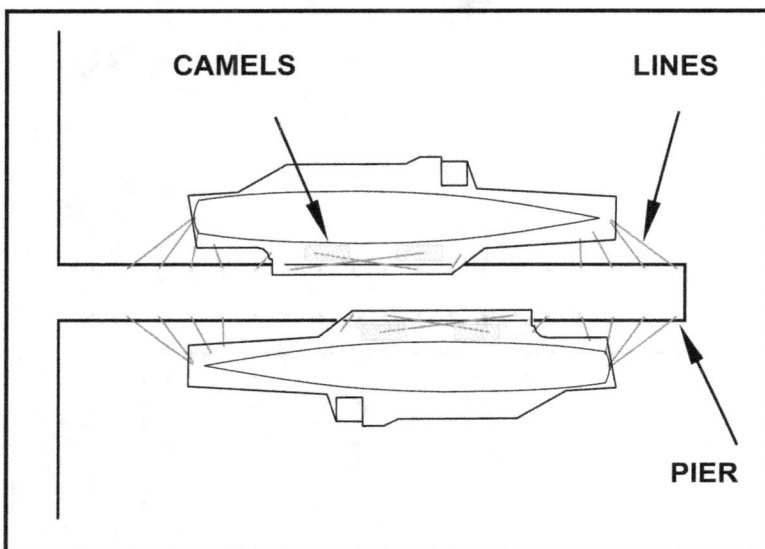

Figure 2-5 Two Ships on One Side of a Pier (plan view)

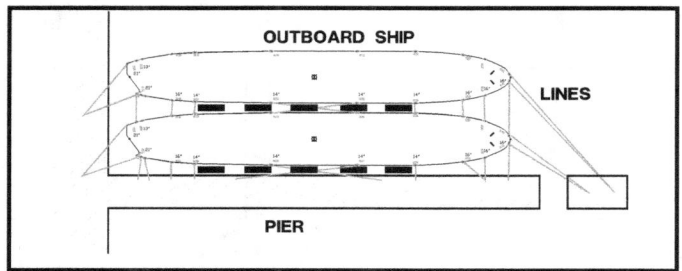

2.3.2. Fleet Moorings

Fleet moorings are pre-existing facilities used to provide temporary berthing for fleet units in ports and harbors where pier space is limited or unavailable. The most common types of fleet moorings consist of one or more buoy systems made up of surface buoys, riser and anchor chain subassemblies, fittings, and anchors. A vessel will moor with its line or anchor chain connected to shackles, ground rings, or other mooring components on top of the buoy.

Several basic types of fleet moorings are discussed below. Most of these are riser or non-riser buoy systems; some do not require a buoy. Special fleet moorings and floating drydock moorings are also discussed briefly, but their maintenance is not further discussed.

Examples of typical fleet moorings are given in Table 2-2 and illustrated in Figure 2-6 through Figure 2-15.

Table 2-2 Examples of Fleet Moorings

a. Vessel Secured at a Single Point

MOORING TYPE	FIGURE NUMBER	DESCRIPTION
At Anchor	Figure 2-6	Typical configuration includes the ship deploying a single drag anchor off the bow. This is usually a temporary mooring used as a last resort in benign conditions. A large amount of harbor room is required for the ship swing watch circle. If the wind changes direction dramatically then the anchor will have to reset. Dynamic fishtailing, even under steady winds and currents, may be a problem. Putting out a second anchor in what is known as a Hammerlock mooring may be required in storm anchoring.
Single Mooring Buoy	Figure 2-7, Figure 2-8, Figure 2-9	A single point mooring (SPM) buoy is secured to the seafloor typically with 1 to 12 ground legs and either drag or plate anchors. The ship moors to the buoy using an anchor chain or hawser. The vessel weathervanes under the action of forcing, which helps to reduce the mooring load. This type of mooring requires much less room than a ship at anchor because the pivot point is much closer to the vessel. A vessel at a mooring buoy is much less prone to fishtailing than a ship at anchor.

b. Vessel Secured at Two Points

MOORING TYPE	FIGURE NUMBER	DESCRIPTION
Bow-Stern Mooring	Figure 2-10	A vessel is moored with one buoy to the bow and another to the stern. This system has a much smaller watch circle than a vessel at a single mooring buoy. Also, two moorings share the load. However, the mooring tension can be much higher if the winds, currents, or waves have a large broadside component to the ship.
Buoy Dolphins	Figure 2-11	Dolphins simulate a longer pier to accommodate larger ships breasted alongside a pier or alongside one or more finger piers by acting as bow and stern moorings. Dolphins may consist of a group of pilings driven close together into the bottom or may employ a buoy system. When buoy systems are used as dolphins they are called buoy dolphins

c. Vessel Secured at Multiple Points

MOORING TYPE	FIGURE NUMBER	DESCRIPTION
Med-Mooring	Figure 2-12	The vessel bow is secured to two mooring buoys and the stern is moored to the end of a pier or wharf. This type of mooring is commonly used for tenders or in cases where available harbor space is limited. Commonly used in the Mediterranean Sea. Hence, the term "Med" Mooring.
Spread Mooring	Figure 2-13	Multiple mooring legs are used to secure a vessel. This arrangement of moorings is especially useful for securing permanently or semi-permanently moored vessels, such as floating drydocks and inactive ships. The ship(s) are usually oriented parallel to the current.

d. Multiple Vessel Moorings

MOORING TYPE	FIGURE NUMBER	DESCRIPTION
Nest	Figure 2-4 Figure 2-5 Figure 2-14 Figure 2-15	Multiple tension members are used to secure several vessels together. Camels or fenders are used to keep the vessels from contacting one another. Nests of vessels are commonly put into spread moorings. Nested vessels may be of similar size (as for inactive ships) or much different size (as a submarine alongside a tender). Advantages of nesting are: a nest takes up relatively little harbor space and forces/moments on a nest may be less than if the ships were moored individually.

Figure 2-6 Ship at Anchor

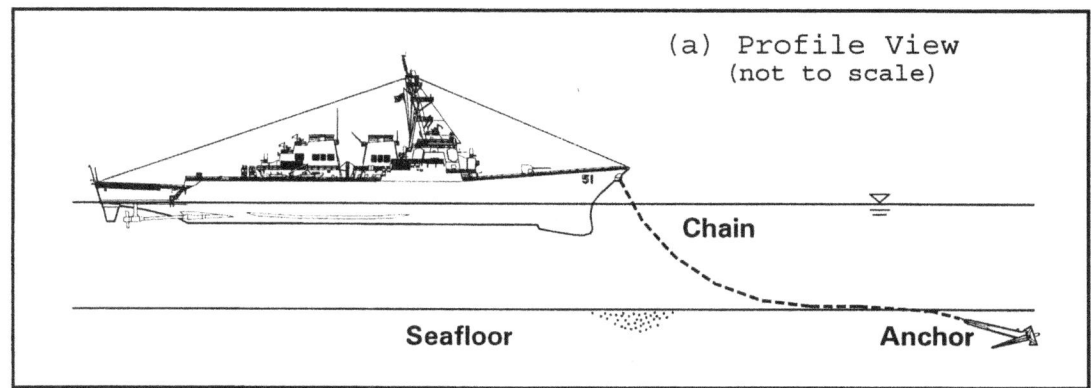

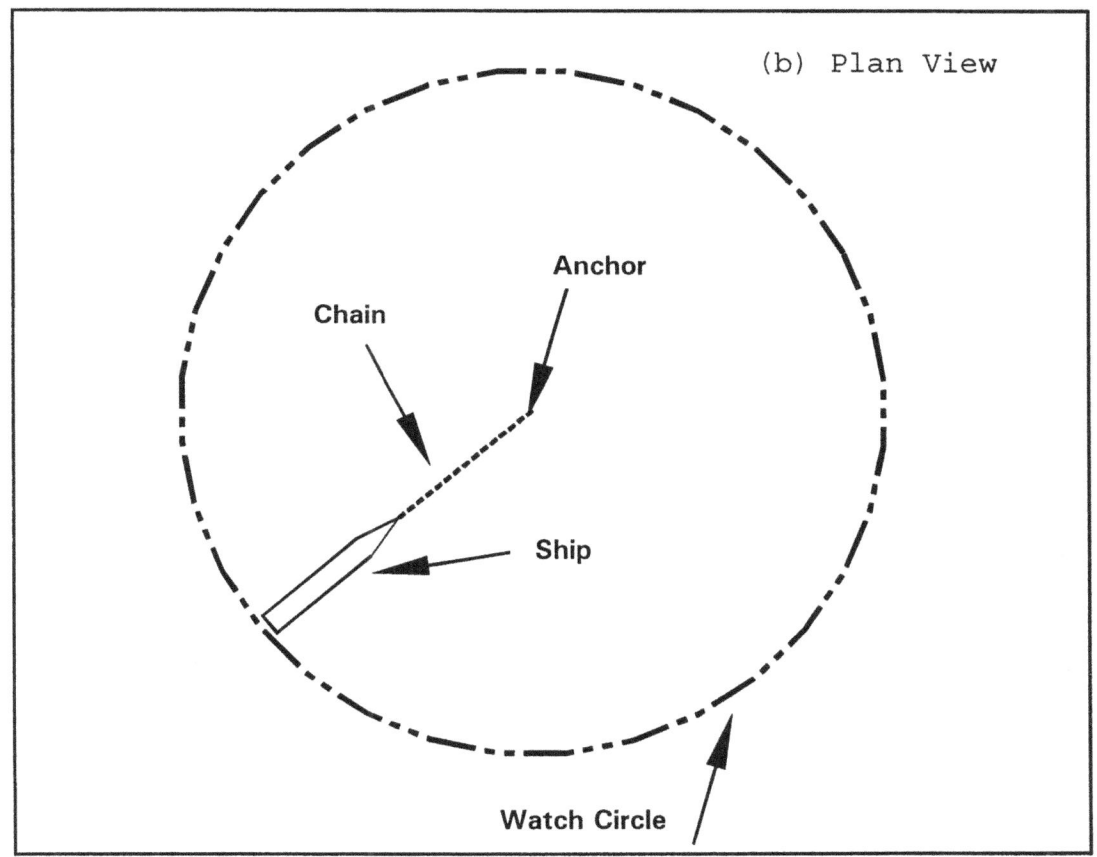

Figure 2-7 Single Point Mooring With Drag Anchors

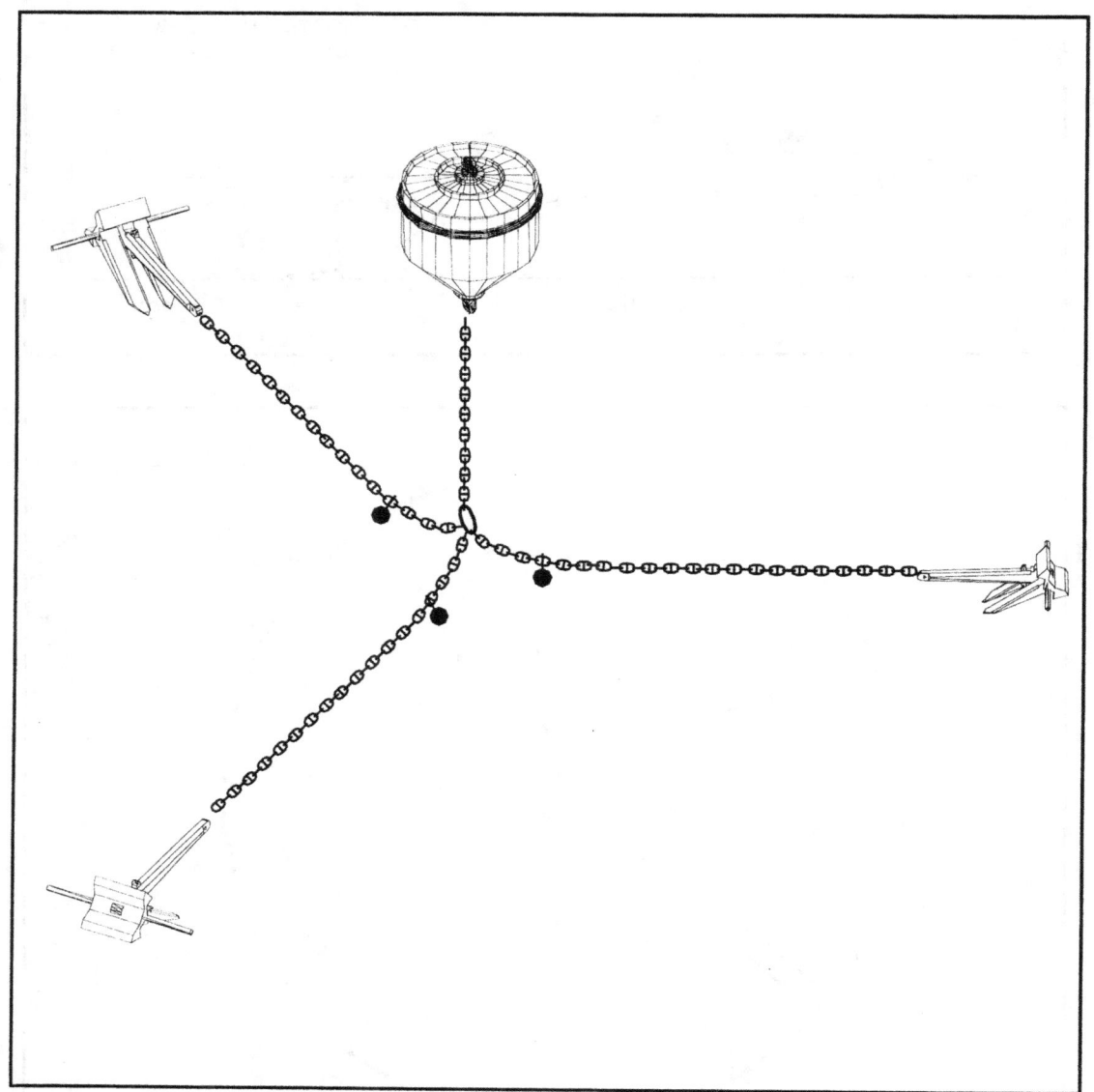

Figure 2-8 Single Point Mooring With a Plate Anchor and a Sinker

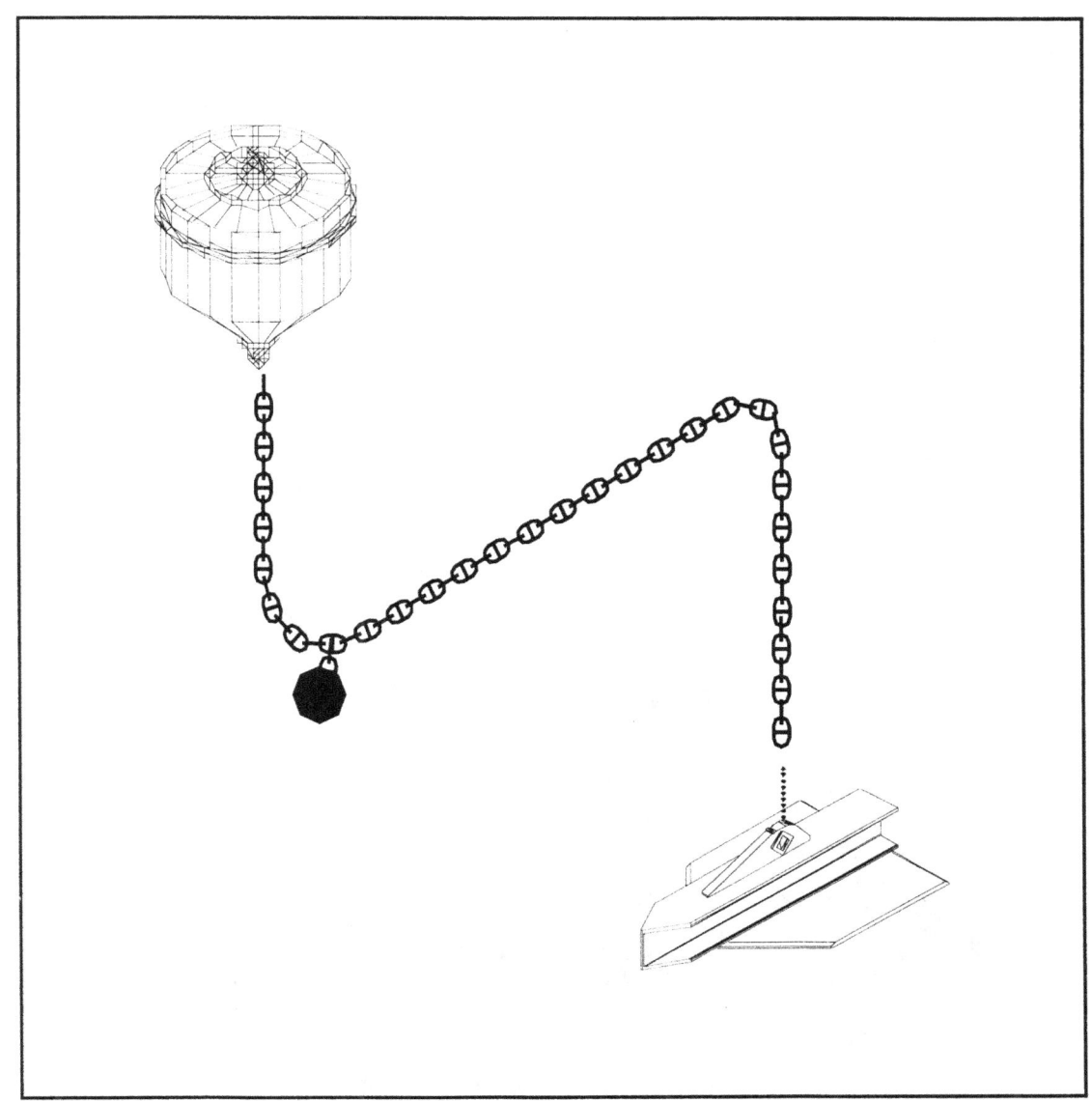

Figure 2-9 Single Point Mooring Example

Figure 2-10 Bow-Stern Mooring

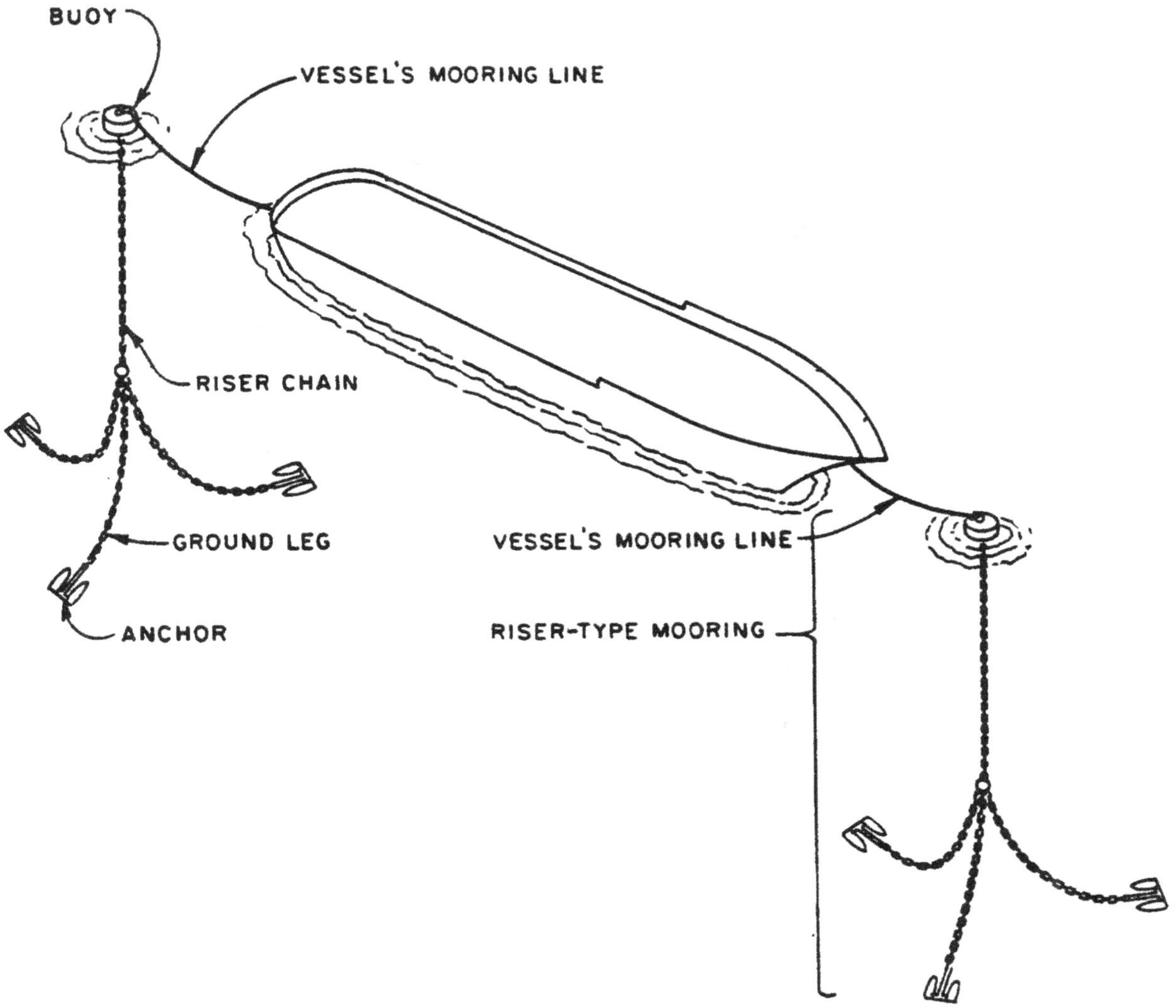

Figure 2-11 Buoy Dolphin Mooring

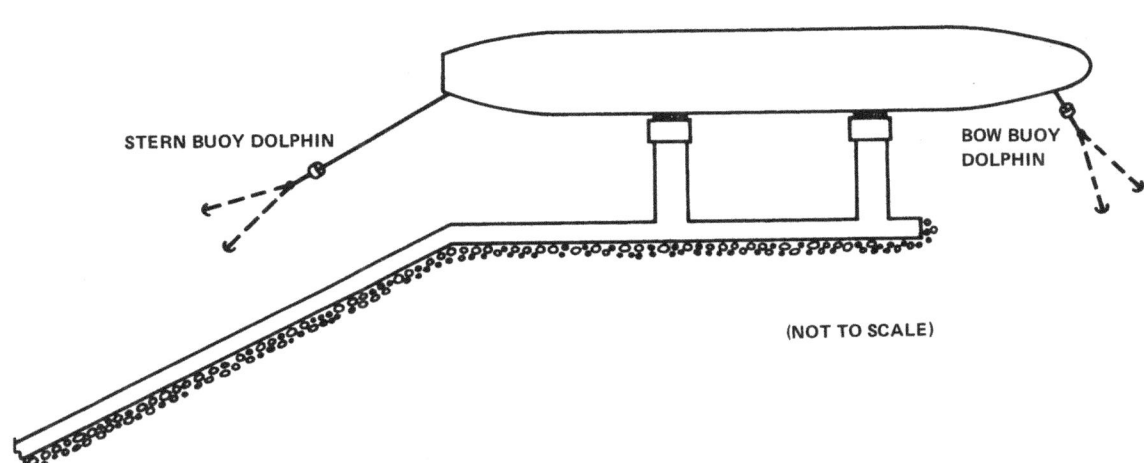

Figure 2-12 Med-Mooring

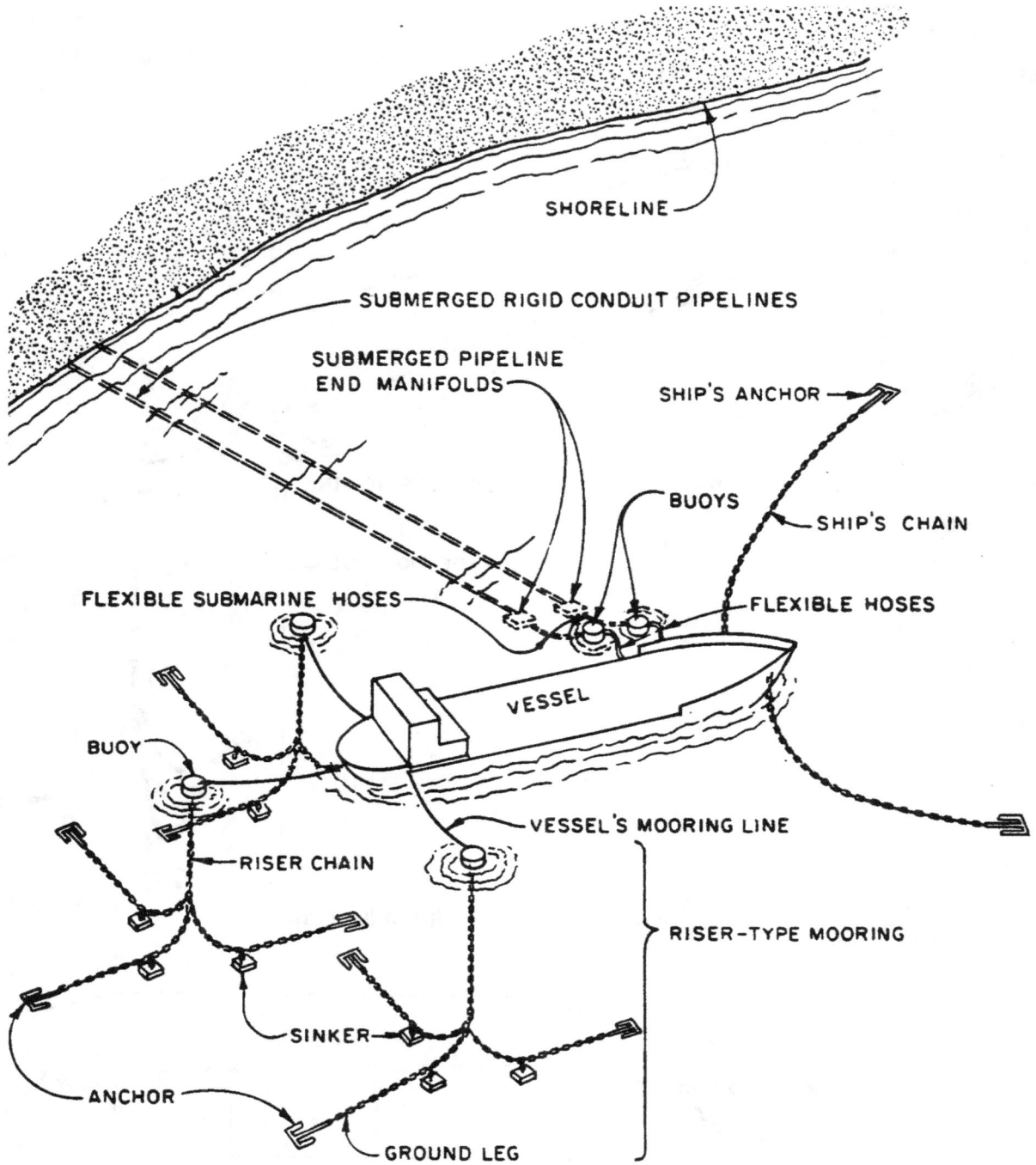

Figure 2-13 Spread Mooring

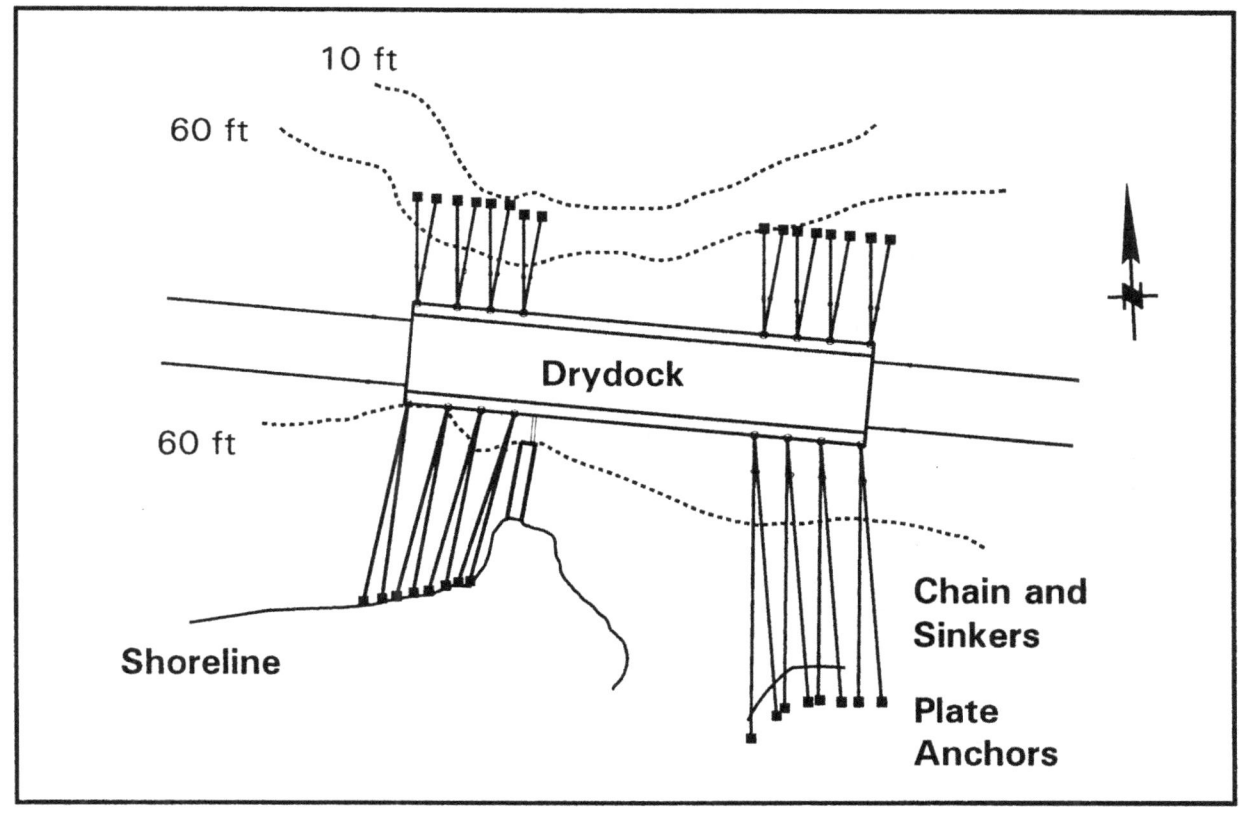

Figure 2-14 Two Inactive Ships Moored at a Wharf

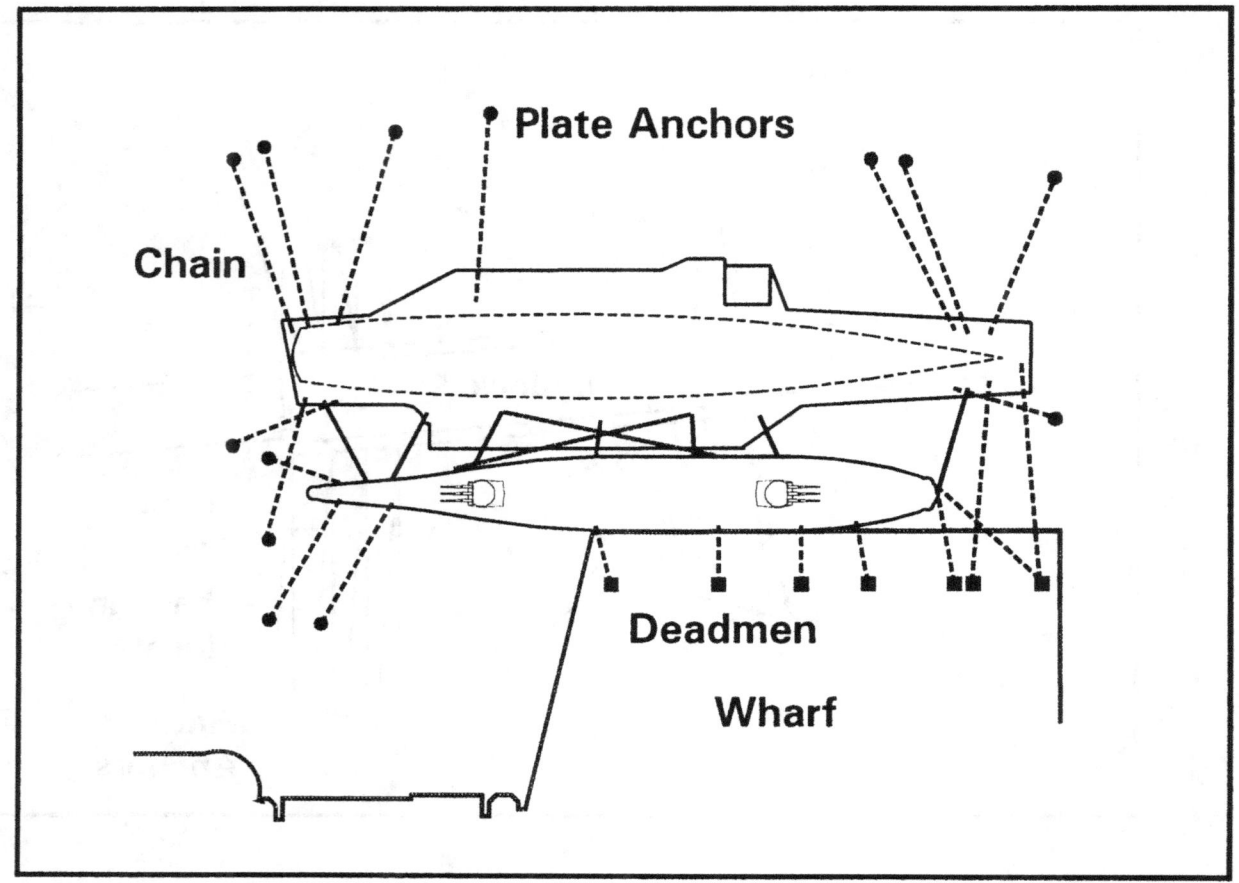

Figure 2-15 Spread Mooring

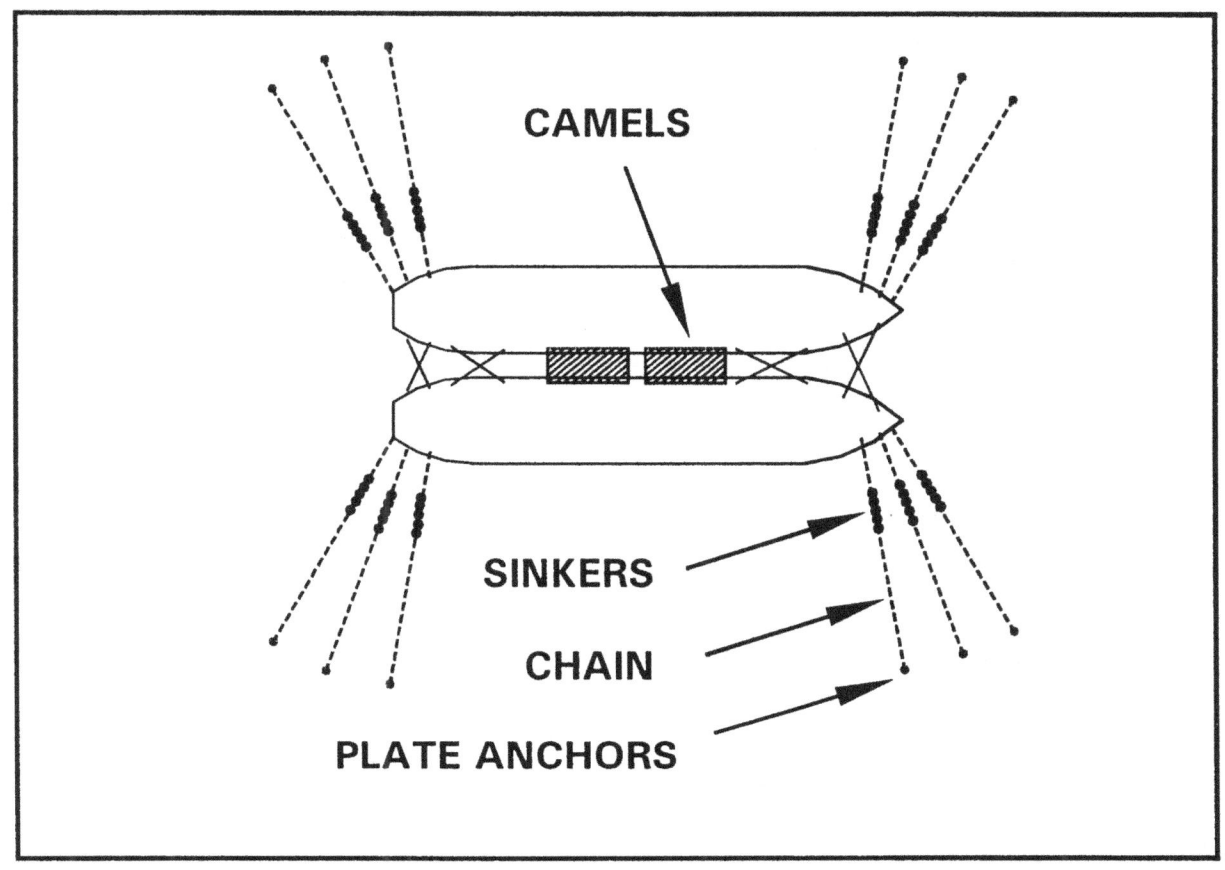

Chapter 3. Mooring Equipment

3.1. Introduction

Equipment most often used in mooring facilities is discussed in this section.

3.2. Key Mooring Components

A mooring is a structure that holds a ship in a position using tension and compression members. The resulting mooring loads are transferred to the earth via anchors or some other members, such as pier piles or a wharf structure.

3.2.1. Tension Members

The most commonly used tension members in moorings are:

- Chain
- Synthetic line
- Wire rope
- Tension bar buoys

3.2.2. Compression Members

The most commonly used compression members in moorings are:

- Marine fenders
- Fenders
- Camels
- Mooring dolphins
- Piers
- Wharves

3.3. Anchors

Anchors are structures used to transmit mooring loads to the earth. Anchors operate on the basis of soil structure interaction, so their behavior can be complex. Fortunately, the U.S. Navy has extensive experience with full-scale testing of a number of different anchor types in a wide variety of soils and conditions (NCEL *Handbook for Marine Geotechnical Engineering*, available from Pile Buck). This experience provides a strong basis for design. However, due to the complex nature of structure/soil interaction, it is strongly recommended that anchors always be pull tested to their design load during installation. Some common anchor types are shown in Figure 3-1.

Figure 3-1 Typical Mooring Anchors

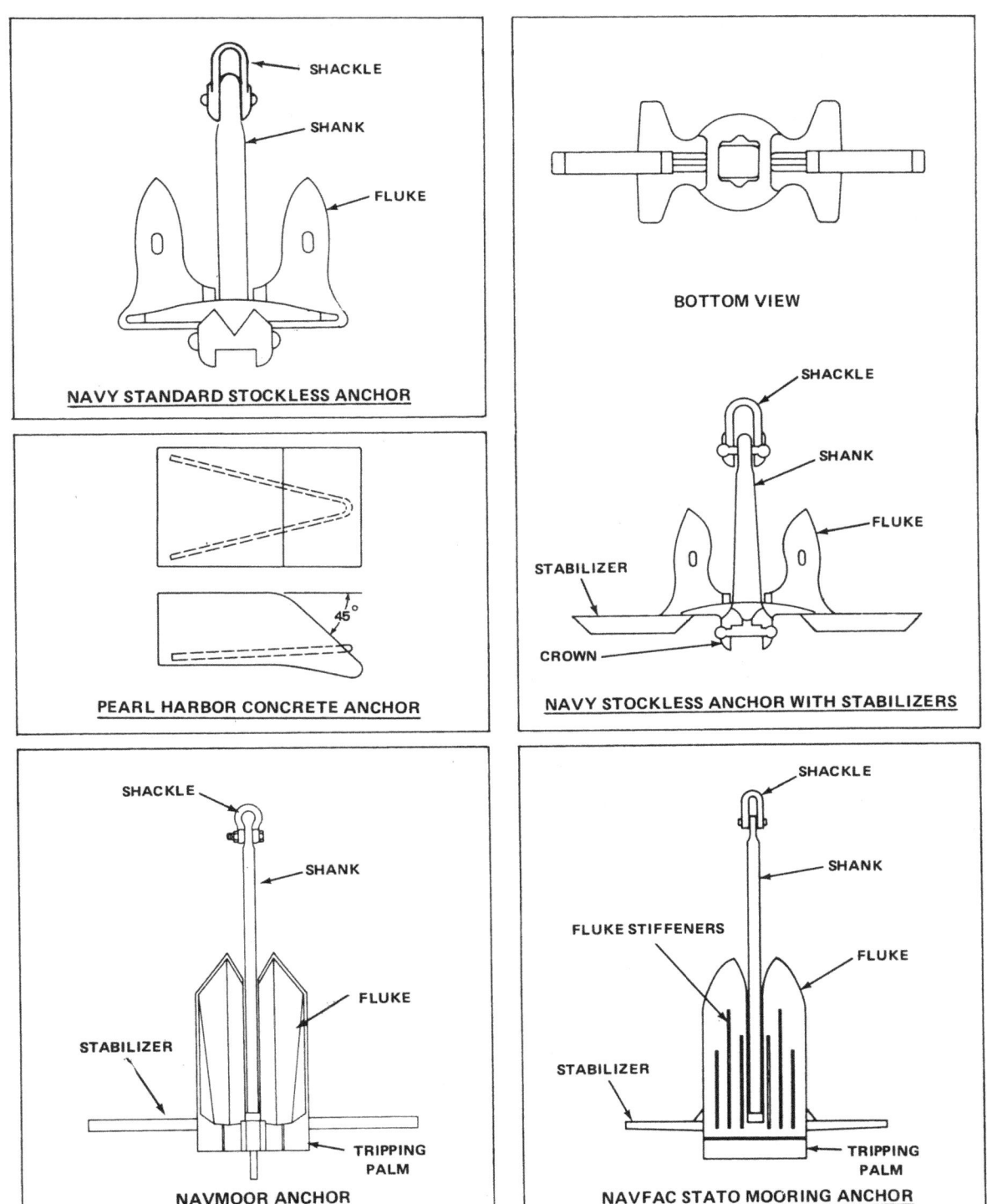

3.3.1. Navy Standard Stockless

These anchors are used extensively in fleet moorings primarily because of their availability. Stockless anchors are not very efficient, but reliable through 'brute force', and there is extensive experience in their use. They are prone to rotate after reaching their maximum holding power, which results in a pullout. To reduce the possibility of rotation, stabilizers are often welded to the anchor. The older type stabilizers consisted of a 1" wall-thickness pipe of variable diameters. A newly designed 1/2" wall-thickness square tube will be used for future stabilizers (see Table 3-1).

Fixing the flukes for the type of soil at the site can also increase efficiency. The flukes should be set and proof tested during installation. Stockless anchors can be used in tandem in various configurations. The vertical angle of the tension member should be approximately zero at the seafloor.

Navy standard stockless anchors range in size from 500 to 40,000 pounds. A summary sheet describing the stockless anchors used by the U.S. Navy is given in Table 3-1.

Table 3-1 Stockless Anchors in the U.S. Navy Fleet Mooring Inventory

ANCHOR IN AIR WEIGHT (1000 lbf)	20	25	30
LENGTH *(inches)*	127.25	137	145.63
STABILIZER EXTENSION *(inches)*	45	48	50
FLUKE LENGTH *(ft)*	7.65	8.24	8.94
FLUKE AREA *(sq. ft)*	35.1	40.7	46.9
SAFE HOLDING CAPACITIES WITH FS = 1.5*			
MUD SEAFLOOR Fluke Angle = 48 deg			
Minimum MUD Thickness (ft)**	22 ft	24 ft	25 ft
Typical Anchor Drag (ft)***	31 ft	33 ft	36 ft
Single Holding (x1000 lbf)	58	71	84
Tandem Holding (x1000 lbf)	116	142	169
SAND SEAFLOOR Fluke Angle = 35 deg			
Minimum SAND Thickness (ft)**	8 ft	8 ft	9 ft
Typical Anchor Drag (ft)***	33 ft	36 ft	39 ft
Single Holding (x1000 lbf)	81	97	112
Tandem Holding (x1000 lbf)	163	194	225

* design mooring properly ** for ultimate holding *** fix flukes open

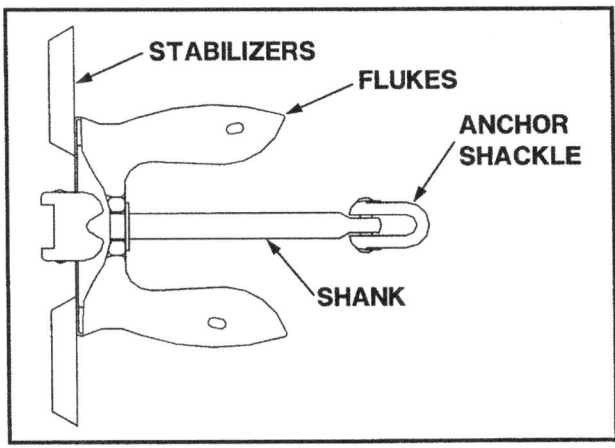

3.3.2. NAVFAC Stato

These anchors were developed for the Naval Facilities Engineering Command as permanent mooring-type anchors. The stabilizers have been designed and tested for maximum stability (see Figure 3-1). The angle between the shank and the flukes is adjustable in the field to a 34° angle or a 50° angle for operations in sand and mud, respectively. STATO anchors range in size from 200 to 15,000 pounds.

3.3.3. NAVMOOR Anchor

NAVMOOR anchors are very efficient, highly reliable and especially designed so it can easily be used in tandem (NCEL TN-1774, *Single and Tandem Anchor Performance of the New Navy Mooring Anchor*). Their performance is excellent in a wide variety of soil conditions. As with stockless anchors, they should be set and proof tested during installation. The vertical angle of tension member should be approximately zero at the seafloor in most cases. Various sizes of NAVMOOR anchors are described in Table 3-2.

Table 3-2 NAVMOOR Anchors in the U.S. Navy Fleet Mooring Inventory

ANCHOR SIZE =	NAVMOOR-12	NAVMOOR-15
IN AIR WEIGHT *(pounds)*	12400	19200
LENGTH OVERALL *(inches)*	192	219
STABILIZER WIDTH *(inches)*	192	219
FLUKE LENGTH *(ft)*	8.54	9.73
FLUKE AREA *(sq. ft)*	38.54	50.07
SAFE HOLDING CAPACITIES FS = 2*		
MUD SEAFLOOR Fluke Angle = 50 deg		
Minimum MUD Thickness (ft)**	38 ft	44 ft
Typical Anchor Drag (ft)***	30-35 ft	35-40 ft
Single Holding (x1000 lbf)	125	168
Tandem Holding (x1000 lbf)	310	420
SAND SEAFLOOR Fluke Angle = 32 deg		
Minimum SAND Thickness (ft)**	9 ft	10 ft
Typical Anchor Drag (ft)***	25 ft	30 ft
Single Holding (x1000 lbf)	160	215
Tandem Holding (x1000 lbf)	400	535

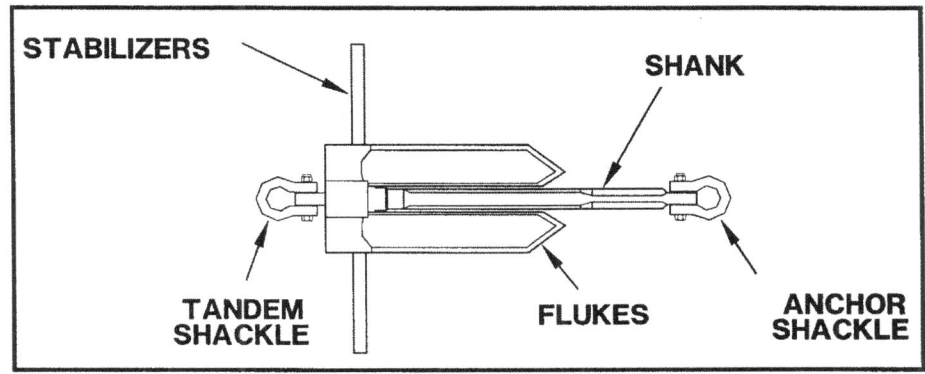

* design mooring properly ** for ultimate holding *** fix flukes open

3.3.4. Stake Piles

This type of anchoring system that consists of wood or steel pilings driven into the bottom can be used in certain moorings as an anchor. The disadvantages of this type system are that the stake pile is fixed and cannot absorb shock energy as well as an anchor, and that a pile-driving rig is required for installation.

3.3.5. "Pearl Harbor" And Other Deadweight Anchors

The "Pearl Harbor" anchor (see Figure 3-1) is essentially a concrete sinker with a wedge shape. It is relatively inexpensive to build and can be fabricated onsite. The size and weight required to obtain adequate holding power, however, make the anchor bulky and difficult to handle.

Most all deadweight anchors have very low efficiency. Full-scale tests (NCEL, *Fleet Mooring Test Program – Pearl Harbor*) show anchor-holding capacity dramatically decreases after anchor starts dragging, just when the anchor capacity required may be most needed. As a result, use of this type of anchor can be dangerous. Deadweight anchors should be used with caution.

3.3.6. Driven Plate Anchor

The driven plate anchor is an efficient and cost effective method of providing permanent moorings. Table 3-3 shows the components of the driven-plate anchor.

Table 3-3 Driven-Plate Anchor Components

COMPONENT	NOTES
Plate	Size the area and thickness of the plate to hold the required working load in the given soils. A plate length-to-width ratio of L/B = 1.5 to 2 is shown by practical experience to give optimum performance.
I-Beam	Size the beam to provide: a driving member; stiffness and strength to the anchor; and to separate the padeye from the plate to provide a moment that helps the anchor key during proof testing.
Padeye	Size this structure as the point where the chain or wire rope is shackled onto the anchor prior to driving.
Follower	Length and size specified so assembly can safely be picked up, driven, and removed.
Hammer	Sized to drive the anchor safely. In most cases it is preferable to use an impact hammer. A vibratory hammer may be used in cohesionless soils or very soft mud. A vibratory hammer may also be useful during follower extraction.
Template	A structure is added to the side of the driving platform to keep the follower in position during setup and driving.

There are three major steps in the installation of a driven-plate anchor (Figure 3-2):

1) Install the anchor itself.

    a) Moor installation platform;

    b) Place anchor in follower;

    c) Shackle anchor to chain;

    d) Place the follower/anchor assembly at the specified anchor location; and

e) Drive the anchor to the required depth in the sediment (record driving blow count), typically 10-40 feet.

2) Remove follower with a crane and/or extractor.

3) Proof-load the anchor. This keys the anchor, proves that the anchor holds the design load, and removes slack from the chain.

Figure 3-2 Major Steps of Driven-Plate Anchor Installation

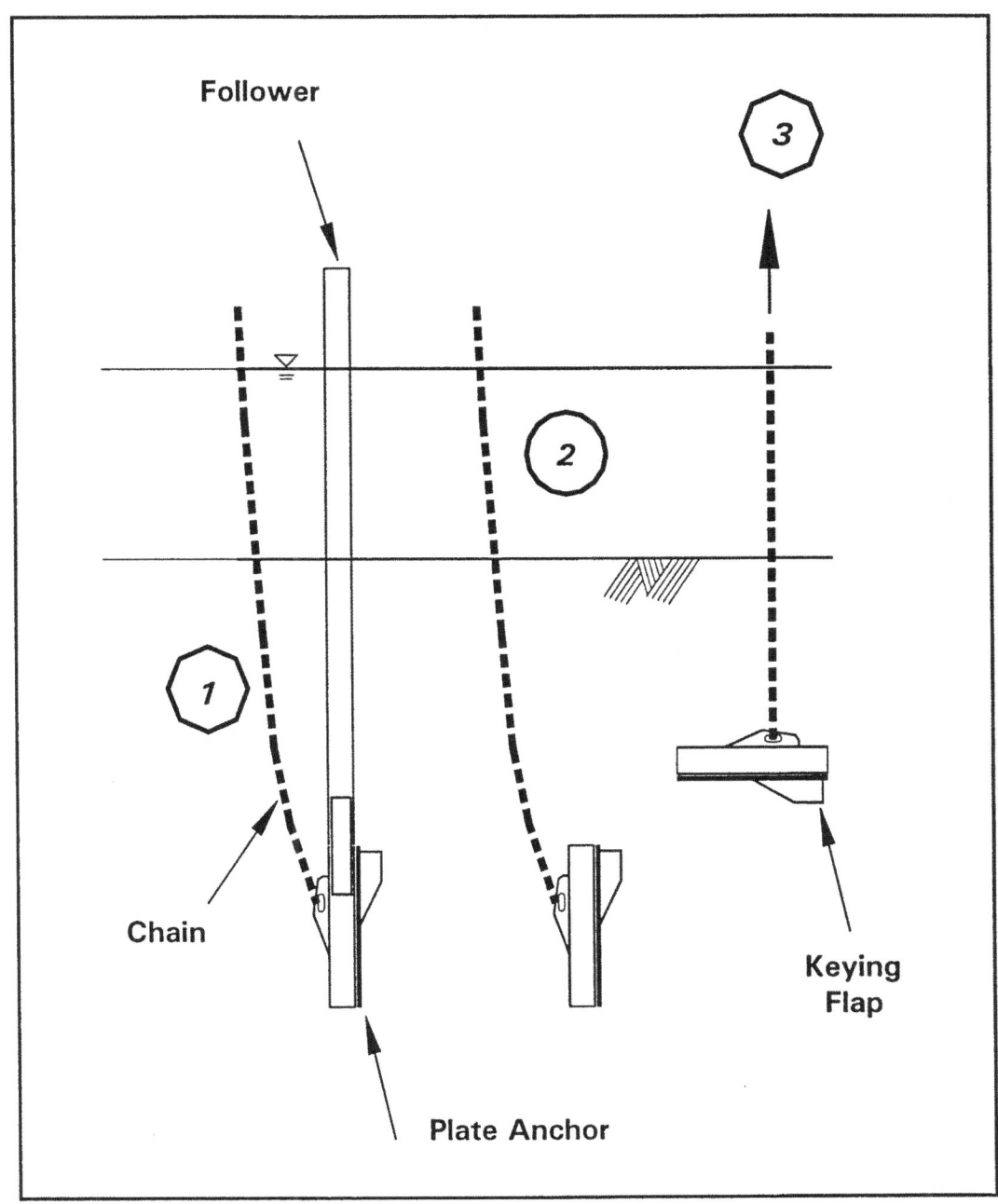

Examples of plate anchors that have been used at various sites are summarized in Figure 3-3.

Figure 3-3 Typical Driven-Plate Anchors

SIZE/LOCATION	SEAFLOOR TYPE	DRIVING DISTANCE INTO COMPOTENT SEDIMENT	PROOF LOAD
0.91 m x 1.22 m (3 ft x 4 ft) Philadelphia, PA	Hard Clay	9 m (30 ft)	670 kN (150 kips) Vertical
0.61 m x 1.22 m (2 ft x 4 ft) San Diego, CA	Sand (Medium)	8 m (27 ft)	890 kN (200 kips) Vertical
1.52 m x 1.83 m (5 ft x 6 ft) Guam	Coral Limestone	12 m (40 ft)	1000 kN (225 kips) Vertical
1.83 m x 3.35 m (6 ft x 11 ft) Pearl Harbor, HI	Mud	21 m (70 ft)	890 kN (200 kips) Horizontal

An alternate to this is the propellant-embedded anchor, as shown in Figure 3-4. An anchor fluke is emplaced in a gun assembly. The entire assembly, consisting of the gun (launch platform), downhaul cable, and anchor fluke (see), is lowered to the bottom. The gun is then fired and the fluke, depending upon the material of the ocean bottom, is driven 10 to 40 feet into the subsurface strata (Figure 3-5.) Attached to the fluke is a wire rope downhaul cable (2 to 3 inches in diameter). The upper end of this downhaul cable is then connected to a wire rope pendant and the anchor chain subassembly or is directly connected to the anchor chain subassembly by swage and shackle fittings. The gun assembly is recovered for re-use.

Figure 3-4 NCEL Propellant-Embedment Anchor

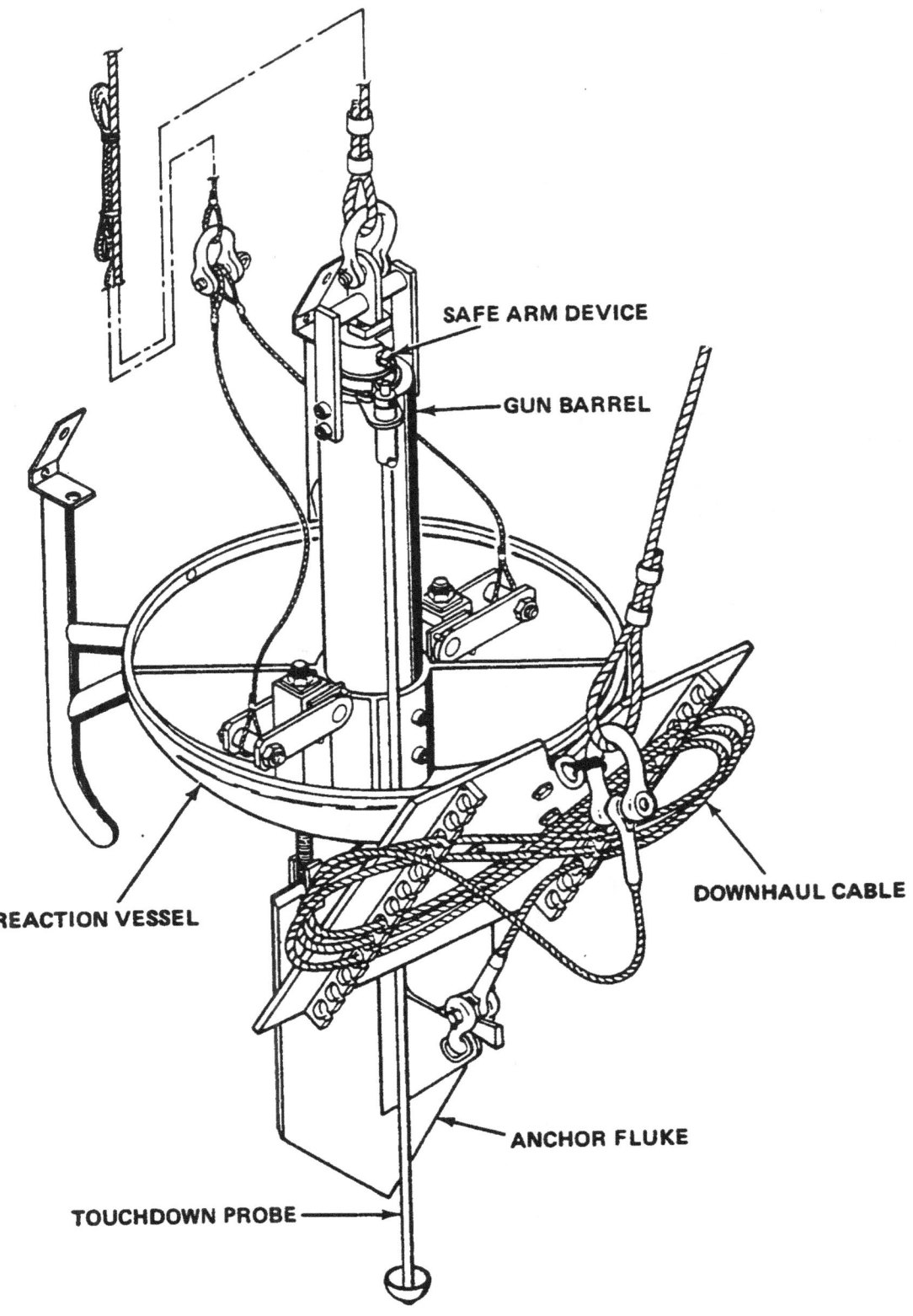

Figure 3-5 Penetration and Keying of a Propellant-Embedment Anchor

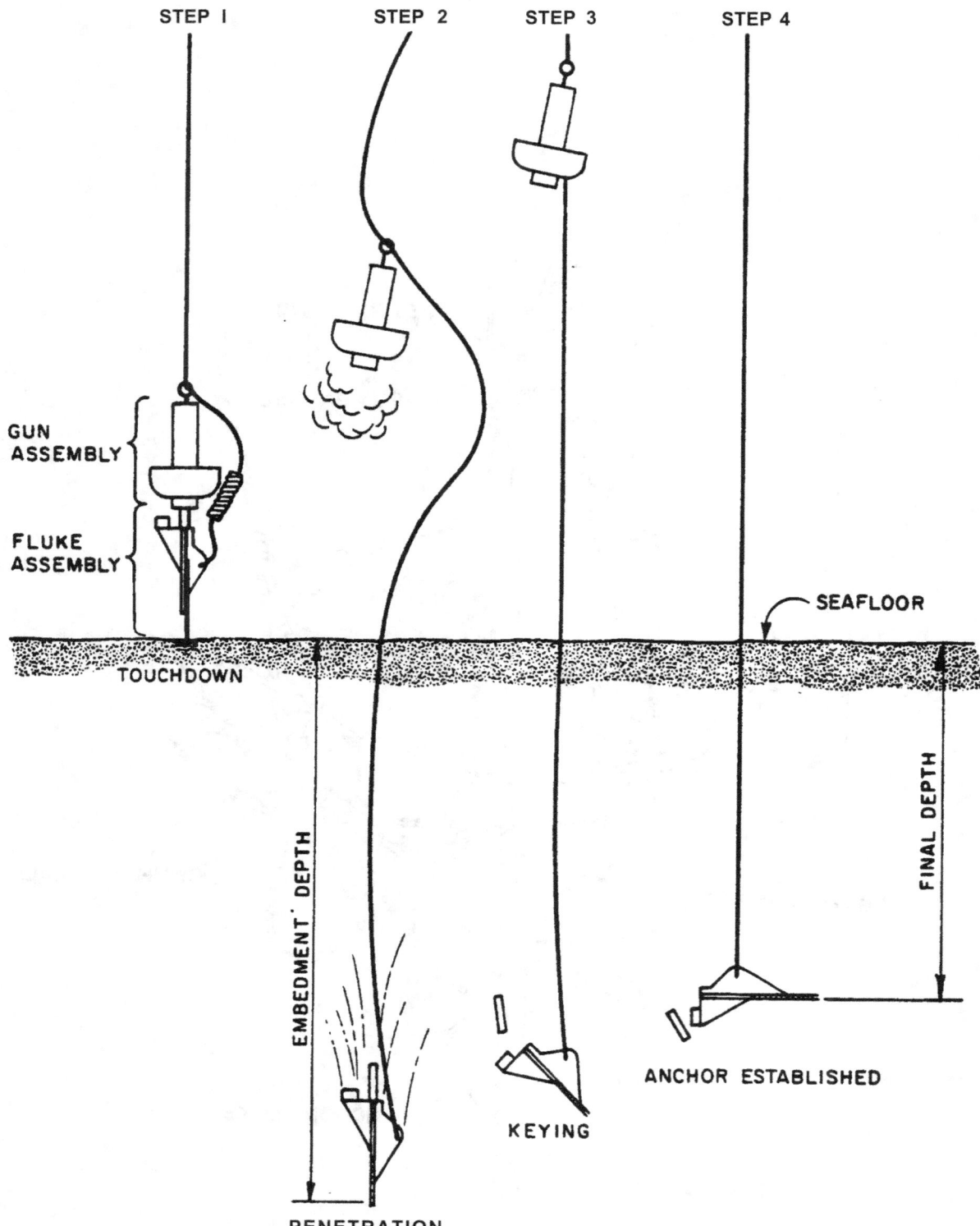

3.3.7. Anchor Pull Test Requirements

This appendix establishes the procedures for setting and pull testing of anchor legs. Tables of predicted anchor system drag distance are included to assist in initial anchor positioning.

3.3.7.1. Setting Drag Anchors

Proof setting of each anchor leg should be accomplished using harbor tugs, Fleet tugs or by pulling diametrically opposite anchor legs against each other. When setting the anchor, it is desirable to have a 10 to 1 scope (see Figure 3-6 for definition) in the mooring line. Adding chain or a work wire may accomplish this. The anchor drag distance to achieve the specified load level must be known to enable proper anchor placement prior to proof loading. Setting shall be accomplished by pulling to the design load of the mooring.

Figure 3-6 Mooring Leg Scope

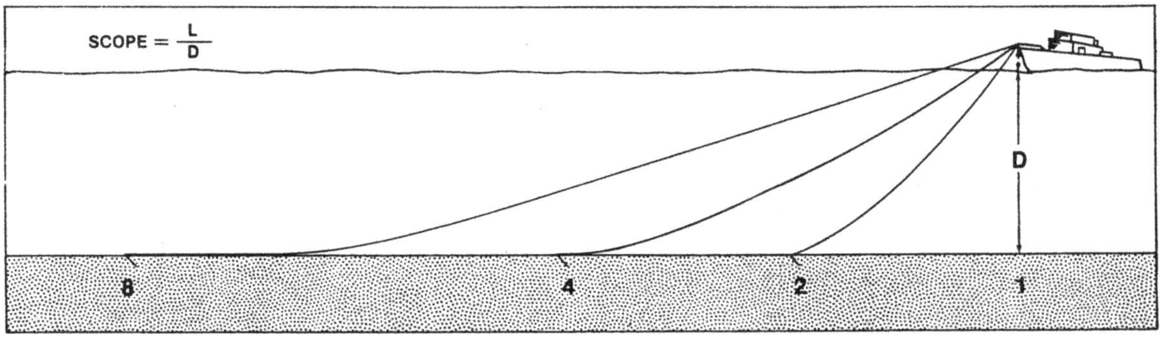

Scope is the ratio of length of rode (L) to depth of water (D), plus allowance for heigth of bow above water. At (1) length of rode equals the depth. At (2) rode length is twice the depth, at (4) four times the depth. Note how the angle between rode and bottom decreases. At (8) the scope is 8:1 and the short length of chain at the anchor lies flat on the bottom.

- Single Anchors. The predicted anchor drag distance for single anchors can be obtained from Table 3-4.

Table 3-4 Predicted Single Anchor Drag Distances

Anchor: Stockless Anchor with Stabilizers and Flukes Fixed at approximately 45°
Seafloor Type: Mud

Horizontal Design Load (Kips)

Anchor Weight (Kips)	25.	50.	75.	100.	125.	150.	175.	200.	225.	250.	275.	300.
6.	54.	*	*	*	*	*	*	*	*	*	*	*
7.	32.	*	*	*	*	*	*	*	*	*	*	*
8.	21.	*	*	*	*	*	*	*	*	*	*	*
9.	13.	*	*	*	*	*	*	*	*	*	*	*
10.	9.	*	*	*	*	*	*	*	*	*	*	*
11.	6.	183.	*	*	*	*	*	*	*	*	*	*
12.	6.	99.	*	*	*	*	*	*	*	*	*	*
13.	6.	64.	*	*	*	*	*	*	*	*	*	*
14.	5.	47.	*	*	*	*	*	*	*	*	*	*
15.	5.	40.	*	*	*	*	*	*	*	*	*	*
16.	5.	33.	*	*	*	*	*	*	*	*	*	*
17.	4.	27.	*	*	*	*	*	*	*	*	*	*
18.	4.	22.	154.	*	*	*	*	*	*	*	*	*
19.	4.	17.	95.	*	*	*	*	*	*	*	*	*
20.	3.	14.	77.	*	*	*	*	*	*	*	*	*
21.	3.	12.	64.	*	*	*	*	*	*	*	*	*
22.	2.	10.	53.	*	*	*	*	*	*	*	*	*
23.	2.	8.	48.	*	*	*	*	*	*	*	*	*
24.	2.	8.	42.	202.	*	*	*	*	*	*	*	*
25.	1.	8.	37.	152.	*	*	*	*	*	*	*	*
26.	1.	8.	33.	104.	*	*	*	*	*	*	*	*
27.	1.	7.	29.	89.	*	*	*	*	*	*	*	*
28.	1.	7.	25.	78.	*	*	*	*	*	*	*	*
29.	1.	7.	21.	68.	*	*	*	*	*	*	*	*
30.	0.	7.	19.	59.	245.	*	*	*	*	*	*	*

Drag Distance (Feet)

Anchor: Stockless Anchor with Stabilizers and Flukes Fixed at approximately 36°
Seafloor Type: Sand

Horizontal Design Load (Kips)

Anchor Weight (Kips)	25.	50.	75.	100.	125.	150.	175.	200.	225.	250.	275.	300.
5.	20.	*	*	*	*	*	*	*	*	*	*	*
6.	19.	*	*	*	*	*	*	*	*	*	*	*
7.	18.	37.	*	*	*	*	*	*	*	*	*	*
8.	17.	33.	*	*	*	*	*	*	*	*	*	*
9.	17.	29.	*	*	*	*	*	*	*	*	*	*
10.	17.	28.	*	*	*	*	*	*	*	*	*	*
11.	17.	27.	46.	*	*	*	*	*	*	*	*	*
12.	17.	26.	43.	*	*	*	*	*	*	*	*	*
13.	17.	26.	39.	*	*	*	*	*	*	*	*	*
14.	17.	25.	37.	*	*	*	*	*	*	*	*	*
15.	17.	24.	35.	*	*	*	*	*	*	*	*	*
16.	17.	24.	34.	52.	*	*	*	*	*	*	*	*
17.	17.	23.	33.	49.	*	*	*	*	*	*	*	*
18.	17.	23.	32.	46.	*	*	*	*	*	*	*	*
19.	18.	23.	32.	44.	*	*	*	*	*	*	*	*
20.	18.	23.	31.	41.	*	*	*	*	*	*	*	*
21.	18.	22.	31.	40.	57.	*	*	*	*	*	*	*
22.	18.	22.	30.	39.	54.	*	*	*	*	*	*	*
23.	18.	22.	30.	38.	52.	*	*	*	*	*	*	*
24.	18.	22.	29.	37.	50.	*	*	*	*	*	*	*
25.	18.	22.	29.	36.	48.	*	*	*	*	*	*	*
26.	17.	22.	28.	36.	46.	62.	*	*	*	*	*	*
27.	17.	23.	28.	36.	45.	60.	*	*	*	*	*	*
28.	17.	23.	28.	35.	43.	58.	*	*	*	*	*	*
29.	17.	23.	28.	35.	43.	56.	*	*	*	*	*	*
30.	17.	23.	27.	35.	42.	54.	*	*	*	*	*	*

Drag Distance (Feet)

Anchor: Stato Anchor with Stabilizers and Flukes Fixed at approximately 50°
Seafloor Type: Mud

Horizontal Design Load (Kips)

Anchor Weight (Kips)	25.	50.	75.	100.	125.	150.	175.	200.	225.	250.	275.	300.
5.	4.	23.	64.	158.	*	*	*	*	*	*	*	*
6.	3.	15.	45.	96.	236.	*	*	*	*	*	*	*
7.	2.	10.	34.	68.	127.	322.	*	*	*	*	*	*
8.	2.	7.	25.	52.	93.	168.	398.	*	*	*	*	*
9.	1.	6.	18.	43.	72.	120.	205.	*	*	*	*	*
10.	1.	6.	14.	35.	57.	94.	148.	280.	*	*	*	*
11.	1.	5.	11.	27.	50.	76.	118.	183.	352.	*	*	*
12.	0.	4.	9.	21.	43.	63.	96.	140.	216.	418.	*	*
13.	0.	4.	8.	18.	36.	56.	81.	118.	169.	266.	*	*
14.	0.	3.	7.	15.	29.	50.	68.	99.	138.	200.	333.	*
15.	0.	3.	7.	13.	24.	43.	61.	85.	118.	158.	229.	395.
16.	0.	3.	6.	10.	21.	37.	55.	73.	102.	138.	188.	265.
17.	0.	2.	6.	10.	18.	32.	50.	66.	90.	120.	156.	216.
18.	0.	2.	5.	9.	16.	27.	44.	61.	78.	106.	138.	179.
19.	0.	2.	5.	8.	14.	24.	39.	56.	71.	94.	122.	156.
20.	0.	2.	5.	8.	12.	21.	34.	51.	66.	83.	109.	139.
21.	0.	1.	4.	7.	11.	19.	30.	46.	61.	75.	98.	124.
22.	0.	1.	4.	7.	10.	17.	26.	41.	56.	70.	88.	112.
23.	0.	1.	4.	7.	10.	15.	24.	36.	52.	66.	79.	102.
24.	0.	1.	3.	6.	9.	13.	22.	32.	47.	61.	74.	92.
25.	0.	1.	3.	6.	9.	12.	20.	28.	43.	57.	70.	83.
26.	0.	0.	3.	5.	8.	11.	18.	26.	39.	53.	66.	78.
27.	0.	0.	2.	5.	8.	11.	16.	24.	35.	49.	62.	74.
28.	0.	0.	2.	5.	8.	10.	15.	22.	31.	44.	58.	70.
29.	0.	0.	2.	5.	7.	10.	13.	21.	28.	41.	54.	66.
30.	0.	0.	2.	4.	7.	10.	12.	19.	26.	37.	50.	63.

Drag Distance (Feet)

Anchor: Stato Anchor with Stabilizers and Flukes Fixed at approximately 30°
Seafloor Type: Sand

Horizontal Design Load (Kips)

Anchor Weight (Kips)	25.	50.	75.	100.	125.	150.	175.	200.	225.	250.	275.	300.
5.	15.	21.	28.	39.	55.	*	*	*	*	*	*	*
6.	15.	20.	27.	36.	45.	63.	*	*	*	*	*	*
7.	15.	20.	27.	32.	42.	51.	77.	*	*	*	*	*
8.	15.	20.	26.	30.	39.	47.	59.	91.	*	*	*	*
9.	15.	20.	25.	30.	36.	45.	53.	67.	103.	*	*	*
10.	15.	21.	24.	30.	33.	42.	50.	58.	74.	116.	*	*
11.	16.	20.	24.	29.	33.	40.	47.	55.	65.	88.	*	*
12.	16.	20.	24.	28.	33.	37.	45.	52.	59.	72.	101.	*
13.	16.	20.	24.	28.	33.	36.	43.	50.	57.	65.	79.	114.
14.	17.	20.	24.	27.	32.	36.	41.	48.	55.	61.	72.	91.
15.	17.	20.	24.	27.	32.	36.	39.	46.	53.	59.	65.	78.
16.	17.	20.	24.	27.	31.	36.	38.	44.	51.	57.	63.	72.
17.	17.	20.	24.	27.	31.	35.	38.	42.	49.	55.	61.	67.
18.	18.	20.	24.	27.	30.	35.	38.	41.	47.	54.	59.	65.
19.	18.	20.	24.	27.	30.	34.	38.	41.	46.	52.	58.	63.
20.	18.	20.	24.	27.	30.	34.	38.	40.	44.	50.	56.	62.
21.	18.	20.	24.	27.	30.	33.	37.	40.	43.	49.	55.	60.
22.	19.	21.	24.	27.	30.	33.	37.	40.	43.	47.	53.	59.
23.	19.	21.	24.	27.	30.	32.	36.	40.	43.	46.	52.	57.
24.	19.	21.	24.	27.	30.	32.	36.	40.	42.	45.	50.	56.
25.	19.	21.	24.	27.	30.	32.	36.	40.	42.	45.	49.	54.
26.	19.	21.	24.	27.	30.	32.	35.	39.	42.	44.	47.	53.
27.	20.	21.	24.	27.	30.	32.	35.	39.	42.	44.	46.	52.
28.	20.	21.	24.	27.	30.	32.	35.	38.	42.	44.	46.	50.
29.	20.	22.	24.	27.	30.	32.	34.	38.	41.	44.	46.	49.
30.	20.	22.	24.	27.	30.	32.	34.	37.	41.	44.	46.	48.

Drag Distance (Feet)

*Exceeds anchor ultimate holding capacity

- Tandem Anchor Systems. The predicted anchor drag distance for tandem anchor systems can be obtained from Table 3-5.

Table 3-5 Predicted Tandem Anchor Drag Distances

Anchor: Tandem Stockless Anchors with Stabilizers and Flukes Fixed at approximately 45°
Seafloor Type: Mud

Anchor Weight (Kips)	\multicolumn{12}{c}{Horizontal Design Load (Kips)}
	25.	50.	75.	100.	125.	150.	175.	200.	225.	250.	275.	300.
5.	6.	*	*	*	*	*	*	*	*	*	*	*
6.	4.	54.	*	*	*	*	*	*	*	*	*	*
7.	4.	32.	*	*	*	*	*	*	*	*	*	*
8.	3.	21.	*	*	*	*	*	*	*	*	*	*
9.	3.	13.	70.	*	*	*	*	*	*	*	*	*
10.	2.	9.	47.	*	*	*	*	*	*	*	*	*
11.	2.	6.	35.	183.	*	*	*	*	*	*	*	*
12.	1.	6.	27.	99.	*	*	*	*	*	*	*	*
13.	1.	6.	21.	64.	*	*	*	*	*	*	*	*
14.	0.	5.	15.	47.	200.	*	*	*	*	*	*	*
15.	0.	5.	12.	40.	128.	*	*	*	*	*	*	*
16.	0.	5.	10.	33.	79.	*	*	*	*	*	*	*
17.	0.	4.	7.	27.	63.	*	*	*	*	*	*	*
18.	0.	4.	7.	22.	50.	154.	*	*	*	*	*	*
19.	0.	4.	7.	17.	44.	95.	*	*	*	*	*	*
20.	0.	3.	7.	14.	38.	77.	*	*	*	*	*	*
21.	0.	3.	6.	12.	33.	64.	179.	*	*	*	*	*
22.	0.	2.	6.	10.	28.	53.	124.	*	*	*	*	*
23.	0.	2.	6.	8.	23.	48.	90.	*	*	*	*	*
24.	0.	2.	6.	8.	19.	42.	77.	202.	*	*	*	*
25.	0.	1.	5.	8.	17.	37.	65.	152.	*	*	*	*
26.	0.	1.	5.	8.	15.	33.	56.	104.	*	*	*	*
27.	0.	1.	5.	7.	13.	29.	51.	89.	224.	*	*	*
28.	0.	1.	5.	7.	11.	25.	46.	78.	177.	*	*	*
29.	0.	0.	4.	7.	10.	21.	42.	68.	132.	*	*	*
30.	0.	0.	4.	7.	9.	19.	38.	59.	101.	245.	*	*

Drag Distance (Feet)

Anchor: Tandem Stockless Anchors with Stabilizers and Flukes Fixed at approximately 36°
Seafloor Type: Sand

Anchor Weight (Kips)	\multicolumn{12}{c}{Horizontal Design Load (Kips)}
	25.	50.	75.	100.	125.	150.	175.	200.	225.	250.	275.	300.
5.	13.	20.	32.	*	*	*	*	*	*	*	*	*
6.	13.	19.	27.	*	*	*	*	*	*	*	*	*
7.	13.	18.	25.	37.	*	*	*	*	*	*	*	*
8.	13.	17.	24.	33.	*	*	*	*	*	*	*	*
9.	13.	17.	23.	29.	42.	*	*	*	*	*	*	*
10.	13.	17.	22.	28.	38.	*	*	*	*	*	*	*
11.	13.	17.	21.	27.	34.	46.	*	*	*	*	*	*
12.	13.	17.	21.	26.	32.	43.	*	*	*	*	*	*
13.	13.	17.	20.	26.	31.	39.	*	*	*	*	*	*
14.	13.	17.	20.	25.	30.	37.	47.	*	*	*	*	*
15.	13.	17.	20.	24.	29.	35.	44.	*	*	*	*	*
16.	13.	17.	20.	24.	29.	34.	41.	52.	*	*	*	*
17.	13.	17.	20.	23.	28.	33.	39.	49.	*	*	*	*
18.	13.	17.	20.	23.	28.	32.	37.	46.	*	*	*	*
19.	13.	18.	20.	23.	27.	32.	36.	44.	53.	*	*	*
20.	13.	18.	20.	23.	27.	31.	35.	41.	50.	*	*	*
21.	13.	18.	20.	22.	26.	31.	35.	40.	48.	57.	*	*
22.	13.	18.	20.	22.	26.	30.	34.	39.	46.	54.	*	*
23.	13.	18.	20.	22.	26.	30.	34.	38.	44.	52.	*	*
24.	14.	18.	20.	22.	25.	29.	33.	37.	42.	50.	58.	*
25.	14.	18.	20.	22.	25.	29.	33.	36.	41.	48.	56.	*
26.	14.	17.	20.	22.	25.	28.	32.	36.	40.	46.	54.	62.
27.	14.	17.	20.	23.	25.	28.	32.	36.	40.	45.	52.	60.
28.	14.	17.	21.	23.	25.	28.	32.	35.	39.	43.	50.	58.
29.	14.	17.	21.	23.	25.	28.	31.	35.	38.	43.	49.	56.
30.	14.	17.	21.	23.	25.	27.	31.	35.	38.	42.	47.	54.

Drag Distance (Feet)

Anchor: Tandem Stato Anchors with Stabilizers and Flukes Fixed at approximately 50°
Seafloor Type: Mud

Anchor Weight (Kips)	\multicolumn{12}{c}{Horizontal Design Load (Kips)}
	25.	50.	75.	100.	125.	150.	175.	200.	225.	250.	275.	300.
5.	.	4.	10.	23.	41.	64.	99.	158.	324.	*	*	*
6.	0.	3.	6.	15.	29.	45.	66.	96.	140.	236.	*	*
7.	0.	2.	5.	10.	19.	34.	48.	68.	94.	127.	182.	322.
8.	0.	2.	5.	7.	14.	25.	39.	52.	70.	93.	122.	168.
9.	0.	1.	4.	6.	11.	18.	30.	43.	55.	72.	93.	120.
10.	0.	1.	3.	6.	8.	14.	22.	35.	46.	57.	74.	94.
11.	0.	1.	3.	5.	7.	11.	18.	27.	39.	50.	60.	76.
12.	0.	0.	2.	4.	7.	9.	15.	21.	32.	43.	53.	63.
13.	0.	0.	2.	4.	6.	8.	12.	18.	25.	36.	46.	56.
14.	0.	0.	1.	3.	5.	7.	10.	15.	21.	29.	40.	50.
15.	0.	0.	1.	3.	5.	7.	9.	13.	18.	24.	34.	43.
16.	0.	0.	1.	3.	4.	6.	8.	10.	16.	21.	28.	37.
17.	0.	0.	0.	2.	4.	6.	8.	10.	13.	18.	24.	32.
18.	0.	0.	0.	2.	4.	5.	7.	9.	11.	16.	21.	27.
19.	0.	0.	0.	2.	3.	5.	7.	8.	10.	14.	19.	24.
20.	0.	0.	0.	1.	3.	5.	6.	8.	10.	12.	17.	21.
21.	0.	0.	0.	1.	3.	4.	6.	7.	9.	11.	15.	19.
22.	0.	0.	0.	1.	2.	4.	5.	7.	9.	10.	13.	17.
23.	0.	0.	0.	1.	2.	4.	5.	7.	8.	10.	11.	15.
24.	0.	0.	0.	1.	2.	3.	5.	6.	8.	9.	11.	13.
25.	0.	0.	0.	0.	1.	3.	4.	6.	7.	9.	10.	12.
26.	0.	0.	0.	0.	1.	3.	4.	5.	7.	8.	10.	11.
27.	0.	0.	0.	0.	1.	2.	4.	5.	7.	8.	9.	11.
28.	0.	0.	0.	0.	1.	2.	4.	5.	6.	8.	9.	10.
29.	0.	0.	0.	0.	1.	2.	3.	5.	6.	7.	9.	10.
30.	0.	0.	0.	0.	1.	2.	3.	4.	6.	7.	8.	10.

Drag Distance (Feet)

Anchor: Tandem Stato Anchor with Stabilizers and Flukes Fixed at approximately 30°
Seafloor Type: Sand

Anchor Weight (Kips)	\multicolumn{12}{c}{Horizontal Design Load (Kips)}
	25.	50.	75.	100.	125.	150.	175.	200.	225.	250.	275.	300.
5.	12.	15.	18.	21.	25.	28.	34.	39.	45.	55.	78.	*
6.	13.	15.	18.	20.	24.	27.	30.	36.	41.	45.	52.	63.
7.	13.	15.	18.	20.	23.	27.	29.	32.	37.	42.	46.	51.
8.	13.	15.	18.	20.	22.	26.	28.	30.	34.	39.	43.	47.
9.	14.	15.	18.	20.	22.	25.	28.	30.	32.	36.	40.	45.
10.	14.	15.	18.	20.	22.	24.	27.	30.	32.	35.	38.	42.
11.	15.	16.	18.	20.	22.	24.	26.	29.	32.	33.	35.	40.
12.	15.	16.	18.	20.	22.	24.	26.	28.	31.	33.	35.	37.
13.	15.	16.	18.	20.	23.	24.	26.	28.	30.	33.	35.	36.
14.	16.	17.	18.	20.	23.	24.	26.	27.	30.	32.	34.	36.
15.	16.	17.	18.	20.	22.	24.	26.	27.	29.	32.	34.	36.
16.	16.	17.	18.	20.	22.	24.	26.	27.	29.	31.	34.	36.
17.	16.	17.	18.	20.	22.	24.	26.	27.	28.	31.	33.	35.
18.	17.	18.	19.	20.	22.	24.	26.	27.	28.	30.	32.	35.
19.	17.	18.	19.	20.	22.	24.	26.	27.	28.	30.	32.	34.
20.	17.	18.	19.	20.	22.	24.	26.	27.	28.	30.	32.	34.
21.	17.	18.	19.	20.	22.	24.	26.	27.	29.	30.	32.	34.
22.	18.	19.	19.	21.	22.	24.	26.	27.	29.	30.	31.	33.
23.	18.	19.	20.	21.	22.	24.	26.	27.	29.	30.	31.	32.
24.	18.	19.	20.	21.	22.	24.	26.	27.	29.	30.	31.	32.
25.	18.	19.	20.	21.	22.	24.	26.	27.	29.	30.	31.	32.
26.	19.	19.	20.	21.	22.	24.	26.	27.	29.	30.	31.	32.
27.	19.	20.	20.	21.	22.	24.	26.	27.	29.	30.	31.	32.
28.	19.	20.	21.	21.	22.	24.	26.	27.	29.	30.	31.	32.
29.	19.	20.	21.	22.	23.	24.	26.	27.	29.	30.	31.	32.
30.	19.	20.	21.	22.	23.	24.	26.	27.	29.	30.	31.	32.

Drag Distance (Feet)

*Exceeds anchor system ultimate holding capacity

While setting the anchor, it is desirable to monitor the loads. A currently calibrated dynamometer should be placed in line to measure the load.

It is expected that the desired holding capacity can be achieved during setting and that the anchor will be within the 40' long by 20' wide allowable anchor area (see Figure 3-7). If after setting, the desired load is not obtained, or if the anchor drags outside the tolerance box, then the anchor must be repositioned and the pull test repeated.

Figure 3-7 Allowable Anchor Area

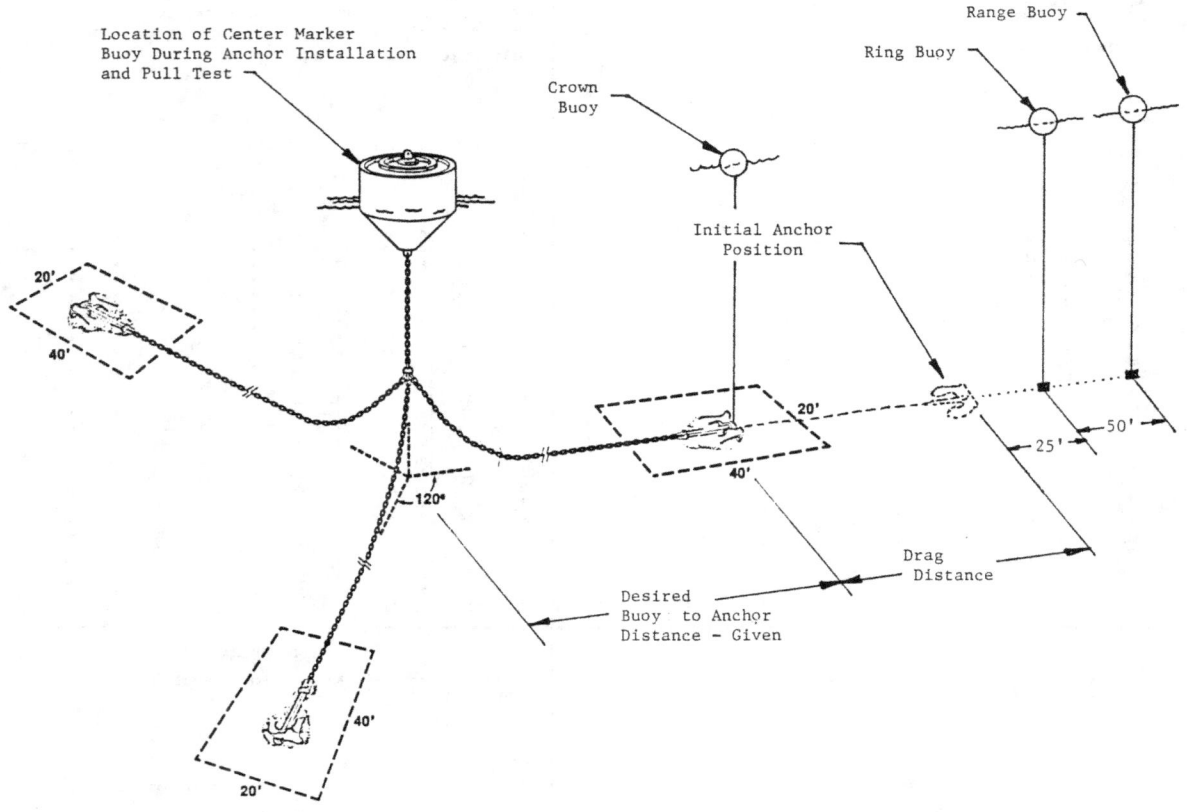

3.3.7.2. Pull Testing

After the anchor is set, the pull test should be conducted. It is important that a length of chain equal to the total length of riser and ground leg be used for the test. Use of shorter chain lengths will create an uplift force on the anchor.

Each anchor leg should be pull tested independently. The vessel performing the pull should gradually build up to the proof test load. Increase the load in 10 kip increments up to the required proof test load. After each 10 kip increase allow the dynamometer reading to stabilize. Once the required pull test load is reached, allow the dynamometer reading to stabilize, and then hold the pull test load for 3 minutes.

1. Anchor Leg Adequate. If the desired load is obtained and the anchor is positioned within the tolerance box, the anchor leg is adequate.

2. Anchor Leg Inadequate. If either the pull test load is not achieved or the anchor is out of position, the anchor must be repositioned and the pull test repeated.

3. Pull test not achieved.

4. If the pull test load cannot be achieved, the anchor system design may not be correct. Several options exist for increasing the anchors capacity:

    - Changing the anchor fluke angle. Usually the fluke is fixed fully open for muds and partially open (approximately 35°) for sands.

- When deploying anchors in silt or clay (muds) it is desirable to allow the anchors to soak. "Soaking" of an anchor is the practice of allowing a newly embedded anchor to rest for a period of time, typically 24 hours, before applying the required proof load.
- Jetting. If a drag anchor does not bury to a sufficient depth to develop the required capacity, it may be possible to use divers to jet the anchor to the required depth. The anchor should be jetted to a depth equal to or greater than the length of the anchor fluke.

5. Anchor out of position.

6. If the anchor is not within the tolerance box, reposition the anchor accordingly, reset and repeat the pull test.[1]

3.4. Chain And Fittings

Mooring chain not only secures the buoy in a predetermined position but its weight also serves to absorb energy caused by the dynamic motion and moored vessels. Chain is usually manufactured in 15-fathom (90', 27.4-meter) lengths, called shots. The chain diameter required for a particular mooring will vary depending on the anticipated maximum load. Mooring chain links normally have center crossbars, called studs, to retain the original shape of the link and to prevent the chain from kinking when it is piled in a heap. The wire diameter of fleet mooring chain links normally varies from 1-3/4 to 4 inches.

Chain is often used in fleet moorings because chain:

- Is easy to terminate
- Can easily be lengthened or shortened
- Is durable
- Is easy to inspect
- Is easy to provide cathodic protection
- Has extensive experience
- Is available
- Is cost effective
- Provides catenary effects

3.4.1. Types of Chain

Chain currently manufactured consists of cast, Di-Loc, or flash butt-welded links. Figure 3-8 contains schematic drawings of these three types of chain links.

---

[1] Ensure that the anchor is not recovered and reset in the furrow or disturbed bottom area caused by the initial pull test.

Figure 3-8 Typical Chain Links

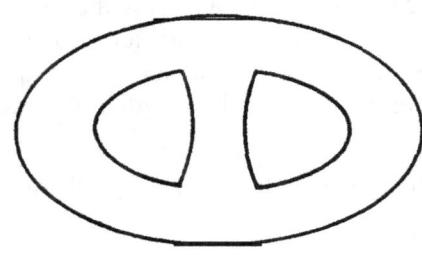

A- CAST CHAIN LINK

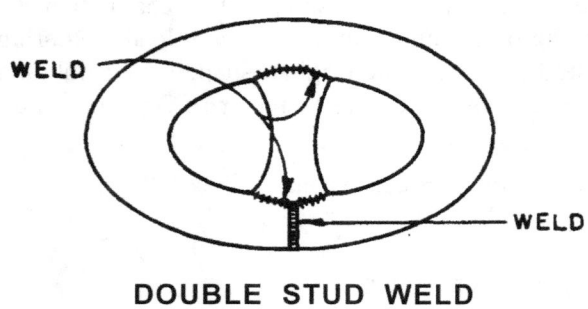

DOUBLE STUD WELD

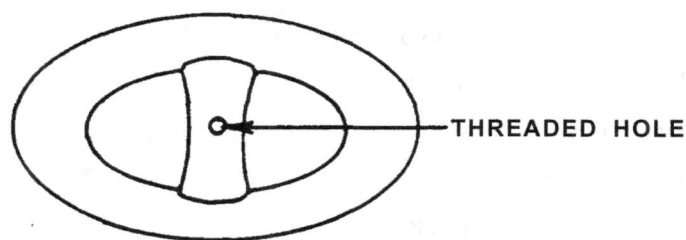

FM3 LINK WITH PRESSED-IN STUD

B-FLASH BUTT-WELDED CHAIN LINKS

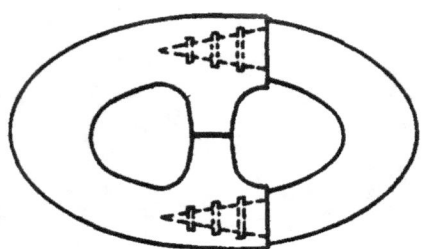

C- DILOK CHAIN LINK

3.4.1.1. Cast Chain

A distinctive feature of this type of chain is that the studs are solid and cast integrally with the links. Cast chain is also made to commercial link standard dimensions.

3.4.1.2. Di-Loc Chain

A Di-Loc link, which is chiefly used in a ship's anchor chain, and not fleet moorings, is made of two forged U-shaped pieces. A forged stem of one piece (the male) contains a series of parallel serrations giving an appearance of concentric rings. The other socket piece (the female) has holes at each end. In joining the two pieces to form a link, the pierced socket section is heated, then the stems of the other section are inserted into the holes. The socket section is then forged with a drop hammer, forcing its material around the indentations in the other piece's stems, while both pieces are held in die blocks (see Figure 3-9). Each chain link is made to commercial link standard dimensions, which are 6 wire diameters long by 3.6 wire diameters wide.

Figure 3-9 Manufacturing Di-Loc Chain

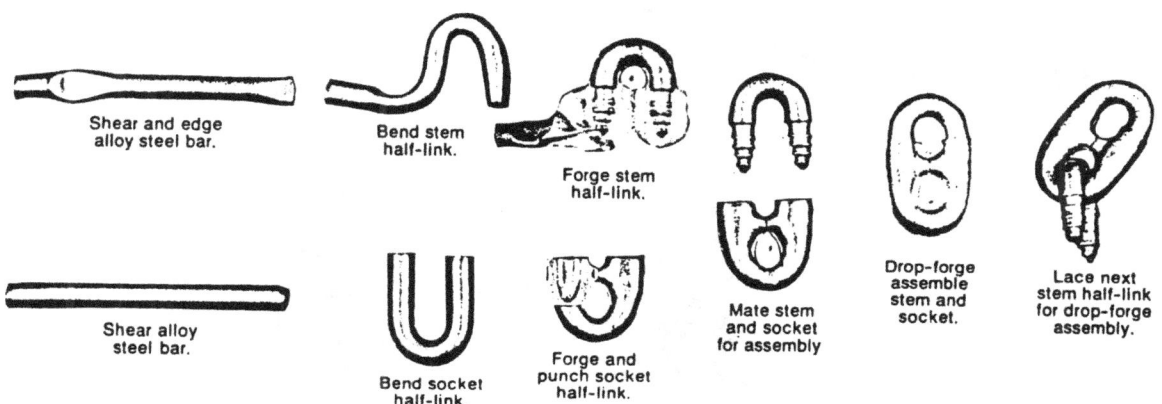

Di-Loc is not recommended for long-term in-water use, because water may seep in between the male and female parts. The resulting corrosion is difficult to inspect.

3.4.1.3. Stud Link Chain

Bending a heated rod into an open chain link shape and flash welding the two butt ends forms each link of this chain. While the link is still hot, a stud is inserted and the link is pressed on both sides to secure the stud. After cooling, one or both ends of the stud may be welded to the link (see Figure 3-10).

Figure 3-10 Producing Flash Butt-Welded Chain

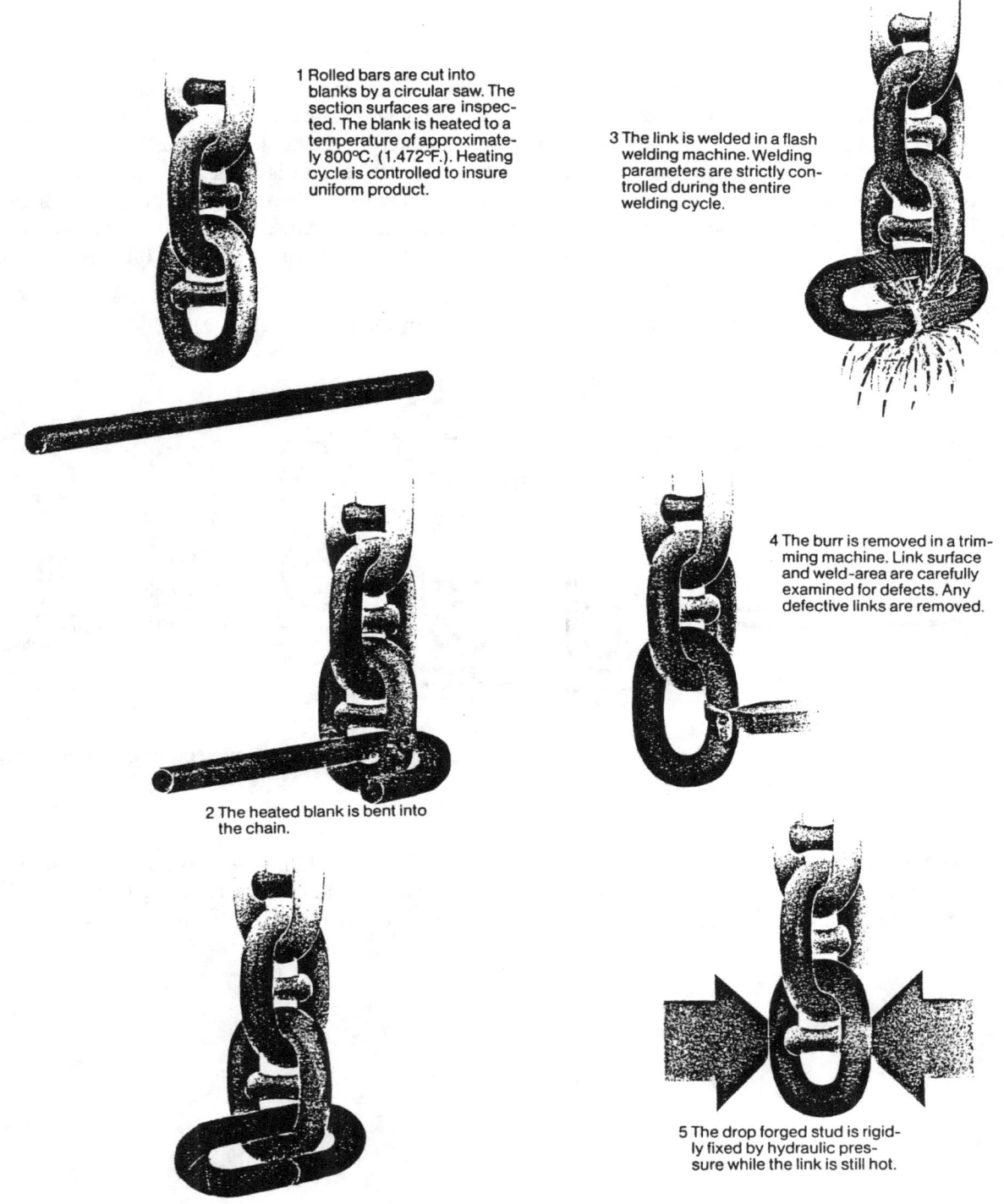

3.4.2. Grades of Chain

The properties of the steel used in the manufacture of chain for the Navy follow specifications that are similar to the rules followed in the manufacture of commercial chain. For commonly available chain, there are three different material qualities designated: Grades 1, 2, and 3, respectively. Grade 3 has the highest strength of the three and is the grade currently being procured by the Navy for fleet mooring use and is singularly suited for application of a cathodic protection system. Grade 3 stud link chain is also specifically designed for long term in-water use (Naval Facilities Engineering Service Center (NFESC), FPO-1-89(PD1), *Purchase Description for Fleet Mooring Chain and Accessories*). This chain is designated as FM3. Properties of FM3 carried in stock are shown in Table 3-6. Catholic protection for chain is discussed in 14.3.

Table 3-6 FM3 Mooring Chain Characteristics

NOMINAL SIZE (inches)	1.75	2	2.25	2.5	2.75	3.5	4
NUMBER OF LINKS PER SHOT	153	133	119	107	97	77	67
LINK LENGTH (inches)	10.6	12.2	13.7	15.2	16.7	21.3	24.3
WEIGHT PER SHOT IN AIR (lbf)	2525	3276	4143	5138	6250	10258	13358
WEIGHT PER LINK IN AIR (lbf)	16.5	24.6	34.8	48	64.4	133.2	199.4
WEIGHT PER FOOT SUB. (lbs/ft)	26.2	33.9	42.6	52.7	63.8	104.1	135.2
BREAKING STRENGTH (kips)	352	454	570	692	826	1285	1632
WORKING STRENGTH (FS=3) (kips)	117.2	151.2	189.8	230.4	275.1	427.9	543.5

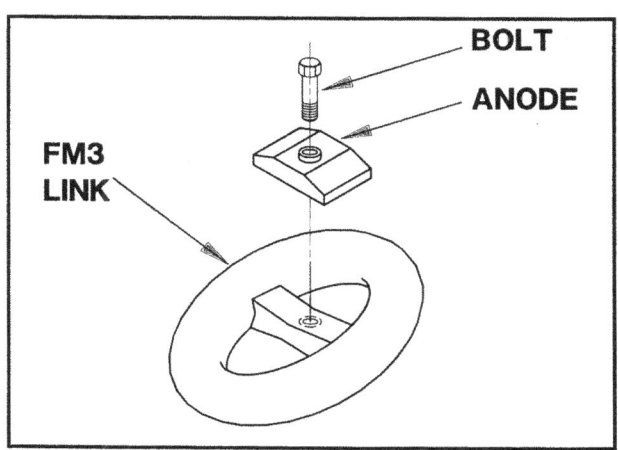

3.4.3. Chain Assembly Accessories

A number of other accessories are used with chain. An overview of some of the various types is shown in Figure 3-11. For example, shots of chain are connected together with chain joining links. Anchor joining links are used to connect chain to anchors. Ground rings provide an attachment point for multiple chains. Buoy swivels are used to connect chain to buoys.

Figure 3-11 Chain Fittings

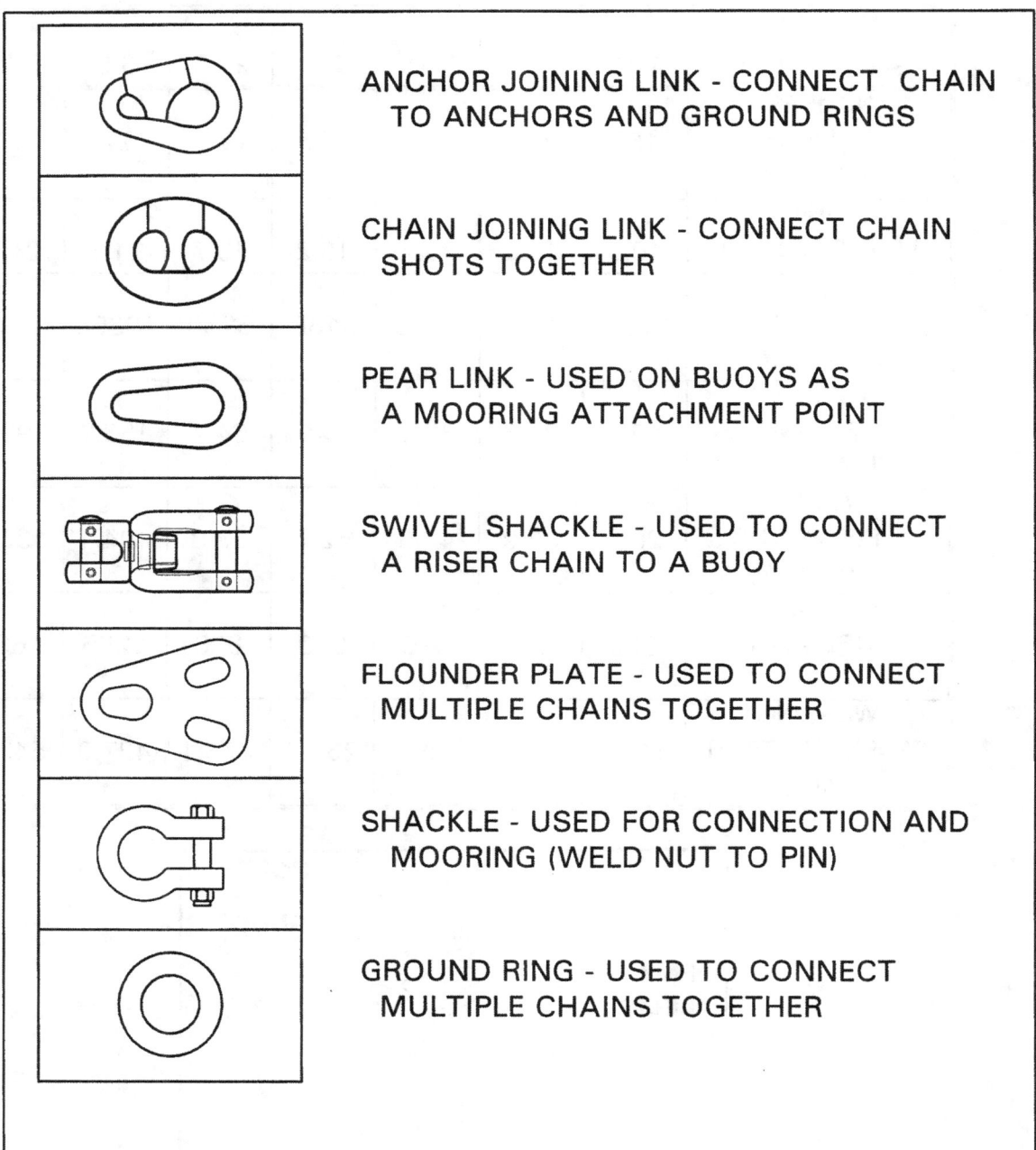

3.4.3.1. *Chain Joining Link*

This link is also called a detachable joining link, a chain-connecting link, a detachable link, a lugless joining shackle, a detachable connecting link, or a Kenter shackle. This link is used to connect two shots of chain, two A-links, a swivel and an A-link, etc. Its wire diameter should be the same size as the mooring components it connects. The two chain joining links commonly found in the Fleet Mooring Inventory are the Baldt and Kenter types, named after the companies that developed their designs. Figure 3-12 and Figure 3-13 provide sketches of and dimensional information concerning the Baldt and Kenter designs, respectively, and marking information for the Ketner design.

Figure 3-12 Chain Joining Link-Baldt Type

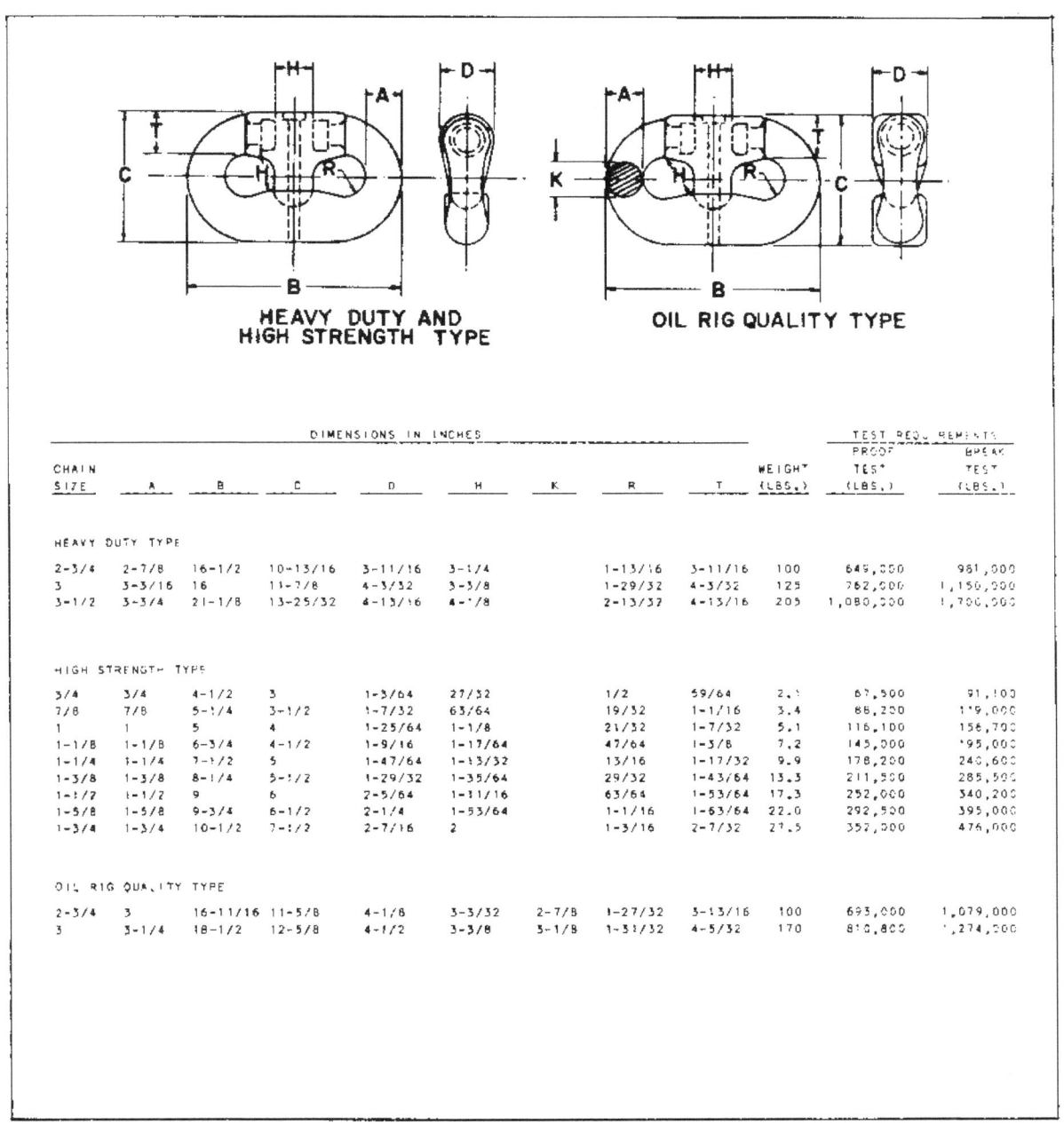

Figure 3-13 Chain Joining Link-Kenter Type

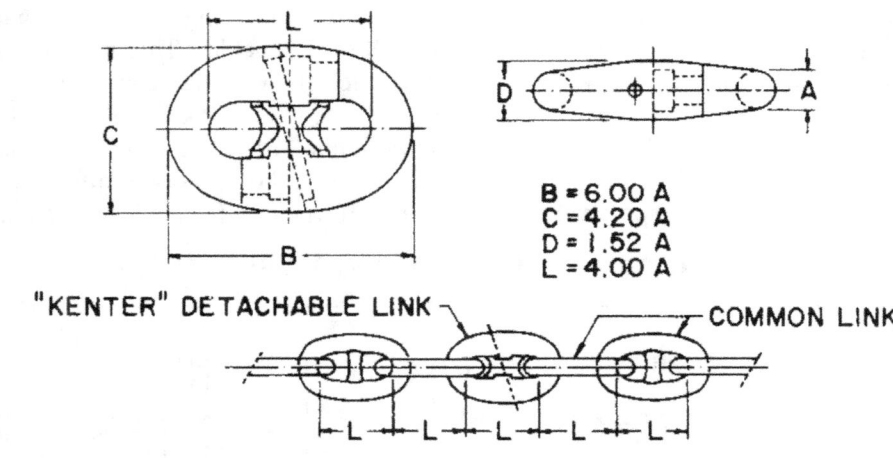

CHAIN SIZE	A	B	C	D	L	WEIGHT (LBS.)
3/4	3/4	4-1/2	3-1/8	1-1/8	3	2.3
13/16	13/16	4-13/16	3-3/8	1-1/4	3-1/4	2.9
7/8	7/8	5-3/16	3-5/8	1-5/16	3-1/2	3.5
15/16	15/16	5-11/16	4	1-7/16	3-3/4	4.6
1	1	6	4-3/16	1-9/16	4	5.7
1-1/16	1-1/16	6-3/8	4-7/16	1-5/8	4-1/4	6.6
1-1/8	1-1/8	6-3/4	4-11/16	1-11/16	4-1/2	7.7
1-3/16	1-3/16	7-1/16	4-15/16	1-13/16	4-3/4	9.3
1-1/4	1-1/4	7-7/16	5-3/16	1-7/8	5	10.6
1-5/16	1-5/16	7-15/16	5-9/16	2	5-1/4	12.8
1-3/8	1-3/8	8-1/4	5-13/16	2-1/8	5-1/2	14.3
1-7/16	1-7/16	8-5/8	6-1/16	2-3/16	5-3/4	16.5
1-1/2	1-1/2	9	6-1/4	2-5/16	6	18.5
1-9/16	1-9/16	9-5/16	6-9/16	2-3/8	6-1/4	20.9
1-5/8	1-5/8	9-13/16	6-7/8	2-1/2	6-1/2	24.2
1-11/16	1-11/16	10-1/8	7-1/8	2-9/16	6-3/4	27.1
1-3/4	1-3/4	10-1/2	7-3/8	2-11/16	7	29.8
1-13/16	1-13/16	10-7/8	7-5/8	2-3/4	7-1/4	33.1
1-7/8	1-7/8	11-1/4	7-7/8	2-7/8	7-1/2	36.4
1-15/16	1-15/16	11-9/16	8-1/8	2-15/16	7-3/4	39.7
2	2	12-1/16	8-7/16	3-1/16	8	45.2
2-1/16	2-1/16	12-3/8	8-11/16	3-1/8	8-1/4	49.1
2-1/8	2-1/8	12-5/4	8-15/16	3-1/4	8-1/2	53.3
2-3/16	2-3/16	13-1/8	9-3/16	3-5/16	8-3/4	58.0
2-1/4	2-1/4	13-7/16	9-7/16	3-7/16	9	62.6
2-5/16	2-5/16	13-13/16	9-11/16	3-1/2	9-1/4	67.9
2-3/8	2-3/8	14-5/16	10	3-5/8	9-1/2	74.9
2-7/16	2-7/16	14-5/8	10-1/4	3-3/4	9-3/4	80.9
2-1/2	2-1/2	15	10-1/2	3-13/16	10	86.8
2-9/16	2-9/16	15-3/8	10-3/4	3-7/8	10-1/4	93.2
2-5/8	2-5/8	15-11/16	11	4	10-1/2	99.8
2-11/16	2-11/16	16-1/16	11-1/4	4-1/16	10-3/4	106.5
2-3/4	2-3/4	16-7/16	11-1/2	4-5/16	11	113.7
2-13/16	2-13/16	16-7/8	11-13/16	4-5/16	11-1/4	123.9
2-7/8	2-7/8	17-1/4	12-1/16	4-3/8	11-1/2	131.8

CHAIN SIZE	A	B	C	D	L	WEIGHT (LBS.)
2-15/16	2-15/16	17-5/8	12-5/16	4-7/16	11-3/4	140
3	3	17-15/16	12-9/16	4-9/16	12	149
3-1/16	3-1/16	18-5/16	12-13/16	4-5/8	12-1/4	158
3-1/8	3-1/8	18-3/4	13-1/8	4-3/4	12-1/2	170
3-3/16	3-3/16	19-1/8	13-3/8	4-7/8	12-3/4	180
3-1/4	3-1/4	19-1/2	13-5/8	4-15/16	13	190
3-5/16	3-5/16	19-13/16	13-7/8	5-1/16	13-1/4	201
3-3/8	3-3/8	20-5/16	14-1/16	5-3/16	13-1/2	215
3-7/16	3-7/16	20-11/16	14-7/16	5-1/4	13-3/4	227
3-1/2	3-1/2	14-3/4	5-5/16		14	238
3-9/16	3-9/16	21-3/8	14-15/16	5-7/16	14-1/4	251
3-5/8	3-5/8	22-7/16	15-3/16	5-1/2	14-1/2	267
3-11/16	3-11/16	22-1/16	15-7/16	5-5/8	14-3/4	276
3-3/4	3-3/4	22-7/16	15-11/16	5-11/16	15	289
3-13/16	3-13/16	22-15/16	16-1/16	5-13/16	15-1/4	309
3-7/8	3-7/8	23-1/4	16-1/4	5-7/8	15-1/2	324
3-15/16	3-15/16	23-5/8	16-9/16	6	15-3/4	339
4	4	24	16-15/16	6-1/16	16	355
4-1/8	4-1/8	24-13/16	17-3/8	6-5/16	16-1/2	395
4-1/4	4-1/4	25-1/2	17-7/8	6-7/16	17	428
4-3/8	4-3/8	26-1/4	18-3/8	6-5/8	17-1/2	465
4-1/2	4-1/2	27-1/16	18-15/16	6-7/8	18	509
4-5/8	4-5/8	27-3/4	19-7/16	7-1/16	18-1/2	551
4-3/4	4-3/4	28-7/16	19-15/16	7-3/16	19	595
4-7/8	4-7/8	29-5/16	20-1/2	7-7/16	19-1/2	648
5	5	30	21	7-5/8	20	696
5-1/8	5-1/8	30-11/16	21-1/2	7-13/16	20-1/2	747
5-1/4	5-1/4	31-9/16	22-1/16	8	21	809
5-3/8	5-3/8	32-1/4	22-9/16	8-3/16	21-1/2	864
5-1/2	5-1/2	33-1/16	23-1/8	8-3/8	22	932
5-5/8	5-5/8	33-3/4	23-5/8	8-9/16	22-1/2	994
5-3/4	5-3/4	34-1/2	24-1/8	8-3/4	23	1058
5-7/8	5-7/8	35-5/16	24-5/8	8-15/16	23-1/2	1133
6	6	36	25-3/16	9-1/8	24	1203

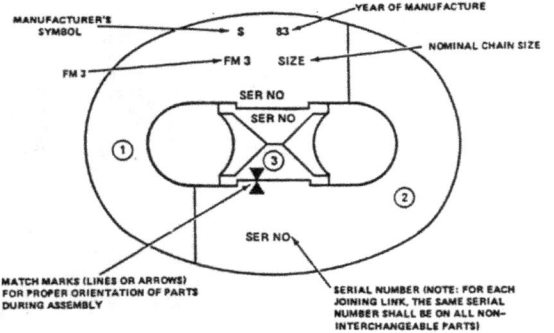

The Kenter-type joining link is of alloy steel an constructed in three parts, one of which is the stud. The two main parts and stud have matching numbers and arrow on the stud, which is lined up with an arrow on the main part for ease of assembly (see the bottom of Figure 3-13). The two main parts are attached to the ends of the chain in the vertical position and then fitted together; the stud slides in place and locks the whole link. The stud is secured by hammering a tapered pin into the hole drilled diagonally through all three parts of the joining link. This hole is tapered, and when the pin is driven home, a small conical recess, called the "Dovetail Chamber" is left clear above its head. A lead pellet is hammered broad end first into this chamber so as to fill it completely and thereby keep the pin in place. During the final stage of hammering the lead pellet into the Dovetail Chamber, precaution must be taken to prevent flat, small pieces of lead from flying off the joining link into the face or eyes. The assembly procedure is depicted in Figure 3-14.

Figure 3-14 Assembly of Ketner Chain Joining Link

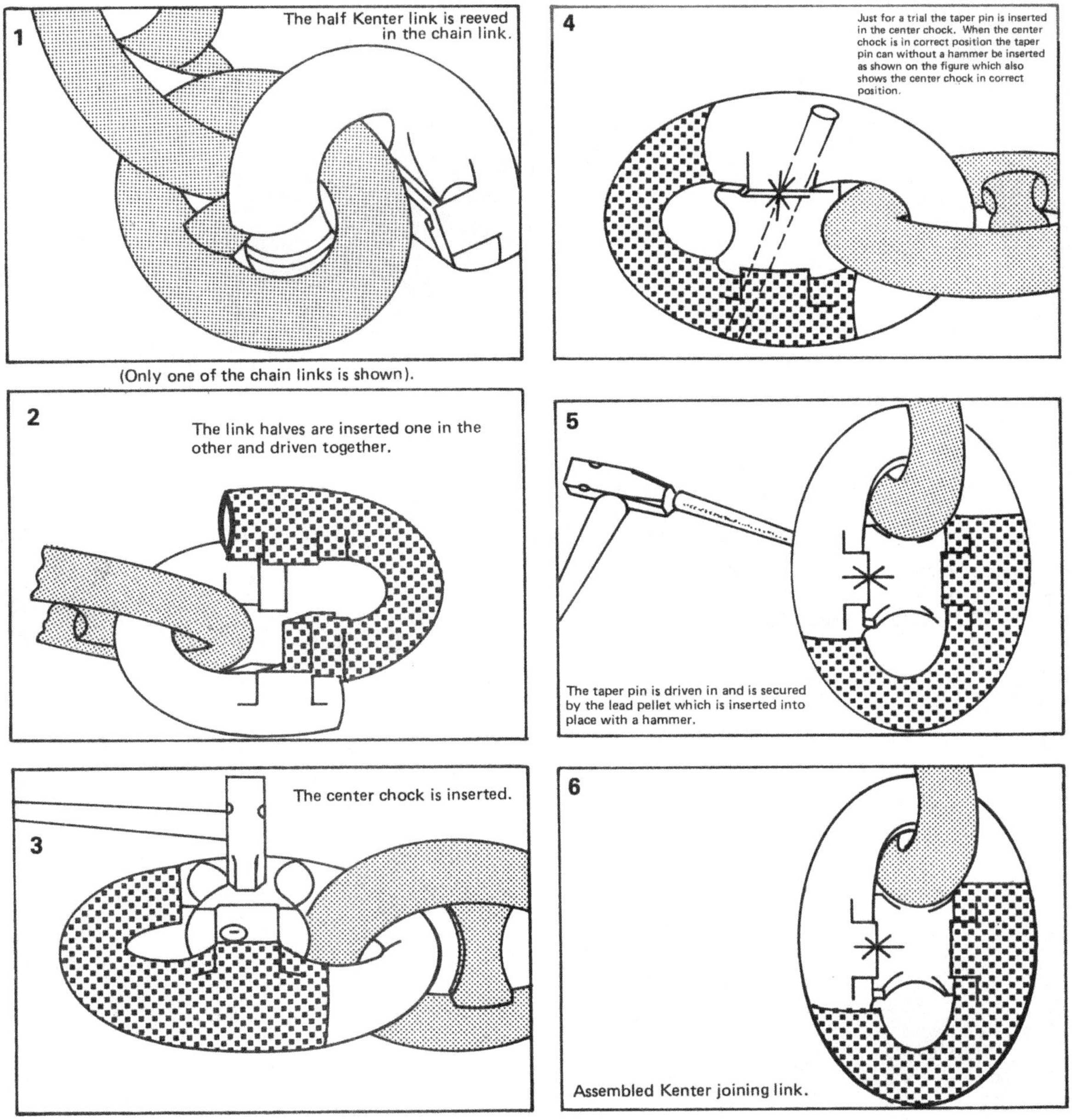

1. The half Kenter link is reeved in the chain link. (Only one of the chain links is shown).
2. The link halves are inserted one in the other and driven together.
3. The center chock is inserted.
4. Just for a trial the taper pin is inserted in the center chock. When the center chock is in correct position the taper pin can without a hammer be inserted as shown on the figure which also shows the center chock in correct position.
5. The taper pin is driven in and is secured by the lead pellet which is inserted into place with a hammer.
6. Assembled Kenter joining link.

Prior to assembly, the internal mating surfaces of a Kenter joining link should be coated with a molybdenum disulphide grease (MIL-G-23549) or an equivalent lithium based grease. When assembling and before inserting a lead pellet, any remaining lead in the Dovetail Chamber must be reamed out with a reamer tool. Failure to do this could result in the new lead pellet working out. After assembly the link is painted with anchor chain paint, MIL-P-24380 (NSN 8010-00-145-0332 and NSN 8010-00-145-0341 for 1 and 5 gallon cans respectively).

When disassembling a Kenter joining link, the locking pin is driven out with a "drift." To part the link, a top swage must always be used between the hammer and link. The swage is shaped to the curvature of the link so that machined surfaces are not damaged (See Figure 3-15).

Figure 3-15 A Top Swage and Reamer

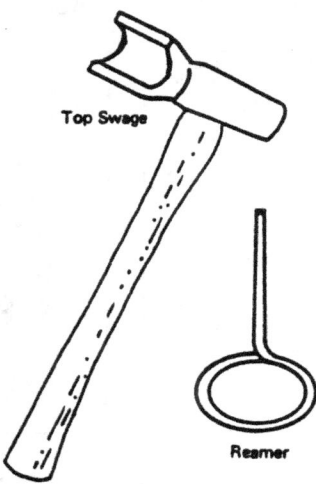

3.4.3.2.   Anchor Joining Link

This link is also called a detachable anchor-connecting link. This component joins the end of a chain composed of common links to any of various enlarged mooring components, such as a ground ring, a buoy padeye, an end link, or an anchor shackle. Figure 3-16 and Figure 3-17 provide sketches of and dimensional information for the Baldt and Kenter types of anchor joining links.

Figure 3-16 Anchor Joining Link-Baldt Type

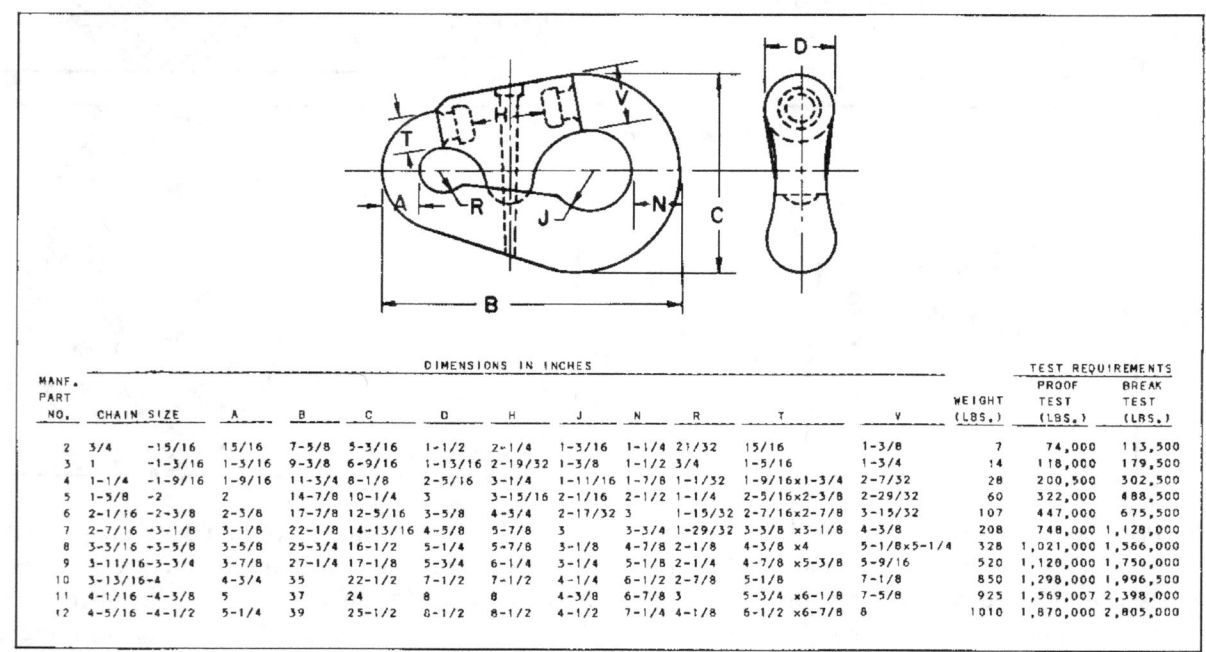

Figure 3-17 Anchor Joining Link-Kenter Type

TYPICAL ANCHOR JOINING LINK - KENTER TYPE

Nominal Diameter	A(max)	B(max)	G(min)	G(max)	H(min)	J(min)	J(max)	K(min)	K(max)	M(min)	N(min)	N(max)	P(min)	P(max)
1 1/4	1.88	1.56	0.75	1.00	1.47	1.50	2.06	1.25	1.56	1.50	1.50	1.75	2.12	2.75
1 3/4	2.63	2.19	1.05	1.40	2.06	2.10	2.89	1.75	2.19	2.10	2.19	2.45	2.98	3.85
2	3.00	2.50	1.20	1.60	2.36	2.40	3.30	2.00	2.50	2.40	2.50	2.80	3.40	4.40
2 1/4	3.38	2.81	1.35	1.80	2.65	2.70	3.71	2.25	2.81	2.70	2.81	3.15	3.83	4.95
2 1/2	3.75	3.13	1.50	2.00	2.95	3.00	4.13	2.50	3.13	3.00	3.13	3.50	4.25	5.50
2 3/4	4.13	3.44	1.65	2.20	3.24	3.30	4.54	2.75	3.44	3.30	3.44	3.85	4.68	6.05
3	4.50	3.75	1.80	2.40	3.54	3.60	4.95	3.00	3.75	3.60	3.75	4.20	5.10	6.60
3 1/2	5.25	4.38	2.10	2.80	4.13	4.20	5.78	3.50	4.38	4.20	4.38	4.90	5.95	7.70
4	6.00	5.00	2.40	3.20	4.72	5.00	6.50	4.00	5.00	4.80	5.00	5.60	6.80	8.80

(All Dimensions in Inches)

NOTE

THE SMALL ENDS OF ALL ANCHOR JOINING LINKS MUST BE COMPATIBLE WITH THE COMMON STUD LINK OF THE SAME NOMINAL CHAIN SIZE.

3.4.3.3. Shackles

Although there are many types of shackles available and in use throughout industry, there are four basic types of shackles used in fleet moorings:

- Joining Shackle. This shackle is also called a D-link, a D-shackle, or a joining shackle, "D" type, and used to connect lengths of chain. The shackle is similar in shape to an anchor joining shackle but smaller in size. Figure 3-18 provides a schematic of a typical joining shackle and a dimensional table of its various sizes.

Figure 3-18 Joining Shackle

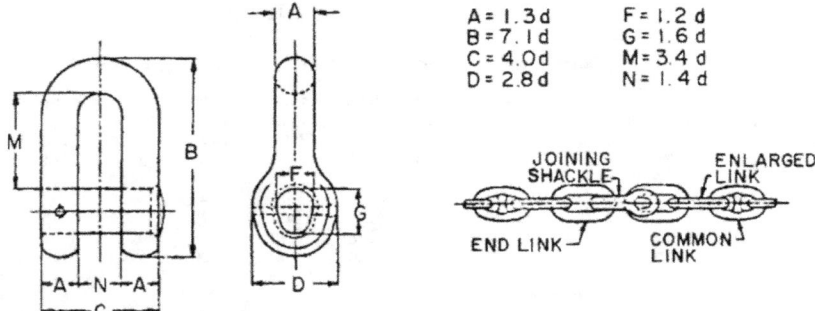

A = 1.3 d   F = 1.2 d
B = 7.1 d   G = 1.6 d
C = 4.0 d   M = 3.4 d
D = 2.8 d   N = 1.4 d

CHAIN SIZE (d)	A	B	C	D	F	G	M	N	WEIGHT (LBS.)
3/4	1	5-5/16	3	2-1/8	15/16	1-3/16	2-9/16	1-1/16	3.7
13/16	1-1/16	5-3/4	3-1/4	2-1/4	1	1-5/16	2-3/4	1-1/8	4.8
7/8	1-1/8	6-3/16	3-1/2	2-7/16	1-1/16	1-3/8	3	1-1/4	5.9
15/16	1-1/4	6-5/8	3-3/4	2-5/8	1-1/8	1-1/2	3-3/16	1-5/16	7.7
1	1-5/16	7-1/8	4	2-13/16	1-3/16	1-5/8	3-3/8	1-3/8	9.2
1-1/16	1-3/8	7-9/16	4-1/4	3	1-1/4	1-11/16	3-5/8	1-1/2	11.0
1-1/8	1-7/16	8	4-1/2	3-3/16	1-3/8	1-13/16	3-13/16	1-9/16	12.8
1-3/16	1-9/16	8-7/16	4-3/4	3-5/16	1-7/16	1-15/16	4-1/16	1-11/16	15.0
1-1/4	1-5/8	8-7/8	5	3-1/2	1-1/2	2	4-1/4	1-3/4	17.2
1-5/16	1-11/16	9-5/16	5-1/4	3-11/16	1-9/16	2-1/8	4-7/16	1-13/16	20.7
1-3/8	1-13/16	9-3/4	5-1/2	3-7/8	1-5/8	2-3/16	4-11/16	1-15/16	23.8
1-7/16	1-7/8	10-3/16	5-3/4	4	1-3/4	2-5/16	4-7/8	2	26.8
1-1/2	1-15/16	10-5/8	6	4-3/16	1-13/16	2-3/8	5-1/8	2-1/8	30.4
1-9/16	2-1/16	11-1/8	6-1/4	4-3/8	1-7/8	2-1/2	5-3/16	2-3/16	34.1
1-5/8	2-1/8	11-9/16	6-1/2	4-9/16	1-15/16	2-5/8	5-1/2	2-1/4	39.4
1-11/16	2-3/16	12	6-3/4	4-3/4	2	2-11/16	5-3/4	2-3/8	44.0
1-3/4	2-1/4	12-7/16	7	4-7/8	2-1/16	2-13/16	5-15/16	2-7/16	48.6
1-13/16	2-3/8	12-7/8	7-1/4	5-1/16	2-3/16	2-7/8	6-3/16	2-9/16	53.7
1-7/8	2-7/16	13-5/16	7-1/2	5-1/4	2-1/4	3	6-3/8	2-5/8	59.2
1-15/16	2-1/2	13-3/4	7-3/4	5-7/16	2-5/16	3-1/8	6-9/16	2-11/16	64.9
2	2-5/8	14-3/16	8	5-5/8	2-3/8	3-3/16	6-13/16	2-13/16	73.3
2-1/16	2-11/16	14-5/8	8-1/4	5-3/4	2-1/2	3-5/16	7	2-7/8	79.9
2-1/8	2-3/4	15-1/16	8-1/2	5-15/16	2-9/16	3-3/8	7-1/4	3	86.9
2-3/16	2-7/8	15-9/16	8-3/4	6-1/8	2-5/8	3-1/2	7-7/16	3-1/16	94.6
2-1/4	2-15/16	16	9	6-5/16	2-11/16	3-5/8	7-5/8	3-1/8	102.3
2-5/16	3	16-7/16	9-1/4	6-1/2	2-3/4	3-11/16	7-7/8	3-1/4	110.7
2-3/8	3-1/16	16-7/8	9-1/2	6-5/8	2-7/8	3-13/16	8-1/16	3-5/16	122.3
2-7/16	3-3/16	17-5/16	9-3/4	6-13/16	2-15/16	3-7/8	8-5/16	3-7/16	131.6
2-1/2	3-1/4	17-3/4	10	7	3	4	8-1/2	3-1/2	141.5
2-9/16	3-5/16	18-3/16	10-1/4	7-3/16	3-1/16	4-1/8	8-11/16	3-9/16	151.8
2-5/8	3-7/16	18-5/8	10-1/2	7-3/8	3-1/8	4-3/16	8-15/16	3-11/16	162.4
2-11/16	3-1/2	19-1/16	10-3/4	7-1/2	3-1/4	4-5/16	9-1/8	3-3/4	173.6
2-3/4	3-9/16	19-1/2	11	7-11/16	3-5/16	4-3/8	9-3/8	3-7/8	185.5
2-13/16	3-11/16	20	11-1/4	7-7/8	3-3/8	4-1/2	9-9/16	3-15/16	202.0
2-7/8	3-3/4	20-7/16	11-1/2	8-7/16	3-7/16	4-5/8	9-3/4	4	214.7
2-15/16	3-13/16	20-7/8	11-3/4	8-1/4	3-1/2	4-11/16	10	4-1/8	228
3	3-7/8	21-5/16	12	8-3/8	3-5/8	4-13/16	10-3/16	4-3/16	242
3-1/16	4	21-3/4	12-1/4	8-9/16	3-11/16	4-7/8	10-7/16	4-5/16	257
3-1/8	4-1/16	22-3/16	12-1/2	8-3/4	3-3/4	5	10-5/8	4-3/8	277
3-3/16	4-1/8	22-5/8	12-3/4	8-15/16	3-13/16	5-1/8	10-13/16	4-7/16	293
3-1/4	4-1/4	23-1/16	13	9-1/8	3-7/8	5-3/16	11-1/16	4-9/16	316
3-5/16	4-5/16	23-1/2	13-1/4	9-5/16	4	5-1/4	11-1/4	4-5/8	328
3-3/8	4-3/8	23-15/16	13-1/2	9-7/16	4-1/16	5-3/8	11-1/2	4-3/4	350
3-7/16	4-1/2	24-7/16	13-3/4	9-5/8	4-1/8	5-1/2	11-11/16	4-13/16	370
3-1/2	4-9/16	24-7/8	14	9-13/16	4-3/16	5-5/8	11-7/8	4-7/8	389
3-9/16	4-5/8	25-5/16	14-1/4	10	4-1/4	5-11/16	12-1/8	5	409
3-5/8	4-11/16	25-3/4	14-1/2	10-1/8	4-3/8	5-13/16	12-5/16	5-1/16	429
3-11/16	4-13/16	26-3/16	14-3/4	10-5/16	4-7/16	5-7/8	12-9/16	5-3/16	451
3-3/4	4-7/8	26-5/8	15	10-1/2	4-1/2	6	12-3/4	5-1/4	473
3-13/16	4-15/16	27-1/16	15-1/4	10-11/16	4-9/16	6-1/8	12-15/16	5-5/16	504
3-7/8	5-1/16	27-1/2	15-1/2	10-7/8	4-5/8	6-3/16	13-3/16	5-7/16	528
3-15/16	5-1/8	27-15/16	15-3/4	11	4-3/4	6-5/16	13-3/8	5-1/2	552
4	5-3/16	28-3/8	16	11-3/16	4-13/16	6-3/8	13-5/8	5-5/8	576
4-1/8	5-3/8	29-5/16	16-1/2	11-9/16	4-15/16	6-5/8	14	5-3/4	640
4-1/4	5-1/2	30-3/16	17	11-7/8	5-1/8	6-13/16	14-7/16	5-15/16	695
4-3/8	5-11/16	31-1/16	17-1/2	12-1/4	5-1/4	7	14-7/8	6-1/8	755
4-1/2	5-7/8	31-15/16	18	12-5/8	5-3/8	7-3/16	15-5/16	6-5/16	829
4-5/8	6	32-13/16	18-1/2	12-15/16	5-9/16	7-3/8	15-3/4	6-1/2	895
4-3/4	6-3/16	33-3/4	19	13-5/16	5-11/16	7-9/16	16-1/8	6-5/8	966
4-7/8	6-5/16	34-5/8	19-1/2	13-5/8	5-7/8	7-13/16	16-9/16	6-13/16	1054
5	6-1/2	35-1/2	20	14	6	8	17	7	1131
5-1/8	6-11/16	36-3/8	20-1/2	14-3/8	6-1/8	8-3/16	17-7/16	7-3/16	1212
5-1/4	6-13/16	37-1/4	21	14-11/16	6-5/16	8-3/8	17-7/8	7-3/8	1313
5-3/8	7	38-1/16	21-1/2	15-1/16	6-5/8	8-5/8	18-1/4	7-1/2	1404
5-1/2	7-1/8	39-1/16	22	15-3/8	6-5/8	8-13/16	18-11/16	7-11/16	1516
5-5/8	7-5/16	39-15/16	22-1/2	15-3/4	6-3/4	9	19-1/8	7-7/8	1615
5-3/4	7-1/2	40-13/16	23	16-1/8	6-7/8	9-3/16	19-9/15	8-1/16	1718
5-7/8	7-5/8	41-11/16	23-1/2	16-7/16	7-1/16	9-3/8	20	8-1/4	1846
6	7-13/16	42-5/8	24	16-13/16	7-3/16	9-5/8	20-3/8	8-3/8	1958

- Anchor Joining Shackle. This shackle is also called a bending shackle, an F-link, an F shackle, or an end shackle. It is an enlarged joining shackle used to connect lengths of chain which have larger sized chain end links.
- Buoy Shackle. This type of shackle is similar to the anchor joining shackle, except that it has a round pin. It is used as top jewelry. A schematic drawing of this shackle and its various dimensions is shown in Figure 3-19.

Figure 3-19 Buoy Shackle

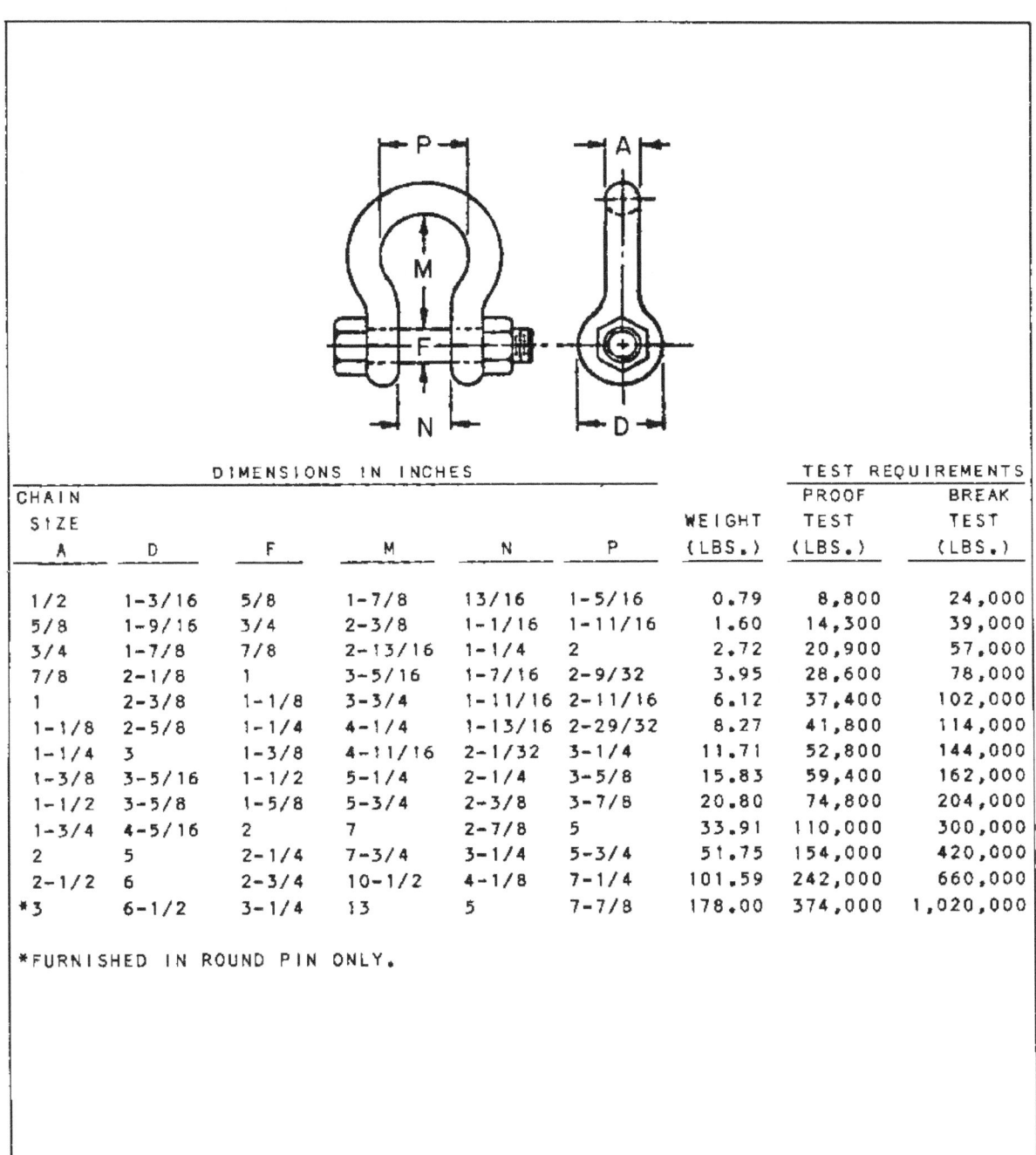

CHAIN SIZE A	D	F	M	N	P	WEIGHT (LBS.)	PROOF TEST (LBS.)	BREAK TEST (LBS.)
1/2	1-3/16	5/8	1-7/8	13/16	1-5/16	0.79	8,800	24,000
5/8	1-9/16	3/4	2-3/8	1-1/16	1-11/16	1.60	14,300	39,000
3/4	1-7/8	7/8	2-13/16	1-1/4	2	2.72	20,900	57,000
7/8	2-1/8	1	3-5/16	1-7/16	2-9/32	3.95	28,600	78,000
1	2-3/8	1-1/8	3-3/4	1-11/16	2-11/16	6.12	37,400	102,000
1-1/8	2-5/8	1-1/4	4-1/4	1-13/16	2-29/32	8.27	41,800	114,000
1-1/4	3	1-3/8	4-11/16	2-1/32	3-1/4	11.71	52,800	144,000
1-3/8	3-5/16	1-1/2	5-1/4	2-1/4	3-5/8	15.83	59,400	162,000
1-1/2	3-5/8	1-5/8	5-3/4	2-3/8	3-7/8	20.80	74,800	204,000
1-3/4	4-5/16	2	7	2-7/8	5	33.91	110,000	300,000
2	5	2-1/4	7-3/4	3-1/4	5-3/4	51.75	154,000	420,000
2-1/2	6	2-3/4	10-1/2	4-1/8	7-1/4	101.59	242,000	660,000
*3	6-1/2	3-1/4	13	5	7-7/8	178.00	374,000	1,020,000

*FURNISHED IN ROUND PIN ONLY.

- Sinker Shackle. This shackle has elongated shanks and is designed to fit over the width of a chain link and attach a sinker bail to the chain. This type shackle is not considered a structural component of a mooring. Figure 3-20 provides a schematic drawing of sinker shackle and its various dimensions.

Figure 3-20 Sinker Shackle

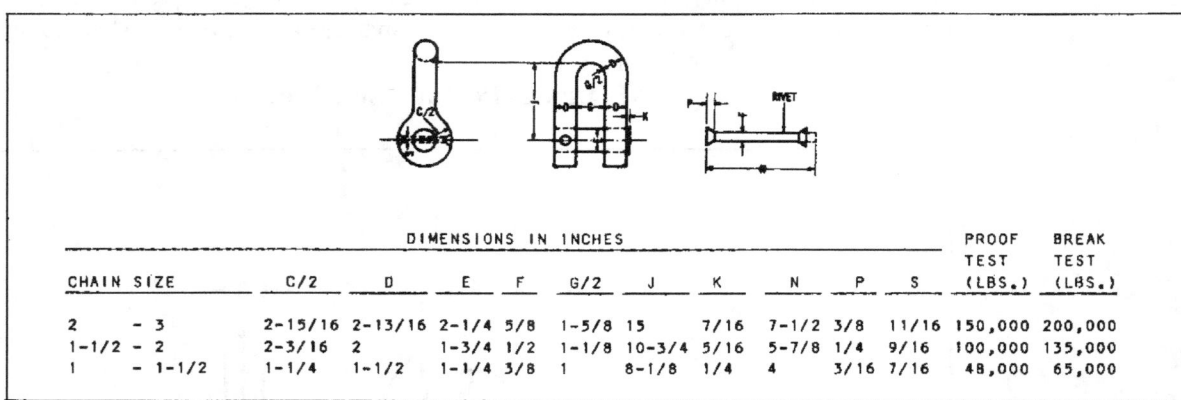

CHAIN SIZE	C/2	D	E	F	G/2	J	K	N	P	S	PROOF TEST (LBS.)	BREAK TEST (LBS.)
2 - 3	2-15/16	2-13/16	2-1/4	5/8	1-5/8	15	7/16	7-1/2	3/8	11/16	150,000	200,000
1-1/2 - 2	2-3/16	2		1-3/4	1/2	1-1/8	10-3/4	5/16	5-7/8	1/4	9/16	100,000	135,000
1 - 1-1/2	1-1/4	1-1/2		1-1/4	3/8	1	8-1/8	1/4	4	3/16	7/16	48,000	65,000

-

3.4.3.4. Common Stud Link Chain

This link is also called a common link, a stud link, or an A-link chain. This chain is the basic component of a fleet mooring and is normally manufactured in 90' (1-shot) lengths. A schematic drawing of a common stud link and its various dimensions is shown in Figure B-13.

Figure 3-21 Common Stud Link Chain

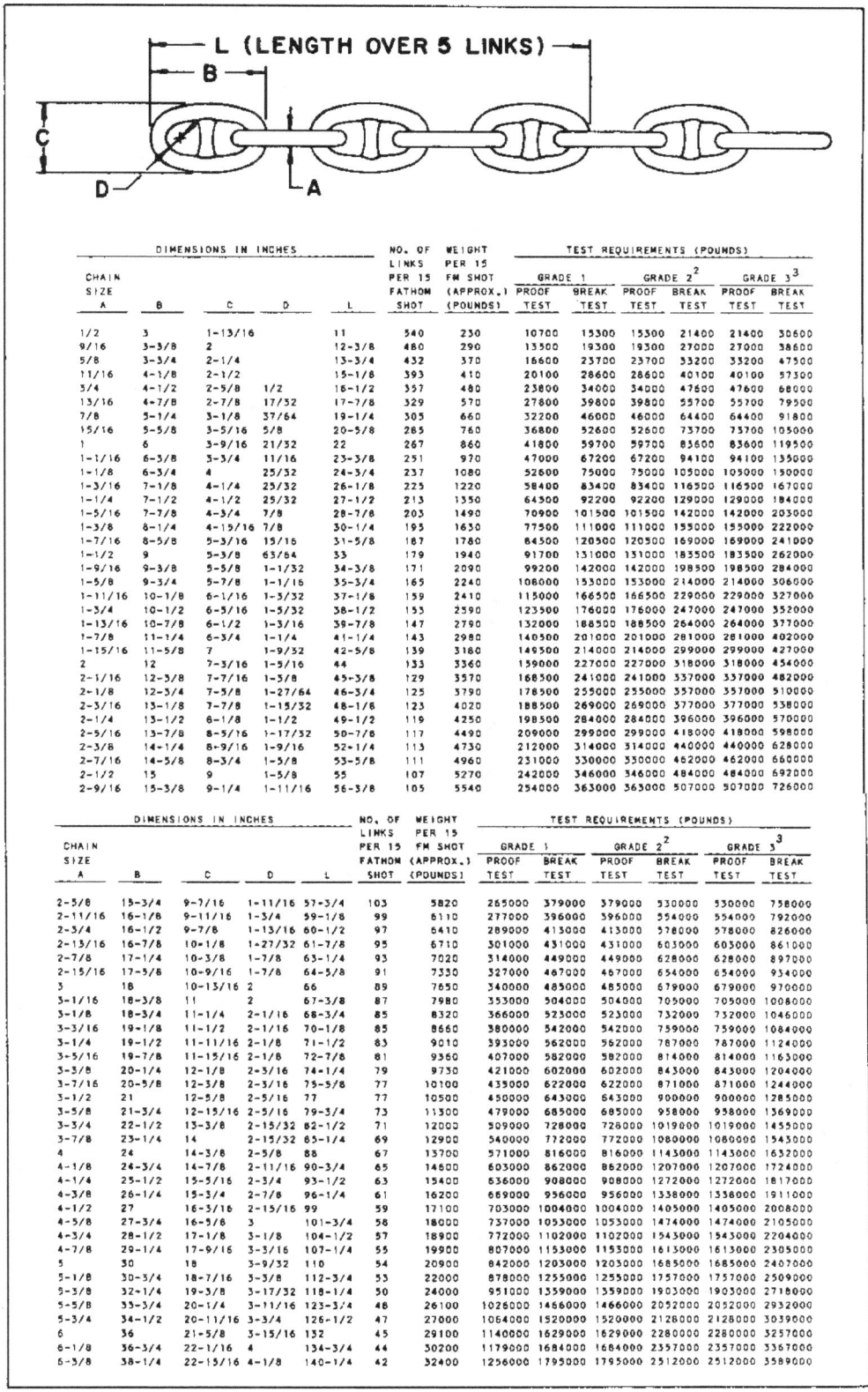

3.4.3.5. Enlarged Link

This link is also called a B-link. It is a large common stud link which acts as an adapter and is used between the last common stud link of a chain and the end link. Figure 3-22 provides a schematic drawing of an enlarged link and its dimensions.

Figure 3-22 Enlarged Link

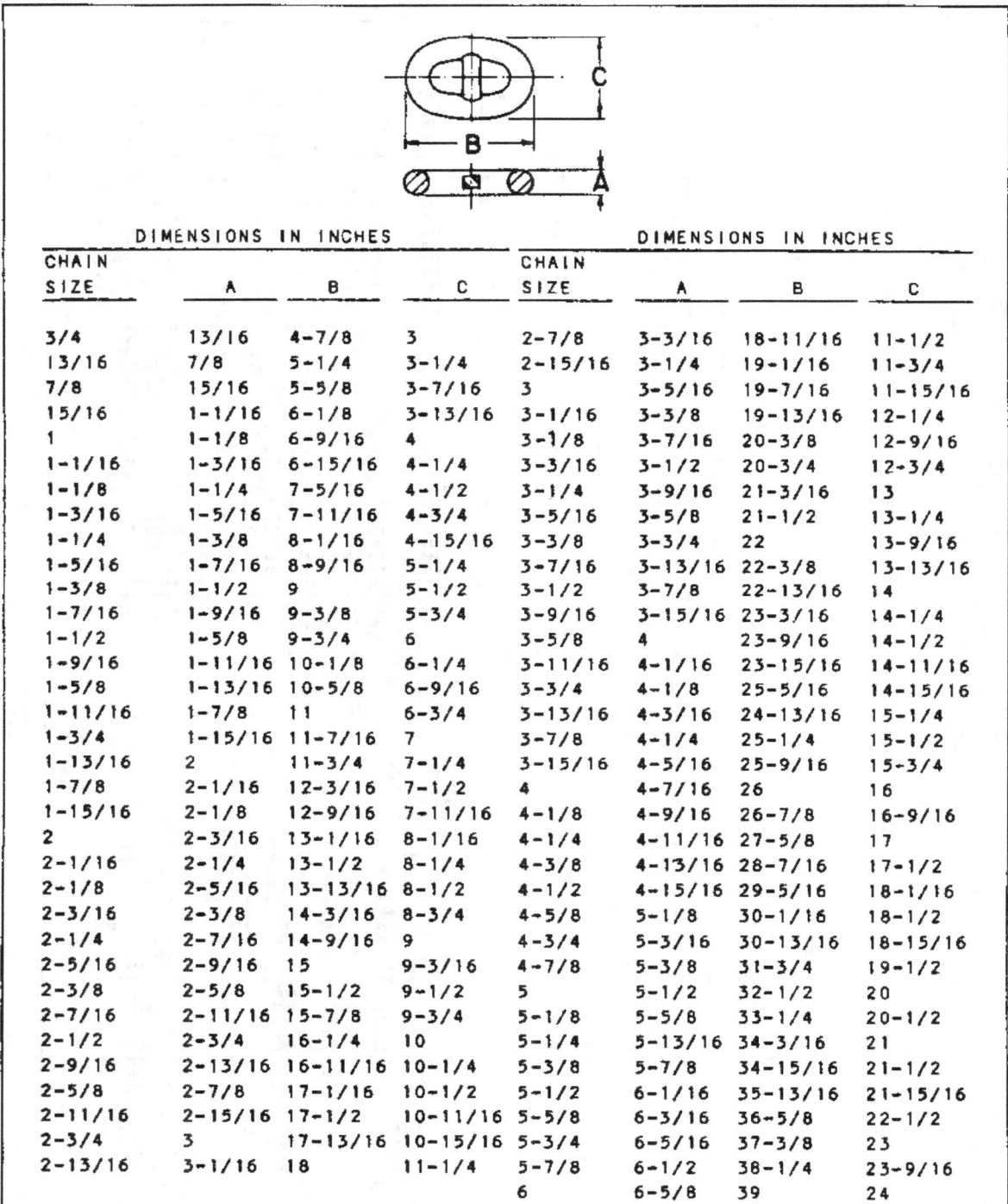

CHAIN SIZE	A	B	C	CHAIN SIZE	A	B	C
3/4	13/16	4-7/8	3	2-7/8	3-3/16	18-11/16	11-1/2
13/16	7/8	5-1/4	3-1/4	2-15/16	3-1/4	19-1/16	11-3/4
7/8	15/16	5-5/8	3-7/16	3	3-5/16	19-7/16	11-15/16
15/16	1-1/16	6-1/8	3-13/16	3-1/16	3-3/8	19-13/16	12-1/4
1	1-1/8	6-9/16	4	3-1/8	3-7/16	20-3/8	12-9/16
1-1/16	1-3/16	6-15/16	4-1/4	3-3/16	3-1/2	20-3/4	12-3/4
1-1/8	1-1/4	7-5/16	4-1/2	3-1/4	3-9/16	21-3/16	13
1-3/16	1-5/16	7-11/16	4-3/4	3-5/16	3-5/8	21-1/2	13-1/4
1-1/4	1-3/8	8-1/16	4-15/16	3-3/8	3-3/4	22	13-9/16
1-5/16	1-7/16	8-9/16	5-1/4	3-7/16	3-13/16	22-3/8	13-13/16
1-3/8	1-1/2	9	5-1/2	3-1/2	3-7/8	22-13/16	14
1-7/16	1-9/16	9-3/8	5-3/4	3-9/16	3-15/16	23-3/16	14-1/4
1-1/2	1-5/8	9-3/4	6	3-5/8	4	23-9/16	14-1/2
1-9/16	1-11/16	10-1/8	6-1/4	3-11/16	4-1/16	23-15/16	14-11/16
1-5/8	1-13/16	10-5/8	6-9/16	3-3/4	4-1/8	25-5/16	14-15/16
1-11/16	1-7/8	11	6-3/4	3-13/16	4-3/16	24-13/16	15-1/4
1-3/4	1-15/16	11-7/16	7	3-7/8	4-1/4	25-1/4	15-1/2
1-13/16	2	11-3/4	7-1/4	3-15/16	4-5/16	25-9/16	15-3/4
1-7/8	2-1/16	12-3/16	7-1/2	4	4-7/16	26	16
1-15/16	2-1/8	12-9/16	7-11/16	4-1/8	4-9/16	26-7/8	16-9/16
2	2-3/16	13-1/16	8-1/16	4-1/4	4-11/16	27-5/8	17
2-1/16	2-1/4	13-1/2	8-1/4	4-3/8	4-13/16	28-7/16	17-1/2
2-1/8	2-5/16	13-13/16	8-1/2	4-1/2	4-15/16	29-5/16	18-1/16
2-3/16	2-3/8	14-3/16	8-3/4	4-5/8	5-1/8	30-1/16	18-1/2
2-1/4	2-7/16	14-9/16	9	4-3/4	5-3/16	30-13/16	18-15/16
2-5/16	2-9/16	15	9-3/16	4-7/8	5-3/8	31-3/4	19-1/2
2-3/8	2-5/8	15-1/2	9-1/2	5	5-1/2	32-1/2	20
2-7/16	2-11/16	15-7/8	9-3/4	5-1/8	5-5/8	33-1/4	20-1/2
2-1/2	2-3/4	16-1/4	10	5-1/4	5-13/16	34-3/16	21
2-9/16	2-13/16	16-11/16	10-1/4	5-3/8	5-7/8	34-15/16	21-1/2
2-5/8	2-7/8	17-1/16	10-1/2	5-1/2	6-1/16	35-13/16	21-15/16
2-11/16	2-15/16	17-1/2	10-11/16	5-5/8	6-3/16	36-5/8	22-1/2
2-3/4	3	17-13/16	10-15/16	5-3/4	6-5/16	37-3/8	23
2-13/16	3-1/16	18	11-1/4	5-7/8	6-1/2	38-1/4	23-9/16
				6	6-5/8	39	24

3.4.3.6. End Link

This link is also called an E-link or an open end link. It is used as the last link on a shot of chain, allowing a joining shackle or other type joining link to connect two shots of chain together. Figure 3-23 provides a schematic drawing of an end link and a dimensional table of its various sizes.

Figure 3-23 End Link

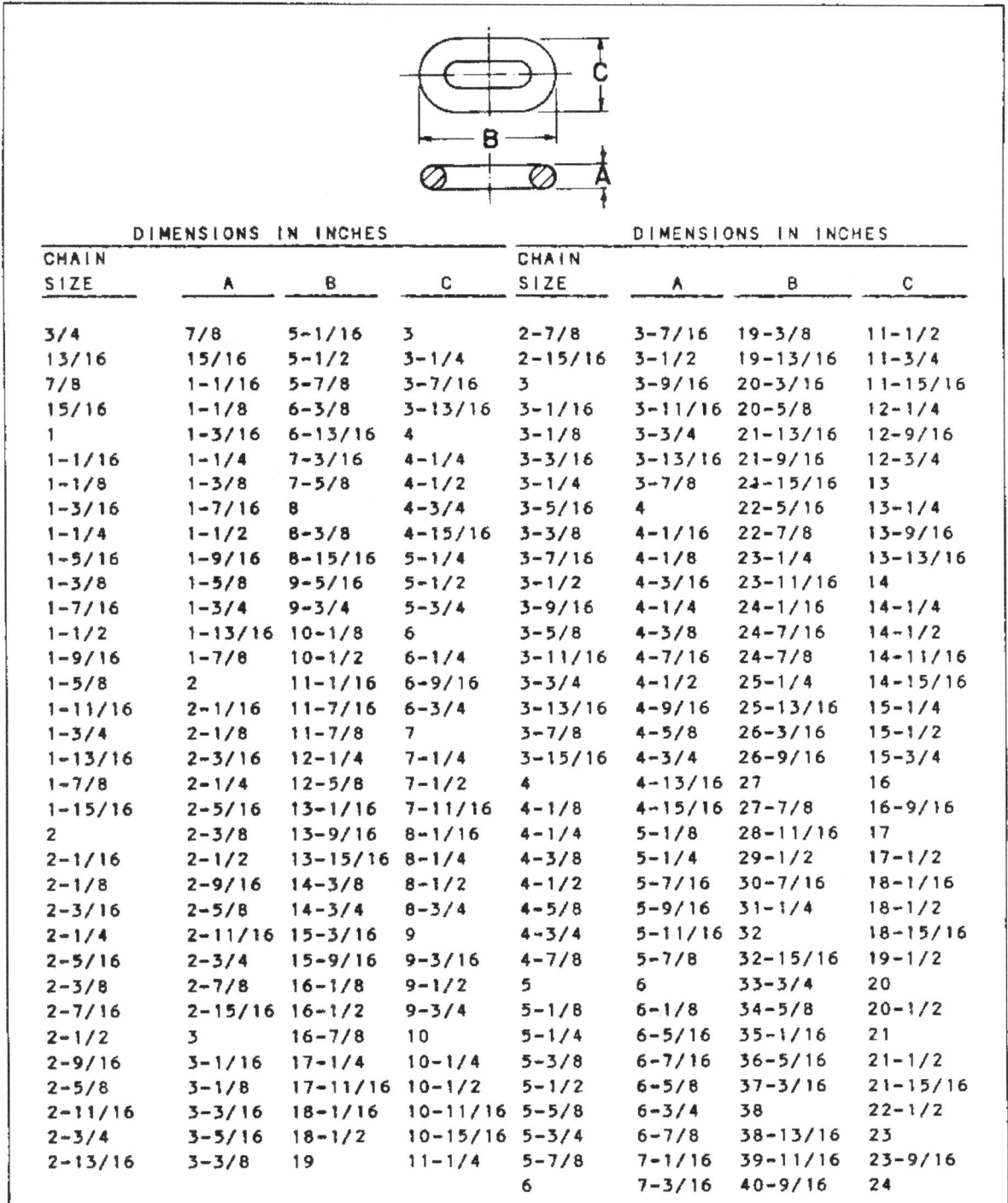

CHAIN SIZE	A	B	C	CHAIN SIZE	A	B	C
3/4	7/8	5-1/16	3	2-7/8	3-7/16	19-3/8	11-1/2
13/16	15/16	5-1/2	3-1/4	2-15/16	3-1/2	19-13/16	11-3/4
7/8	1-1/16	5-7/8	3-7/16	3	3-9/16	20-3/16	11-15/16
15/16	1-1/8	6-3/8	3-13/16	3-1/16	3-11/16	20-5/8	12-1/4
1	1-3/16	6-13/16	4	3-1/8	3-3/4	21-13/16	12-9/16
1-1/16	1-1/4	7-3/16	4-1/4	3-3/16	3-13/16	21-9/16	12-3/4
1-1/8	1-3/8	7-5/8	4-1/2	3-1/4	3-7/8	21-15/16	13
1-3/16	1-7/16	8	4-3/4	3-5/16	4	22-5/16	13-1/4
1-1/4	1-1/2	8-3/8	4-15/16	3-3/8	4-1/16	22-7/8	13-9/16
1-5/16	1-9/16	8-15/16	5-1/4	3-7/16	4-1/8	23-1/4	13-13/16
1-3/8	1-5/8	9-5/16	5-1/2	3-1/2	4-3/16	23-11/16	14
1-7/16	1-3/4	9-3/4	5-3/4	3-9/16	4-1/4	24-1/16	14-1/4
1-1/2	1-13/16	10-1/8	6	3-5/8	4-3/8	24-7/16	14-1/2
1-9/16	1-7/8	10-1/2	6-1/4	3-11/16	4-7/16	24-7/8	14-11/16
1-5/8	2	11-1/16	6-9/16	3-3/4	4-1/2	25-1/4	14-15/16
1-11/16	2-1/16	11-7/16	6-3/4	3-13/16	4-9/16	25-13/16	15-1/4
1-3/4	2-1/8	11-7/8	7	3-7/8	4-5/8	26-3/16	15-1/2
1-13/16	2-3/16	12-1/4	7-1/4	3-15/16	4-3/4	26-9/16	15-3/4
1-7/8	2-1/4	12-5/8	7-1/2	4	4-13/16	27	16
1-15/16	2-5/16	13-1/16	7-11/16	4-1/8	4-15/16	27-7/8	16-9/16
2	2-3/8	13-9/16	8-1/16	4-1/4	5-1/8	28-11/16	17
2-1/16	2-1/2	13-15/16	8-1/4	4-3/8	5-1/4	29-1/2	17-1/2
2-1/8	2-9/16	14-3/8	8-1/2	4-1/2	5-7/16	30-7/16	18-1/16
2-3/16	2-5/8	14-3/4	8-3/4	4-5/8	5-9/16	31-1/4	18-1/2
2-1/4	2-11/16	15-3/16	9	4-3/4	5-11/16	32	18-15/16
2-5/16	2-3/4	15-9/16	9-3/16	4-7/8	5-7/8	32-15/16	19-1/2
2-3/8	2-7/8	16-1/8	9-1/2	5	6	33-3/4	20
2-7/16	2-15/16	16-1/2	9-3/4	5-1/8	6-1/8	34-5/8	20-1/2
2-1/2	3	16-7/8	10	5-1/4	6-5/16	35-1/16	21
2-9/16	3-1/16	17-1/4	10-1/4	5-3/8	6-7/16	36-5/16	21-1/2
2-5/8	3-1/8	17-11/16	10-1/2	5-1/2	6-5/8	37-3/16	21-15/16
2-11/16	3-3/16	18-1/16	10-11/16	5-5/8	6-3/4	38	22-1/2
2-3/4	3-5/16	18-1/2	10-15/16	5-3/4	6-7/8	38-13/16	23
2-13/16	3-3/8	19	11-1/4	5-7/8	7-1/16	39-11/16	23-9/16
				6	7-3/16	40-9/16	24

3.4.3.7. C-Link

This link is similar to an end link except that it has an off-center stud. Offsetting the stud provides sufficient space for the lugs of a shackle to pass through the larger opening.

3.4.3.8. Pear Link

This link is also called a pear-shaped link, a pear-shaped ring, or a pear-shaped end link. This is an end link with one end larger than the other (see Figure 3-24). When cast/forged onto a ground ring, as shown in Figure 3-24, it is used as an adapter to connect the ground ring to an anchor joining link (see Figure 3-16 and Figure 3-28).

Figure 3-24 Pear Link

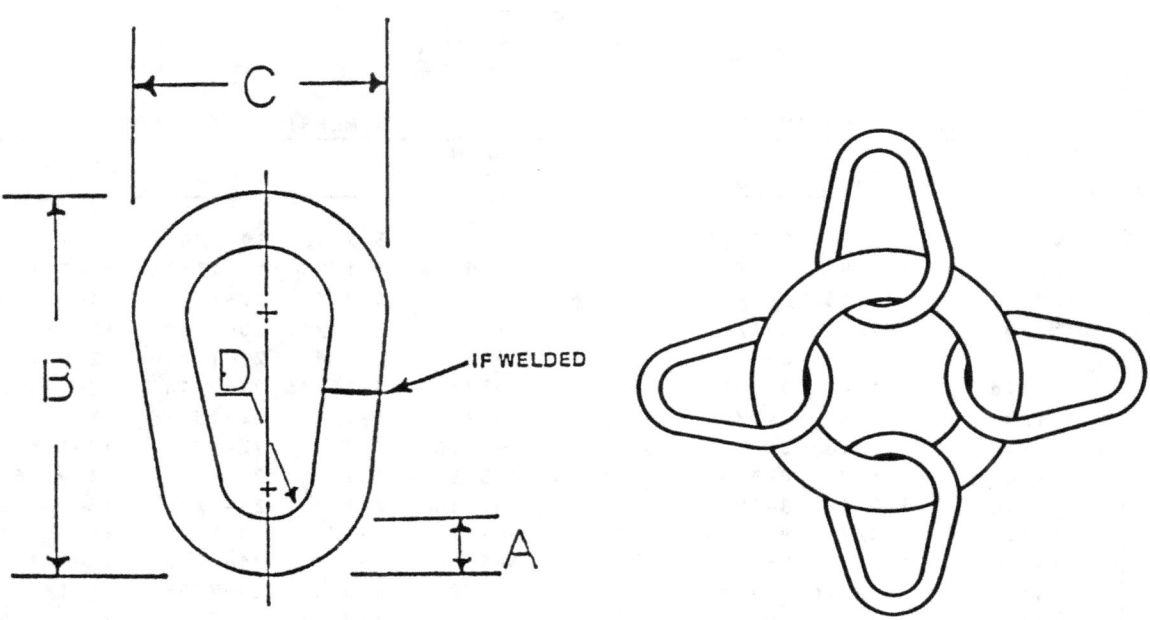

(All Dimensions in Inches)

Nominal Chain Size	A (min)	A (max)	B (min)	B (max)	C (max)	C (min)	D (max)	D (min)
1 1/4	1.28	1.34	10.70	10.76	5.55	5.61	.95	1.01
1 3/4	1.84	1.91	15.29	15.35	7.94	8.00	1.38	1.44
2	2.20	2.30	18.33	18.43	9.51	9.61	1.64	1.74
2 1/4	2.33	2.42	19.35	19.45	10.04	10.14	1.73	1.83
2 1/2	2.58	2.67	21.40	21.50	11.11	11.21	1.92	2.02
2 3/4	2.83	2.92	23.44	23.54	12.17	12.27	2.11	2.21
3	3.06	3.19	25.27	25.39	13.22	13.34	2.28	2.40
3 1/2	3.82	3.94	31.60	31.72	16.41	16.53	2.85	2.97
4	4.19	4.31	34.66	34.78	18.00	18.12	3.13	3.25

3.4.3.9. Swivel

A swivel is used to prevent twisting of the riser chain subassembly of a riser-type, free-swinging mooring, and is also used in each anchor chain subassembly (near the anchor) to prevent the chain from twisting during installation (see Figure 3-25). This standard swivel is normally manufactured with a similar sized common link on each end.

Figure 3-25 Swivel

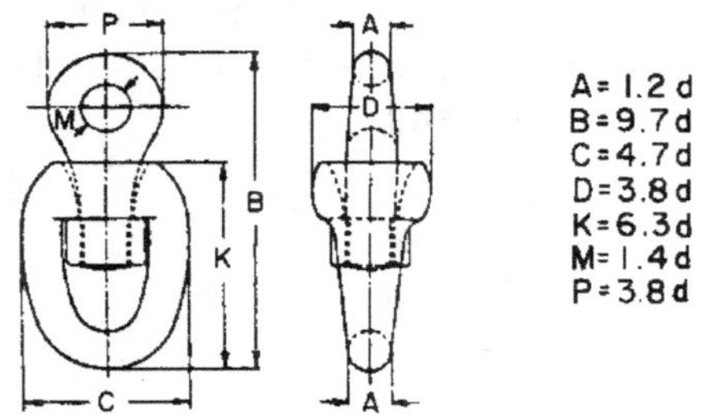

A = 1.2 d
B = 9.7 d
C = 4.7 d
D = 3.8 d
K = 6.3 d
M = 1.4 d
P = 3.8 d

DIMENSIONS IN INCHES

CHAIN SIZE	A	B	C	D	K	M	P	WEIGHT (LBS.)
2-15/16	3-1/2	28-1/2	13-13/16	11-3/16	18-1/2	4-1/8	11-3/16	373
3	3-5/8	29-1/8	14-1/8	11-3/8	19-1/16	4-3/16	11-3/8	395
3-1/16	3-11/16	29-11/16	14-3/8	11-5/8	19-5/16	4-5/16	11-5/8	419
3-1/8	3-3/4	30-5/16	14-11/16	11-7/8	19-11/16	4-3/8	11-7/8	452
3-3/16	3-13/16	30-15/16	15	12-1/8	20-1/16	4-7/16	12-1/8	478
3-1/4	3-7/8	31-1/2	15-1/4	12-3/8	20-1/2	4-9/16	12-3/8	505
3-5/16	4	32-1/8	15-9/16	12-9/16	20-7/8	4-5/8	12-9/16	534
3-3/8	4-1/16	32-3/4	15-7/8	12-13/16	21-1/4	4-3/4	12-13/16	571
3-7/16	4-1/8	33-3/8	16-1/8	13-1/16	21-11/16	4-13/16	13-1/16	602
3-1/2	4-3/16	33-15/16	16-7/16	13-5/16	22-1/16	4-7/8	13-5/16	633
3-9/16	4-1/4	34-9/16	16-3/4	13-9/16	22-7/16	5	13-9/16	666
3-5/8	4-3/8	35-3/16	17-1/16	13-3/4	22-13/16	5-1/16	13-3/4	699
3-11/16	4-7/16	35-3/4	17-5/16	14	23-1/4	5-3/16	14	734
3-3/4	4-1/2	36-3/8	17-5/8	14-1/4	23-5/8	5-1/4	14-1/4	769
3-13/16	4-9/16	37	17-15/16	14-1/2	24	5-5/16	14-1/2	820
3-7/8	4-5/8	37-9/16	18-3/16	14-3/4	24-7/16	5-7/16	14-3/4	856
3-15/16	4-3/4	38-3/16	18-1/2	14-15/16	24-13/16	5-1/2	14-15/16	897
4	4-13/16	38-13/16	18-13/16	15-3/16	25-3/16	5-5/8	15-3/16	939
4-1/8	4-15/16	40	19-3/8	15-11/16	26	5-3/4	15-11/16	1083
4-1/4	5-1/8	41-1/4	20	16-3/16	26-3/4	5-15/16	16-1/8	1131
4-3/8	5-1/4	42-7/16	20-9/16	16-5/8	27-9/16	6-1/8	16-5/8	1228
4-1/2	5-3/8	43-5/8	21-1/8	17-1/8	28-3/8	6-5/16	17-1/8	1347
4-5/8	5-9/16	44-7/8	21-3/4	17-9/16	29-1/8	6-1/2	17-9/16	1455
4-3/4	5-11/16	46-1/16	22-5/16	18-1/16	29-15/16	6-5/8	18-1/16	1570
4-7/8	5-7/8	47-5/16	22-15/16	18-1/2	30-11/16	6-13/16	18-1/2	1711
5	6	48-1/2	23-1/2	19	31-1/2	7	19	1839
5-1/8	6-1/8	49-11/16	24-1/16	19-1/2	32-5/16	7-3/16	19-1/2	1971
5-1/4	6-5/16	50-15/16	24-11/16	19-15/16	33-1/16	7-3/8	19-15/16	2136
5-3/8	6-7/16	52-1/8	25-1/4	20-7/16	33-7/8	7-1/2	20-7/16	2283
5-1/2	6-5/8	53-3/8	25-7/8	20-7/8	34-5/8	7-11/16	20-7/8	2463
5-5/8	6-3/4	54-9/16	26-7/16	21-3/8	35-7/16	7-7/8	21-3/8	2623
5-3/4	6-7/8	55-3/4	27	21-7/8	36-1/4	8-1/16	21-7/8	2793
5-7/8	7-1/16	57	27-5/8	22-5/16	37	8-1/4	22-5/16	2998
6	7-3/16	58-3/16	28-3/16	22-13/16	37-13/16	8-3/8	22-13/16	3181

DIMENSIONS IN INCHES

CHAIN SIZE (d)	A	B	C	D	K	M	P	WEIGHT (LBS.)
3/4	15/16	7-1/8	3-1/2	2-7/8	4-3/4	1-1/16	2-7/8	6.2
13/16	1	7-7/8	3-13/16	3-1/16	5-1/8	1-1/8	3-1/16	7.7
7/8	1-1/16	8-1/2	4-1/8	3-5/16	5-1/2	1-1/4	3-5/16	9.7
15/16	1-1/8	9-1/16	4-7/16	3-9/16	5-15/16	1-5/16	3-9/16	12.3
1	1-3/16	9-11/16	4-11/16	3-3/16	6-5/16	1-3/8	3-13/16	15.0
1-1/16	1-1/4	10-5/16	5	4-1/16	6-11/16	1-1/2	4-1/16	17.6
1-1/8	1-3/8	10-15/16	5-5/16	4-1/4	7-1/16	1-9/16	4-1/4	20.7
1-3/16	1-7/16	11-1/2	5-9/16	4-1/2	7-1/2	1-11/16	4-1/2	24.3
1-1/4	1-1/2	12-1/8	5-7/8	4-3/4	7-7/8	1-3/4	4-3/4	28.0
1-5/16	1-9/16	12-3/4	6-3/16	5	8-1/4	1-13/16	5	33.7
1-3/8	1-5/8	13-5/16	6-7/16	5-1/4	8-11/16	1-15/16	5-1/4	38.6
1-7/16	1-3/4	13-15/16	6-3/4	5-7/16	9-1/16	2	5-7/16	43.7
1-1/2	1-13/16	14-9/16	7-1/16	5-11/16	9-7/16	2-1/8	5-11/16	49.6
1-9/16	1-7/8	15-3/16	7-3/8	5-15/16	9-7/8	2-3/16	5-15/16	55.1
1-5/8	1-15/16	15-3/4	7-5/8	6-3/16	10-1/4	2-1/4	6-3/16	63.9
1-11/16	2	16-3/8	7-15/16	6-7/16	10-5/8	2-3/8	6-7/16	71.7
1-3/4	2-1/8	17	8-1/4	6-5/8	11	2-7/16	6-5/8	79.4
1-13/16	2-3/16	17-9/16	8-1/2	6-7/8	11-7/16	2-9/16	6-7/8	87.1
1-7/8	2-1/4	18-3/16	8-13/16	7-1/8	11-13/16	2-5/8	7-1/8	95.9
1-15/16	2-5/16	18-13/16	9-1/8	7-3/8	12-3/16	2-11/16	7-3/8	105.8
2	2-3/8	19-3/8	9-3/8	7-5/8	12-5/8	2-13/16	7-5/8	119.0
2-1/16	2-1/2	20	9-11/16	7-13/16	13	2-7/8	7-13/16	130.1
2-1/8	2-9/16	20-5/8	10	8-1/16	13-3/8	3	8-1/16	141.1
2-3/16	2-5/8	21-1/4	10-3/16	8-5/16	13-13/16	3-1/16	8-5/16	153.2
2-1/4	2-11/16	21-13/16	10-9/16	8-9/16	14-3/16	3-1/8	8-9/16	166.4
2-5/16	2-3/4	22-7/16	10-7/8	8-13/16	14-9/16	3-1/4	8-13/16	179.7
2-3/8	2-7/8	23-1/16	11-3/8	9	14-15/16	3-5/16	9	195.4
2-7/16	2-15/16	23-5/8	11-7/16	9-1/4	15-3/8	3-7/16	9-1/4	214
2-1/2	3	24-1/4	11-3/4	9-1/2	15-3/4	3-1/2	9-1/2	230
2-9/16	3-1/16	24-7/8	12-1/16	9-3/4	16-1/8	3-9/16	9-3/4	247
2-5/8	3-1/8	25-7/16	12-5/16	10	16-9/16	3-11/16	10	269
2-11/16	3-1/4	26-1/16	12-5/8	10-3/16	16-15/16	3-3/4	10-3/16	282
2-3/4	3-5/16	26-11/16	10-7/16	12-15/16	17-5/16	3-7/8	10-7/16	302
2-13/16	3-3/8	27-5/16	13-1/4	10-11/16	17-3/4	3-15/16	10-11/16	328
2-7/8	3-7/16	27-7/8	13-1/2	10-15/16	18-1/8	4	10-15/16	351

3.4.3.10. Swivel Shackle (Chain)

This component can be used in place of the standard swivel. Both ends of the swivel shackle are required to fit a common link of a specified nominal chain size and are procured without attached common links (see Figure 3-26).

Figure 3-26 Typical Chain Swivel Shackle

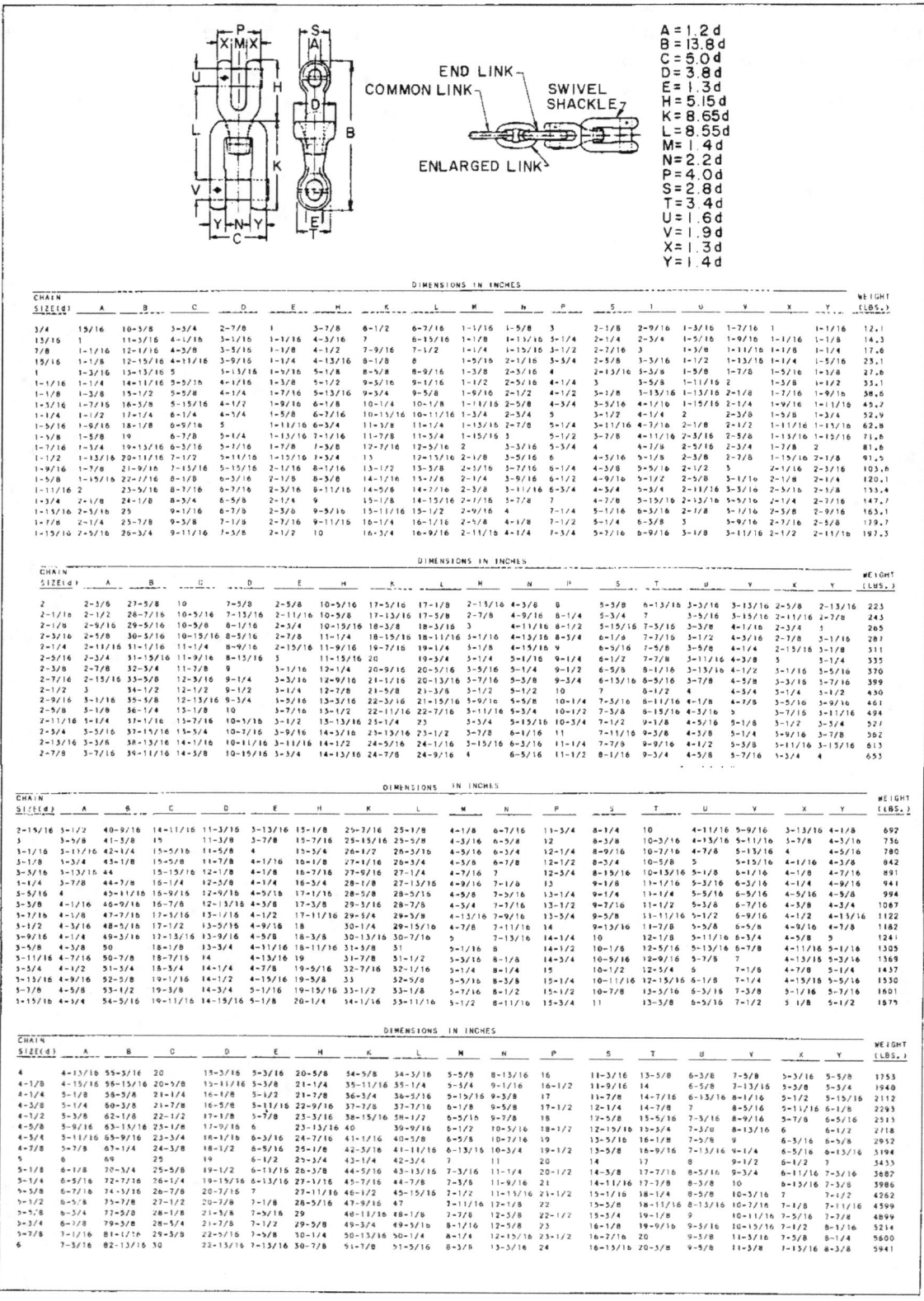

3.4.3.11. Modified Swivel Shackle (Riser)

This modified swivel shackle has two different jaw sizes and is designed to be used in lieu of the standard swivel in a riser. The top end of the riser swivel shackle is required to fit a shackle eye plate of the tension bar while the other end will be sized to fit the upper riser link to which it will attach (see Figure 3-27).

Figure 3-27 Typical Riser Swivel Shackle

TYPICAL RISER SWIVEL SHACKLE
(All Dimensions in Inches)

Nominal Chain Size	A (min)	A (max)	B (min)	C (min)	C (max)	G (max)
1 1/4	1.61	1.69	5.62	3.19	3.50	5.25
1 3/4	2.22	2.34	5.62	3.19	3.50	5.25
2	2.57	2.71	5.62	3.19	3.50	5.25
2 1/4	2.84	2.98	5.62	3.19	3.50	5.25
2 1/2	3.23	3.39	5.62	3.19	3.50	5.25
2 3/4	3.23	3.39	5.62	3.19	3.50	5.25

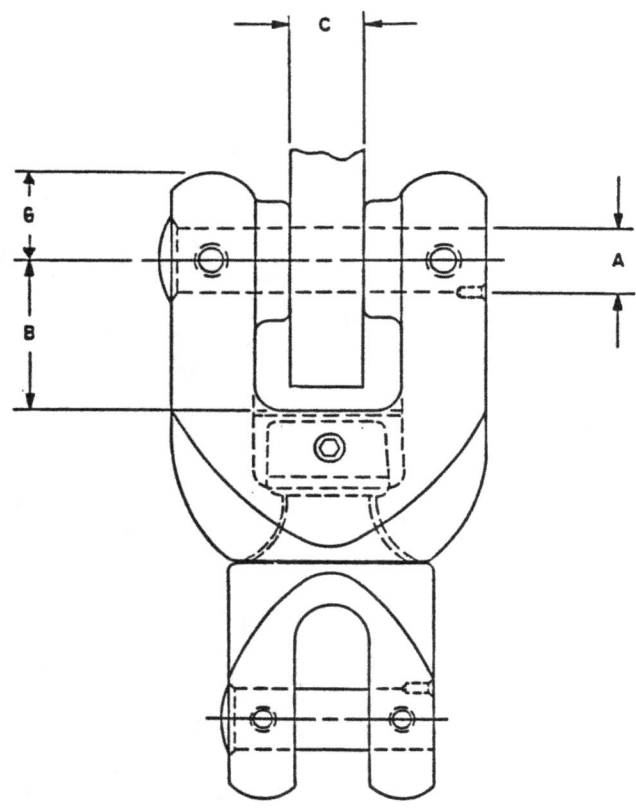

NOTE: THE TOP END OF A RISER SWIVEL SHACKLE IS REQUIRED TO FIT A SHACKLE EYE PLATE OF A FLEET MOORING BUOY. THE BOTTOM END OF THE RISER SWIVEL SHACKLE WILL BE THE SAME SIZE AS THE CHAIN SWIVEL SHACKLE AND WILL BE REQUIRED TO FIT A COMMON LINK OF THE SPECIFIED NOMINAL CHAIN SIZE, BUT THE COMMON LINK SHALL NOT BE PROVIDED WITH THE RISER SWIVEL SHACKLE. RISER SWIVEL SHACKLES SHALL MEET ALL CHEMICAL, MECHANICAL, PHYSICAL AND TESTING REQUIREMENTS SPECIFIED FOR THE SWIVEL OF THE SAME NOMINAL CHAIN SIZE.

3.4.3.12. Ground Ring

This is a large steel ring in a riser-type mooring which joins a riser chain subassembly to three or more anchor chain subassemblies. Figure 3-28 shows a schematic drawing of FM3 ground rings. The size of the ring for withstanding a particular proof load may vary widely due to different manufacturing processes.

Figure 3-28 Ground Ring

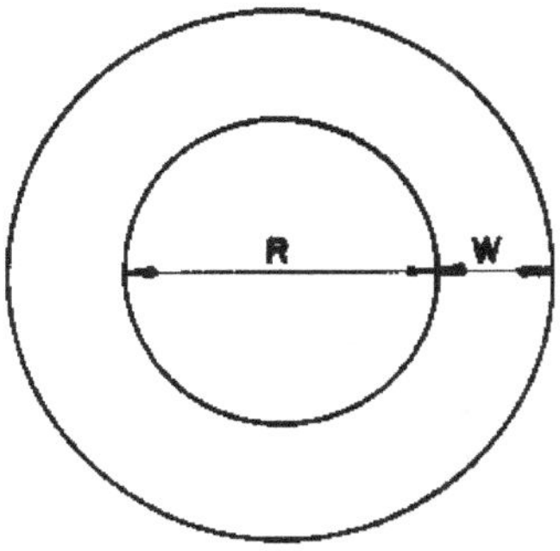

CHAIN SIZE	R (in)	W (in)	PROOF TEST (LBS.)	BREAK TEST (LBS.)
*4-1/2	21	7-1/4	1,405,000	2,008,000
4	18	6-1/2	840,000	1,176,000
3-1/2	18	6-1/2	840,000	1,176,000
3	16	6	495,000	693,000
2-3/4	15	5-1/2	420,660	588,930
2-1/2	14	4-3/4	351,560	492,190
2-1/4	14	4-1/2	287,930	403,100
2	14	4	230,000	322,000
1-3/4	12	3-1/2	178,000	249,210
1-1/4	10	2-3/4	92,910	130,070
3/4	10	1-7/8	34,680	48,550

*PROPOSED FOR USE IN THE AAA AND BBB CLASS MOORINGS (NOT STOCKED).

3.4.3.13. Spider Plate

This is a steel plate or casting, triangular in shape, that has three or more holes for joining several chains together (see Figure 3-29).

Figure 3-29 Spider Plate

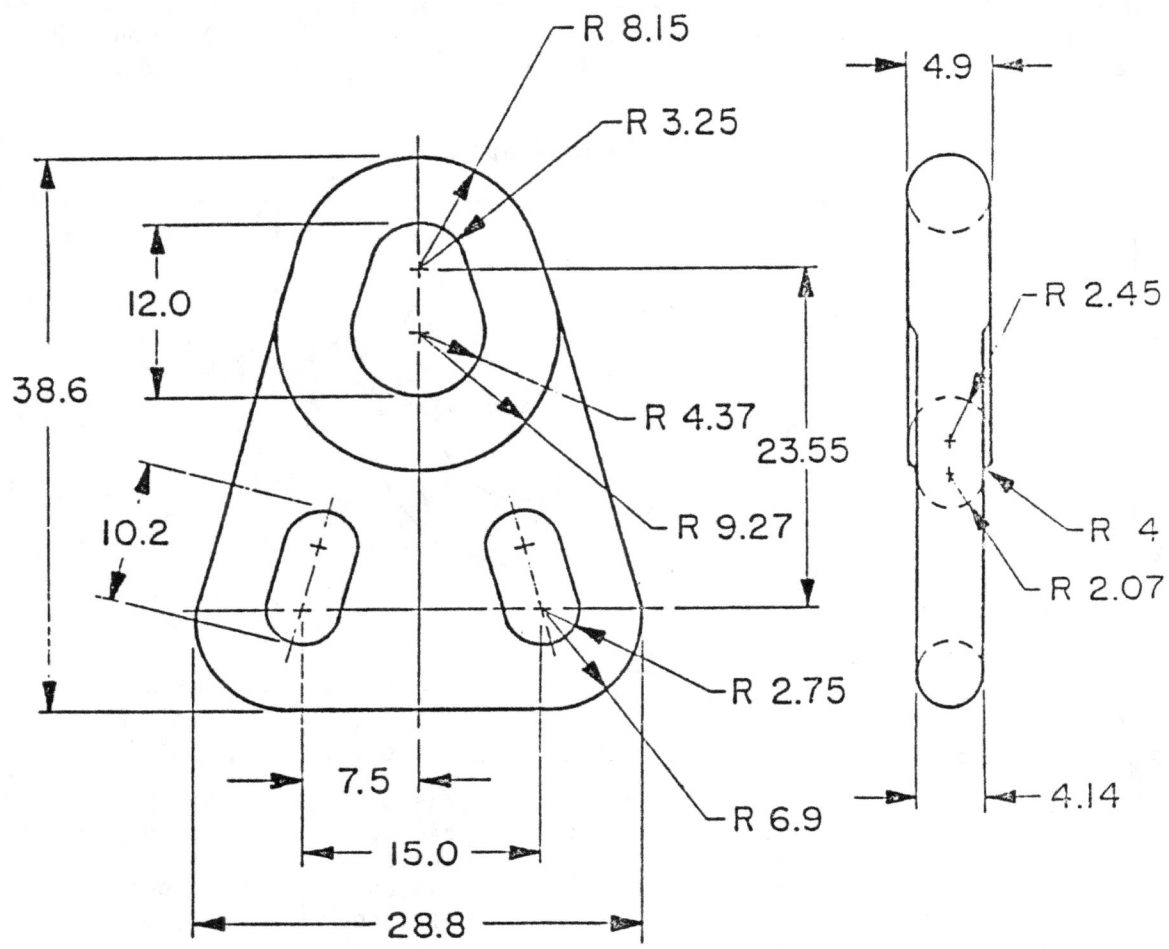

(All Dimensions in Inches)

3.4.3.14. Chain Equalizer

The equalizer is a half-rounded type of fitting (similar to a nonmovable pulley) with wide jaws through which an anchor chain subassembly is passed and then connected to two separate anchors. When tension is applied to the mooring, the chain has free movement to slide through the equalizer until the loads are equal in both subassembly legs (see Figure 3-30).

Figure 3-30 Chain Equalizer and Its Use

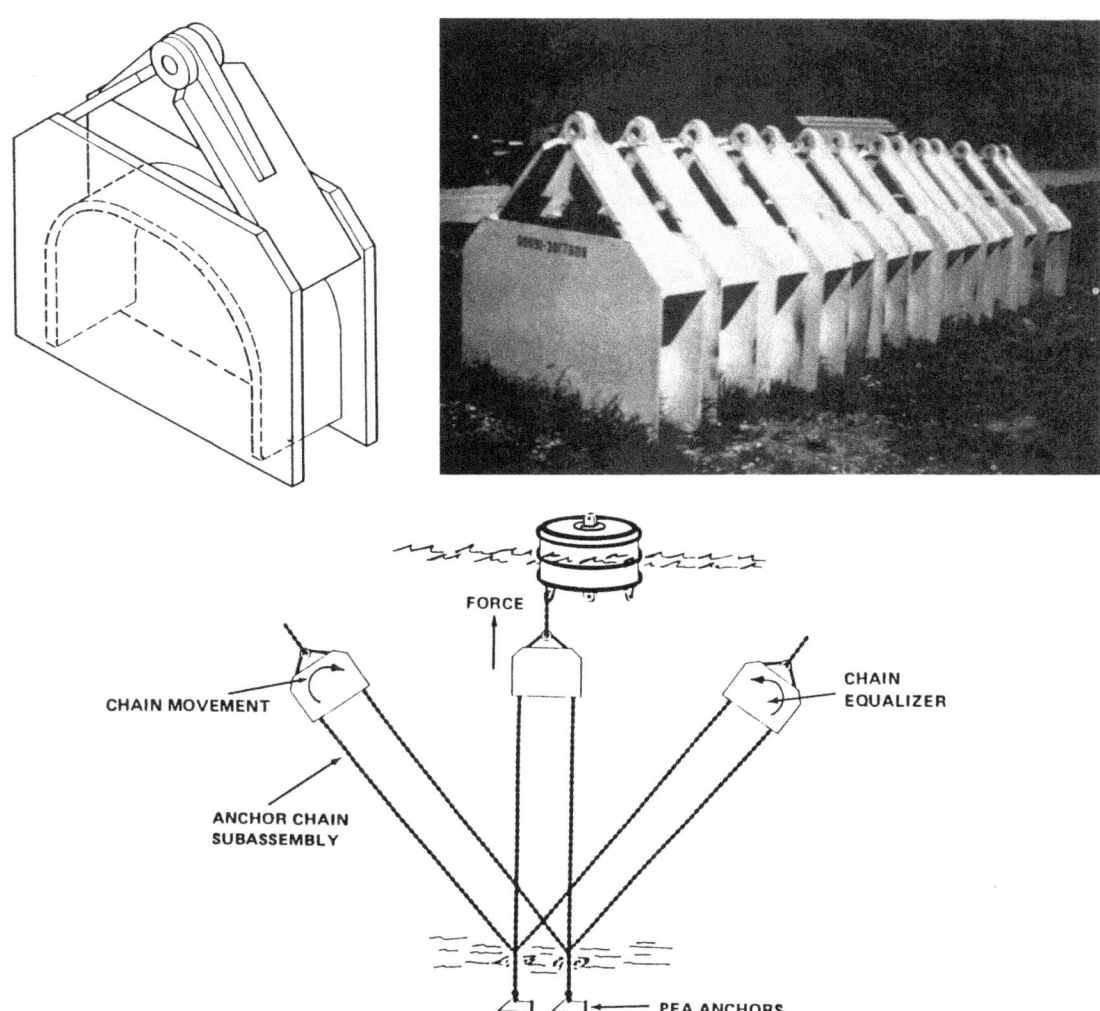

3.5. Buoys

The size of the buoy used in a particular fleet mooring is dependent on the depth of the water and the weight of the chain suspended in the water column. Figure 3-31 shows the four types of buoys commonly used in fleet moorings.

Figure 3-31 Commonly Used Fleet Mooring Buoys

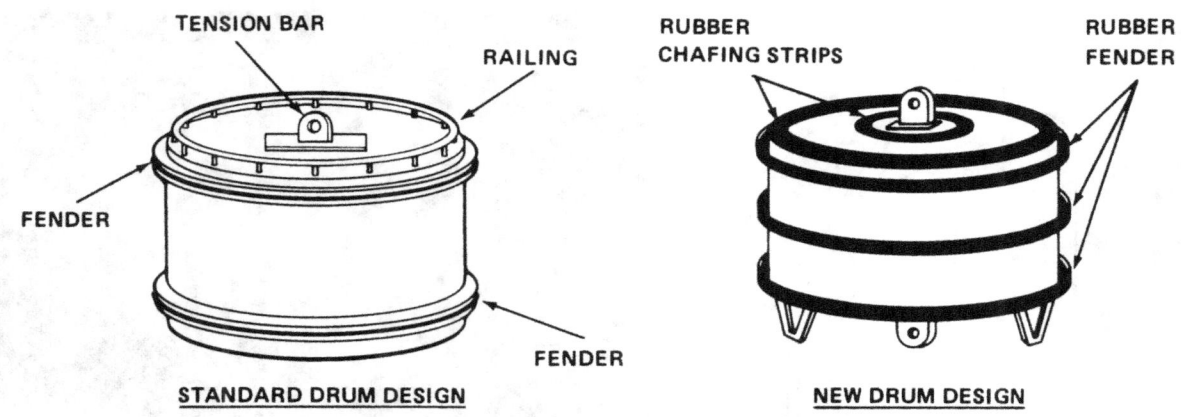

A. DRUM RISER-TYPE BUOY WITH TENSION BAR

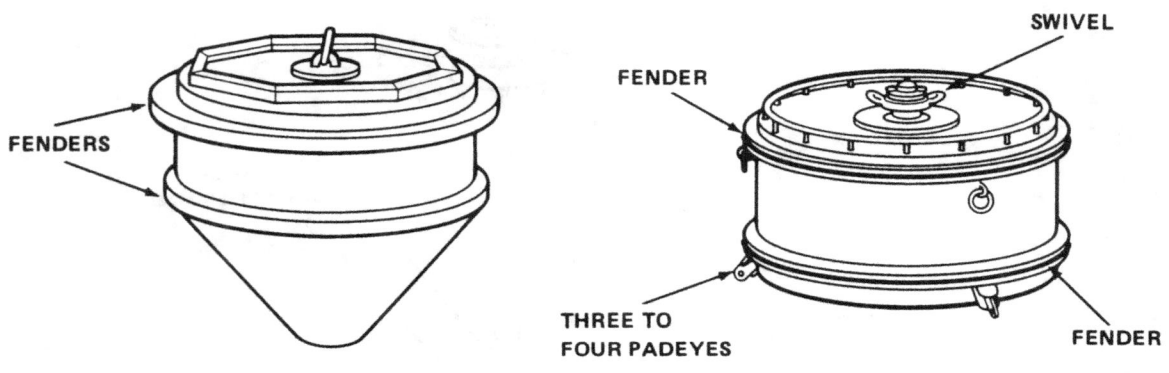

B. PEG-TOP MARK II RISER-TYPE BUOY

C. DRUM NON-RISER TYPE (TELEPHONE) BUOY

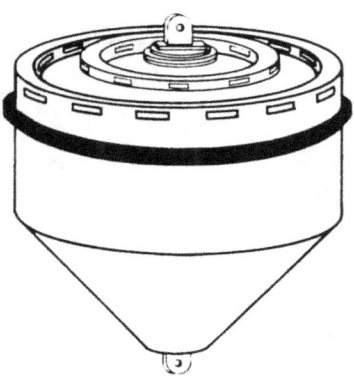

D. NEW FOAM BUOY

Buoy systems fall into two categories, riser and non-riser-types, depending upon the configuration of the ground tackle securing them to the sea floor. Riser-type systems have a single point at the center of the bottom of the buoy to which a riser chain is attached. The non-riser-type system has a larger buoy with three to four padeyes attached to its circular bottom edge. Each of these padeyes is attached to one or more anchor leg subassemblies.

3.5.1. Steel Buoys

These have been used for many years; however, their high maintenance cost has led to their phase-out. Information is included for reference.

3.5.1.1. Drum Buoy

This buoy is normally made of steel and is available in a variety of sizes. The shape of the buoy resembles a drum, and it is primarily used in smaller classes of moorings. The older standard design has a paint or fiberglass coating while the new design has a polyurethane coating. This type of buoy contains either a tension bar or a hawsepipe (see Figure 3-31).

- Tension Bar. A tension bar is a steel bar that passes through the center of the buoy with padeyes on top and bottom. The new foam buoy has a 10" diameter pipe used as a tension bar. The lower padeye is connected to the upper link of the riser chain. A moored vessel can be connected to the upper padeye by its anchor chain, wire rope, or mooring hawsers.
- Hawsepipe. A hawsepipe is a cylindrical tube passing through the center of the buoy. The riser chain is passed through this tube and its upper link is held on the top of the buoy with a slotted chain plate. The lower portion of the riser chain within the hawsepipe is protected from wear by a rubbing casting which encircles the chain and greatly reduces the probability of chain abrasion. A moored vessel ties directly to the riser chain. The buoy is nonstructural in that the mooring load passes through the buoy via the riser chain.

3.5.1.2. Peg Top Buoy

This buoy is also used to support riser-type moorings and includes either a tension bar or hawsepipe. Peg top buoys are conically shaped, with the top deck area considerable larger than the bottom surface (see Figure 3-31).

3.5.1.3. Non-Riser-Type Buoy

These buoys are larger than those used in riser-type moorings since they have the additional weight of three or more ground legs to support in the water column. These buoys have a swivel at the top to which the moored vessel's anchor chain or hawser is attached. Three or four padeyes, to which the anchor leg assemblies are connected, are equally spaced around the buoy's hull (see Figure 3-31).

3.5.2. Foam Buoys

The foam buoy consists of a tension bar encompassed by a rigid closed-cell interior foam which is covered by a flexible cross-linked polyethylene foam adhered to the rigid interior foam. The overall foam buoy is encased within a minimum of a 3/4" thick aliphatic urethane elastomer shell (see Figure 3-31).

This buoy is vastly superior to the older steel formed buoy in that:

- If its outer hull is punctured, it will not flood and sink.
- It is considerably lighter than the steel buoy required to support a comparable weight in the water column.
- It requires minimum preventative and corrective maintenance.
- Its resilient construction greatly reduces the probability of damage caused by collision with mooring ships; i.e., they are self-fendering.

The projected operational maintenance costs are considerably less than those of the standard steel buoy. There are two foam buoys commonly used on U.S. Navy Fleet moorings: an 8' diameter buoy and a 12' diameter buoy. Properties of these buoys are given in Table 3-7. An example of a foam buoy being secured is shown in Figure 3-32.

Table 3-7 Foam-Filled Polyurethane Coated Buoys

PARAMETERS	8' BUOY	12' BUOY
Weight in Air	4,500 lbs	10,400 lbs
Net Buoyancy	15,000 lbs	39,000 lbs
Working Buoyancy (24" FB)	6,150 lbs	20,320 lbs
Proof Load on Bar (0.6 F_y)	300 kips	600 kips
Working Load of Bar (0.3F_y)	150 kips	300 kips
Diameter Overall (w/fenders)	8 ft 6 in	12 ft
Diameter of Hull	8 ft	11 ft 6 in
Length of Hull Overall	7 ft 9 in	8 ft 9 in
Length of Tension Bar	11 ft 4 In	13 ft I in
Height of Cylindrical Portion	4 ft 4 in	5 ft 7 in
Height of Conical Portion	3 ft 5 in	3 ft 2 in
Bar Thickness (top/bottom)	4.5/3 in	5/3.5 in
Top Padeye ID (top/bottom)	3.5/3.5 in	4.5/5 in
Shackle on Top	3 inch	4 inch
Maximum Chain Size	2.75 inch	4 inch
Min. Recommended Riser Wt	1,068 lbs	7,500 lbs
Riser Wt for 24" freeboard	8,850 lbs	18,680 lbs
Max. Recommended Riser Wt	7,500 lbs	21,264 lbs
Moment to Heel 1°:		
Min Riser Wt	108 ft-lbs	1,183 ft-lbs
Max Riser Wt	648 ft-lbs	2,910 ft-lbs

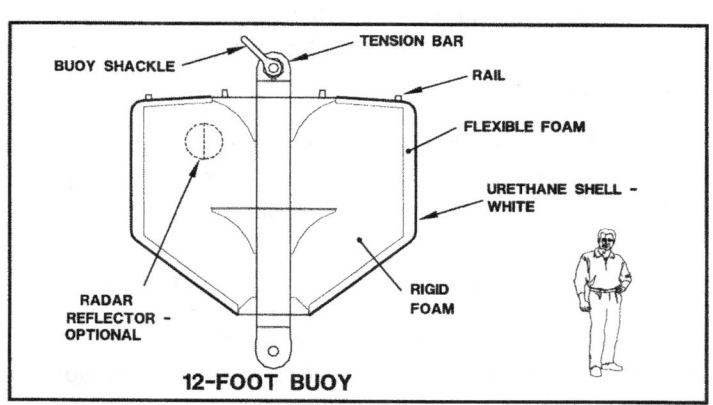

Figure 3-32 Example of Foam Buoy Being Secured

3.6. Sinkers

Sinkers are placed on fleet moorings to tune the static and dynamic behavior of a mooring. Sinkers are usually made of concrete or low cost metal. Key sinker parameters that can be specified in design include:

- Mass
- Weight
- Location
- Number
- Size
- Design

3.7. Mooring Lines

The most common tension member lines used are synthetic fiber ropes and wire rope. Synthetic lines have the advantage of easy handling and some types have stretch, which can be used to fine tune static and dynamic mooring behavior and aid in load sharing between tension members. Wire rope has the advantage of durability. The size and type of line specified in a given design will depend upon parameters such as the following:

- Safety
- Break strength
- Diameter
- Weight
- Buoyancy and hydrodynamic properties
- Ease of handling
- Equipment to be used
- Stretch/strain properties
- Load sharing between lines
- Dynamic behavior
- Reliability
- Durability
- Fatigue
- Exposure
- Chaffing/abrasion
- Wet vs. dry condition
- Experience
- Ability to splice
- Ability to provide terminations
- Inspection

- ✓ Cost
- ✓ Availability

3.7.1. Synthetic Fiber Ropes

Weaving a number of strands together to form a composite tension member forms mooring lines. Lines are made of different types of fiber and various constructions. Stretch/strain properties of selected lines are shown in Table 3-8 and Figure 3-33. Engineering characteristics of some double braided nylon and polyester lines are given in Table 3-9 and Table 3-10. Additional information is provided in NFESC TR-6014-OCN, *Mooring Design Physical and Empirical Data*. An example of fiber ropes is shown in Figure 3-34.

Table 3-8 Stretch of Synthetic Lines

% Break Strength (T/Tb)	SPECTRA BRAID[3] % Stretch (1)	DOUBLE BRAIDED POLYTESTER[4] % Stretch (2)	DOUBLE BRAIDED NYLON[5] % Stretch (3)	DOUBLE BRAIDED NYLON[6] % Stretch (4)	ULTRA-STRONG[2] % Stretch (5)
0	0	0	0	0	0
5	0.38	0.8	5.8	2.04	
10	0.58	1.5	9.1	3.89	1.68
15	0.72	2.1	11.5	5.46	
20	0.87	2.6	13.2	6.85	3.23
25	0.92	3.1	14.6	8.13	
30	0.98	3.6	15.7	9.26	4.7
35	1.07	4	16.6	10.28	
40	1.135	4.7	17.6	11.3	6.02
45	1.196	5.2	18.5	12.1	
50	1.25	5.7	19.3	12.8	7.58
55	1.305	6.2	20.1	(no data)	
60	1.354	6.8	20.9		9.05
65	1.412	7.3	21.7		
70	1.448	7.8	22.5		10.51
75	1.492	8.3	23.3		
80	1.535	8.8	24.1		12.08
85	1.578	9.3	24.9		
90	1.617	9.8	25.7		13.73
95	1.655	10.3	26.6		
100	1.693	10.9	27.4		15.35

---

[2] SPECTRON 12; Sampson
[3] 2-IN-1 STABLE BRAID; Sampson (engineering data sheet); cyclic loading
[4] DOUBLE BRAIDED; Sampson (engineering data sheet); cyclic loading; WET
[5] DOUBLE BRAIDED; Sampson (engineering data sheet); cyclic loading; DRY
[6] BLENDED ROPE; Sampson (engineering data sheet); cycled 50 times

Figure 3-33 Synthetic Line Stretch

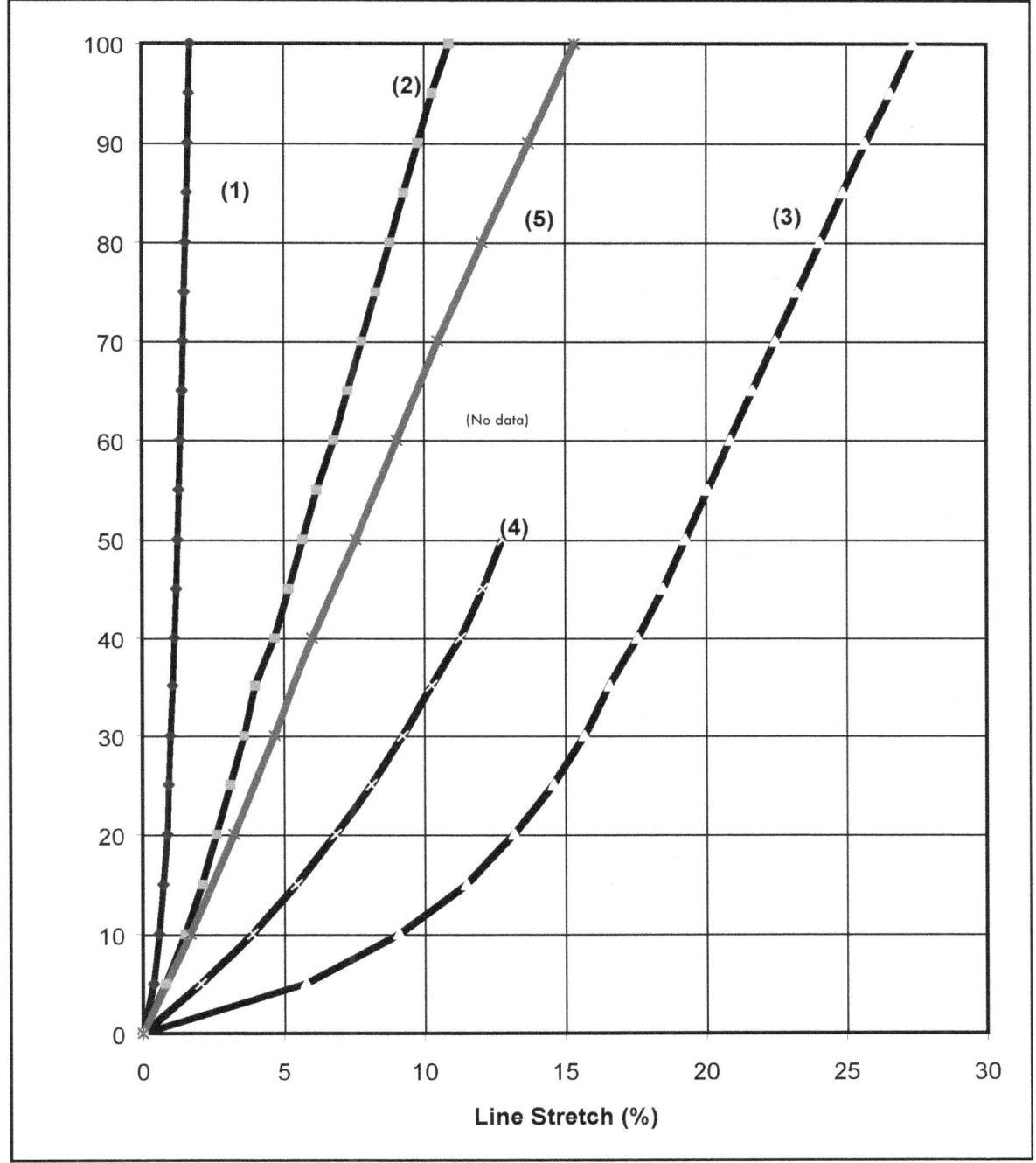

Table 3-9 Double Braided Nylon Line[7]

DIA. (in.)	CIR. (in)	SINGLE LINE				THREE PARTS LINE			
		Av Fb (kips)	Av Fb (E5 N)	AE (kips)	AE (E5 N)	Av Fb (kips)	Av Fb (E5 N)	AE (kips)	AE (E5 N)
1.0	3	33.6	1.495	118.9	5.29	100.8	4.48	356.8	15.87
1.1	3.5	45	2.002	159.3	7.09	135	6.01	477.9	21.26
1.2	3.75	52	2.313	184.1	8.19	156	6.94	552.2	24.56
1.3	4	59	2.624	208.8	9.29	177	7.87	626.5	27.87
1.4	4.5	74	3.292	261.9	11.65	222	9.88	785.8	34.96
1.6	5	91	4.048	322.1	14.33	273	12.14	966.4	42.99
1.8	5.5	110	4.893	389.4	17.32	330	14.68	1168.1	51.96
1.9	6	131	5.827	463.7	20.63	393	17.48	1391.2	61.88
2.1	6.5	153	6.806	541.6	24.09	459	20.42	1624.8	72.27
2.2	7	177	7.873	626.5	27.87	531	23.62	1879.6	83.61
2.4	7.5	202	8.985	715.0	31.81	606	26.96	2145.1	95.42
2.5	8	230	10.231	814.2	36.22	690	30.69	2442.5	108.65
2.7	8.5	257	11.432	909.7	40.47	771	34.30	2729.2	121.40
2.9	9	285	12.677	1008.8	44.88	855	38.03	3026.5	134.63
3.2	10	322	14.323	1139.8	50.70	966	42.97	3419.5	152.11
3.5	11	384	17.081	1359.3	60.46	1152	51.24	4077.9	181.39
3.8	12	451	20.061	1596.5	71.01	1353	60.18	4789.4	213.04
4.1	13	523	23.264	1851.3	82.35	1569	69.79	5554.0	247.05
4.5	14	599	26.645	2120.4	94.32	1797	79.93	6361.1	282.95
4.8	15	680	30.248	2407.1	107.07	2040	90.74	7221.2	321.22

---

[7] After Sampson, dry, cyclic loading; reduce nylon lines by 15% for wet conditions.
- Dia. = diameter
- Cir. = circumference
- Av Fb = average break strength
- AE = cross-sectional area times modulus of elasticity (this does not include the highly nonlinear properties of nylon, shown in Table 3-9.)

Table 3-10 Double Braided Polyester Lines[8]

DIA. (in.)	CIR. (in)	SINGLE LINE				THREE PARTS LINE			
		Av Fb (kips)	Av Fb (E5 N)	AE (kips)	AE (E5 N)	Av Fb (kips)	Av Fb (E5 N)	AE (kips)	AE (E5 N)
1.0	3	37.2	1.655	316.6	14.08	111.6	4.96	949.8	42.25
1.1	3.5	45.8	2.037	389.8	17.34	137.4	6.11	1169.4	52.02
1.2	3.75	54.4	2.420	463.0	20.59	163.2	7.26	1388.9	61.78
1.3	4	61.5	2.736	523.4	23.28	184.5	8.21	1570.2	69.85
1.4	4.5	71.3	3.172	606.8	26.99	213.9	9.51	1820.4	80.98
1.6	5	87.2	3.879	742.1	33.01	261.6	11.64	2226.4	99.03
1.8	5.5	104	4.626	885.1	39.37	312	13.88	2655.3	118.11
1.9	6	124	5.516	1055.3	46.94	372	16.55	3166.0	140.83
2.1	6.5	145	6.450	1234.0	54.89	435	19.35	3702.1	164.68
2.2	7	166	7.384	1412.8	62.84	498	22.15	4238.3	188.53
2.4	7.5	190	8.452	1617.0	71.93	570	25.35	4851.1	215.79
2.5	8	212	9.430	1804.3	80.26	636	28.29	5412.8	240.77
2.7	8.5	234	10.409	1991.5	88.59	702	31.23	5974.5	265.76
2.9	9	278	12.366	2366.0	105.24	834	37.10	7097.9	315.73
3.2	10	343	15.257	2919.1	129.85	1029	45.77	8757.4	389.55
3.5	11	407	18.104	3463.8	154.08	1221	54.31	10391.5	462.24
3.8	12	470	20.907	4000.0	177.93	1410	62.72	12000.0	533.79
4.1	13	533	23.709	4536.2	201.78	1599	71.13	13608.5	605.34
4.5	14	616	27.401	5242.6	233.20	1848	82.20	15727.7	699.60
4.8	15	698	31.049	5940.4	264.24	2094	93.15	17821.3	792.73

A comparison of the physical characteristics of natural and synthetic ropes is given in Table 3-11.

---

[8] After Sampson, dry, cyclic loading.
- Dia. = diameter
- Cir. = circumference
- Av Fb = average break strength
- AE = cross-sectional area times modulus of elasticity

Table 3-11 Comparison of Physical Characteristics of Natural and Synthetic Wire Rope

ROPE MATERIAL:	Manila-3 Strand		Sisal-3 Strand		Nylon-3 Strand		Kevlar-12 Strand	
Manufactured or derived from:	Abaca Plant-3' to 6' leaf fibers		Agave Sisalana plant -2' to 4' leaf fibers		Hexamethelene diamine and adipic acid derived from coke, air, water, and petroleum products		Generic name aramid-aromatic polyamide fibers	
ROPE DIAMETER:	1"	3"	1"	3"	1"	3"	1"	2"
Strength Characteristics:								
1. Tensile strength dry	8,100	57,600	6,480	46,080	22,500	180,000	63,000	214,000
2. Recommended factor of safety	6^2	6^2	6^2	6^2	9^2	9^2	9^2	9^2
3. Working strength^2	$1,350^2$	$9,600^2$	$1,080^2$	$7,680^2$	$2,500^2$	$20,000^2$	$7,000^2$	$23,800^2$
4. Wet strength versus dry strength (percent)	up to 120%	up to 120%	up to 120%	up to 120%	90-95%	90-95%	95%	95%
5. Strength per unit of weight (tensile strength/lbs per foot)	30,000	23,800	24,000	19,000	86,500	85,700	210,000	181,000
6. Rope shock load absorption ability	poor	poor	poor	poor	excellent	excellent	poor	poor
7. Repeat loading capabilities	poor	poor	poor	poor	very good	very good	very good	very good
8. Individual filament or fiber strength (grams per denier)	5.0-7.0	5.0-7.0	4.0-5.0	4.0-5.0	8.0-9.0	8.0-9.0	18.0	18.0
Weight & Density Characteristics:								
1. Pounds per 100 feet	27	242	27	242	26	210	30	118
2. Specific gravity of fiber	1.5	1.5	1.4	1.4	1.14	1.14	1.44	1.44
3. Ability to float	no	no	no	no	no	no	no	no
Elasticity - Stretch:								
1. Permanent elongation at working loads (break-in)	5%	5%	5%	5%	8%	8%	-	-
2. Working elasticity at working load (recoverable stretch under load)	5%	5%	5%	5%	16%	16%	14% Includes permanent elongation	14% Includes permanent elongation
3. Elongation at 100% load (at break) used rope	13%	13%	13%	13%	35%	35%	over 21%	over 21%
4. Creep (elongation under sustained load)	very low	very low	very low	very low	moderate	moderate	very low	very low
Surface Characteristics:								
1. Rope feeling to touch	some harshness due to hairs, after use considerable harshness due to broken fiber ends		same as Manila		smooth, after use becomes fuzzy with a softer feel		smooth and somewhat sticky but not slippery, after use becomes fuzzier than nylon or polyester	
2. Rendering qualities (ability of rope to ease out smoothly under load around bitts or capstan heads)	excellent		good		fair		very poor	

[1] Data provided by Wall, Columbian, American, Tubbs/Jackson, and Samson Rope Companies and E.I. Dupont.

[2] Working loads and factors of safety are as recommended by the Cordage Institute and from manufacturers' minimum catalog breaking strength values. These values are based on a percentage of the minimum breaking test of new and unused rope of current manufacture as follows: nylon and kevlar - 11%; polyester, polypropylene, polyethylene, manila, and sisal-17%. These values are based on normal service conditions and do not cover exceptional conditions such as shock loads, sustained loads, etc., nor do they cover conditions where life, limb, or valuable property are involved. In these cases a lower working load or higher factor of safety may be advisable. A higher working load or lower factor of safety should be used only with an expert knowledge of conditions or professional estimates of risk. The factors of safety given are for 3/4 inch dia. rope and larger. Smaller ropes should have larger safety factors; up to 12 for nylon and kevlar and up to 10 for the remainder.

Table 3-11 Comparison of Physical Characteristics of Natural and Synthetic Wire Rope (continued)

ROPE MATERIAL:	Polyester 3 strand		Polypropylene monofilament 3 strand		Polypropylene multifilament 3 strand		Polyethylene 3 strand	
Manufactured or derived from:	Ethylene Glycol & Terephthalic acid		Propane		Propane		Polymers and copolymers of ethylene	
ROPE DIAMETER:	1"	3"	1"	3"	1"	3"	1"	3"
Strength Characteristics:								
1. Tensile strength dry	19,800	157,000	12,600	103,000	11,470	93,750	11,470	93,750
2. Recommended factor of safety	6^2	6^2	6^2	6^2	6^2	6^2	6^2	6^2
3. Working strength^2	$3,300^2$	$26,170^2$	$2,100^2$	$17,200^2$	$1,900^2$	$15,600^2$	$1,900^2$	$15,600^2$
4. Wet strength versus dry strength (percent)	100%	100%	102-105%	102-105%	102-105%	102-105%	102-105%	102-105%
5. Strength per unit of weight (tensile strength/lbs per foot)	66,000	60,900	70,000	67,300	58,800	57,200	62,000	58,900
6. Rope shock load absorption ability	good	good	very good	very good	very good	very good	fair	fair
7. Repeat loading capabilities	excellent	excellent	very good	very good	very good	very good	fair	fair
8. Individual filament or fiber strength (grams per denier)	7.5-9.0	7.5-9.0	6.0-7.5	6.0-7.5	5.0-6.5	5.0-6.5	5.5-7.0	5.5-7.0
Weight & Density Characteristics:								
1. Pounds per 100 feet	30	258	18	153	19-1/2	164	18-1/2	159
2. Specific gravity of fiber	1.38	1.38	0.91	0.91	0.91	0.91	0.95	0.95
3. Ability to float	no	no	yes	yes	yes	yes	yes	yes
Elasticity - Stretch:								
1. Permanent elongation at working loads (break-in)	6%	6%	4%	4%	7-1/2%	7-1/2%	6%	6%
2. Working elasticity at working load (recoverable stretch under load)	6%	6%	9%	9%	10-1/2%	10-1/2%	6%	6%
3. Elongation at 100% load (at break) used rope	20%	20%	24%	24%	36%	36%	22%	22%
4. Creep (elongation under sustained load)	low	low	high	high	high	high	high	high
Surface Characteristics:								
1. Rope feeling to touch	smooth & hard, not slippery, after use becomes fuzzy with a softer feel		smooth but not slippery, after use becomes harsh due to broken fiber ends		smooth and soft with some natural fuzziness, remains same after use		smooth and very slippery, after use becomes slightly harsh due to broken fiber ends	
2. Rendering qualities (ability of rope to ease out smoothly under load around)	good		poor		fair		good but requires extra wraps	

[1] Data provided by Wall, Columbian, American, Tubbs/Jackson, and Samson Rope Companies and E.I. Dupont.

[2] Working loads and factors of safety are as recommended by the Cordage Institute and from manufacturers' minimum catalog breaking strength values. These values are based on a percentage of the minimum breaking test of new and unused rope of current manufacture as follows: nylon and kevlar - 11%; polyester, polypropylene, polyethylene, manila, and sisal - 17%. These values are based on normal service conditions and do not cover exceptional conditions such as shock loads, sustained loads, etc., nor do they cover conditions where life, limb, or valuable property are involved. In these cases a lower working load or higher factor of safety may be advisable. A higher working load or lower factor of safety should be used only with an expert knowledge of conditions or professional estimates of risk. The factors of safety given are for 3/4 inch dia. rope and larger. Smaller ropes should have larger safety factors; up to 12 for nylon and kevlar and up to 10 for the remainder.

Table 3-11 Comparison of Physical Characteristics of Natural and Synthetic Wire Rope (conclusion)

ROPE MATERIAL:	Manila 3 strand	Sisal 3 Strand	Nylon 3 Strand	Kevlar 12 Strand
Water Absorption: (some water will be held between fibers of all ropes)	up to 100% of weight of rope	up to 100% of weight of rope	up to 9%	less than 1%
Flexure:				
1. Dry	good	good	superior	excellent
2. Wet	good	good	excellent	excellent
Resistance to Rot, Mildew, & Attack by Marine Organisms:	poor	very poor	excellent	excellent
Wear:				
1. Resistance to surface abrasion	good	fair	very good	poor
2. Resistance to internal wear from flexing	good	very good	excellent	poor to fair
3. Resistance to cutting (toughness)	good	poor	excellent	excellent
High & Low Temperature Properties:				
1. Melting point	progressive strength loss above 160°F, chars at 275°F	same as Manila	480°F. progressive strength loss above 300°F	does not melt, carbonizes at about 800°F
2. High temperature working limit	200°F	200°F	300°F	400°F
3. Low temperature working limit	-100°F	-100°F	-70°F	-320°F
4. Flammability	burns like wood	burns like wood	burns with difficulty	does not burn
Deterioration:				
1. Due to aging (properly stored ropes)	about 1% per year	about 1% per year	zero	zero
2. Due to exposure to sunlight (ultraviolet rays)	some slight	some slight	some slight	poor
Chemical Resistance:				
1. To acids	very poor	very poor	fair, except to concentrated sulphuric & hydrochloric acids.	fair, except to concentrated sulphuric and nitric acids
2. To alkalis	very poor	very poor	excellent	very good to excellent, slight deterioration from sodium hydroxide
3. To organic solvents	good but hydrocarbons will remove protective rope lubricants	same as Manila	good, soluble in some phenolic compounds and in 90% formic acid	excellent

[1] Data provided by Wall, Columbian, American, Tubbs/Jackson, and Samson Rope Companies and E.I. Dupont.

[2] Working loads and factors of safety are as recommended by the Cordage Institute and from manufacturers' minimum catalog breaking strength values. These values are based on a percentage of the minimum breaking test of new and unused rope of current manufacture as follows: nylon and kevlar - 11%; polyester, polypropylene, polyethylene, manila, and sisal-17%. These values are based on normal service conditions and do not cover exceptional conditions such as shock loads, sustained loads, etc., nor do they cover conditions where life, limb, or valuable property are involved. In these cases a lower working load or higher factor of safety may be advisable. A higher working load or lower factor of safety should be used only with an expert knowledge of conditions or professional estimates of risk. The factors of safety given are for 3/4 inch dia. rope and larger. Smaller ropes should have larger safety factors; up to 12 for nylon and kevlar and up to 10 for the remainder.

Figure 3-34 Example of Fiber Ropes

3.7.2. Wire Ropes

Wire rope is composed of three parts: wires, strands, and a core. The basic unit is the wire. A predetermined number of wires of the proper size are fabricated in a uniform geometric arrangement of definite pitch or lay to form a strand of the required diameter. The required number of strands are then laid together symmetrically around a core to form the rope. Typical properties of marine wire rope are given in Table 3-12.

Table 3-12 Typical Properties of Marine Wire Rope

3.8. Fenders

Fendering is used between ships and compression structures, such as piers and wharves, in fixed moorings. Fenders act to distribute forces on ship hull(s) and minimize the potential for damage. Fendering is also used between moored ships. A wide variety of types of fenders are used including:

- Wooden piles
- Cylindrical marine fenders
- Hard rubber fenders
- Mooring dolphins
- Specially designed structures

The pressure exerted on ships hulls is a key factor to consider when specifying fenders. Allowable hull pressures on ships are discussed in NFESC TR-6015-OCN, *Foam-Filled Fender Design to Prevent Hull Damage*.

Behaviors of some common types of cylindrical marine fenders are shown in Figure 3-35 and Figure 3-36.

Figure 3-35 SEA-GUARD Fender Information

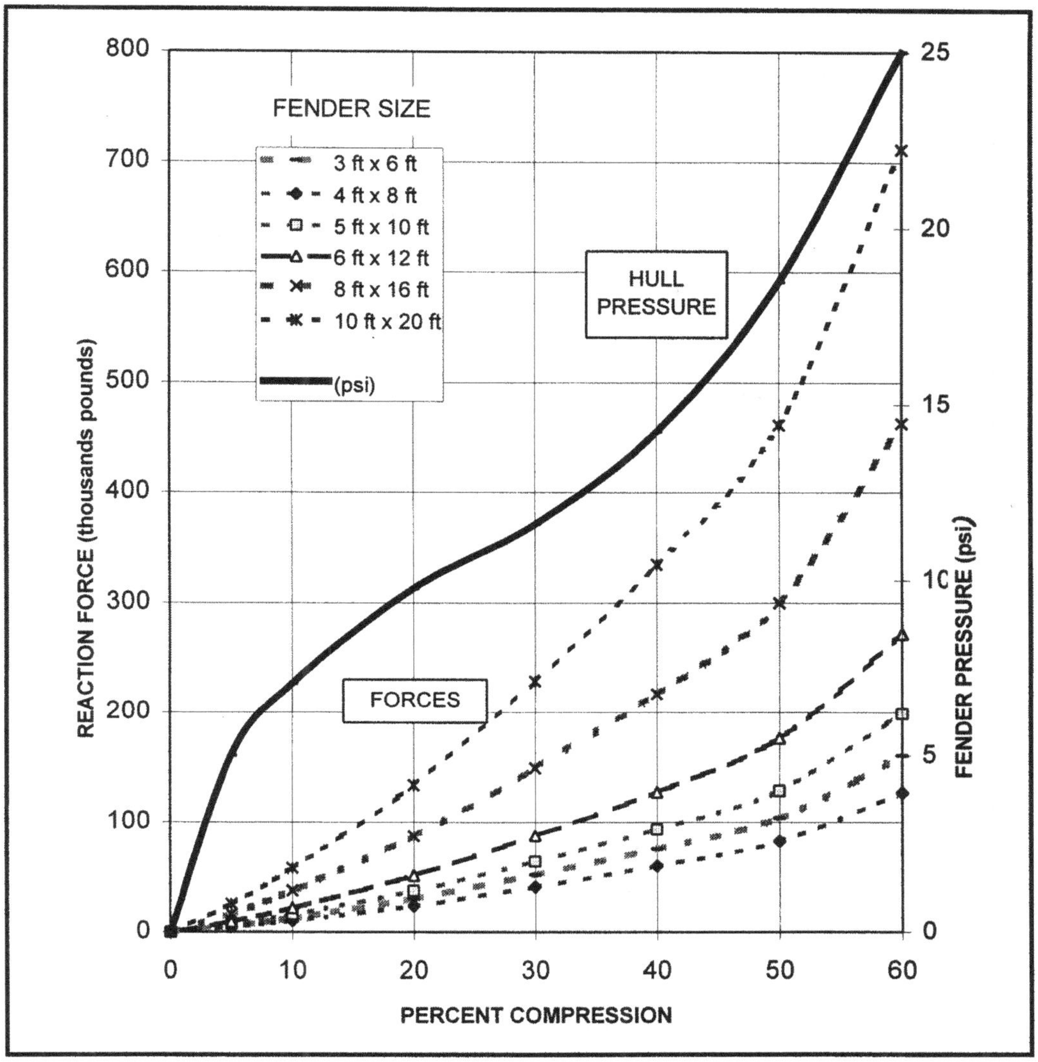

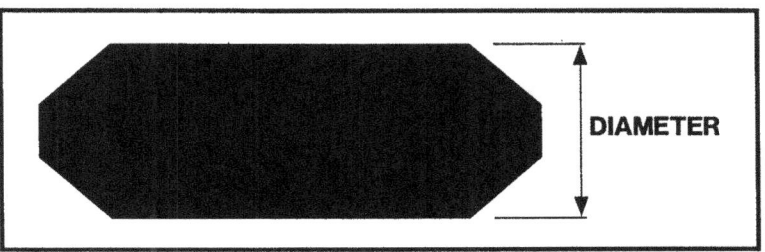

Figure 3-36 SEA-CUSHON Fender Performance

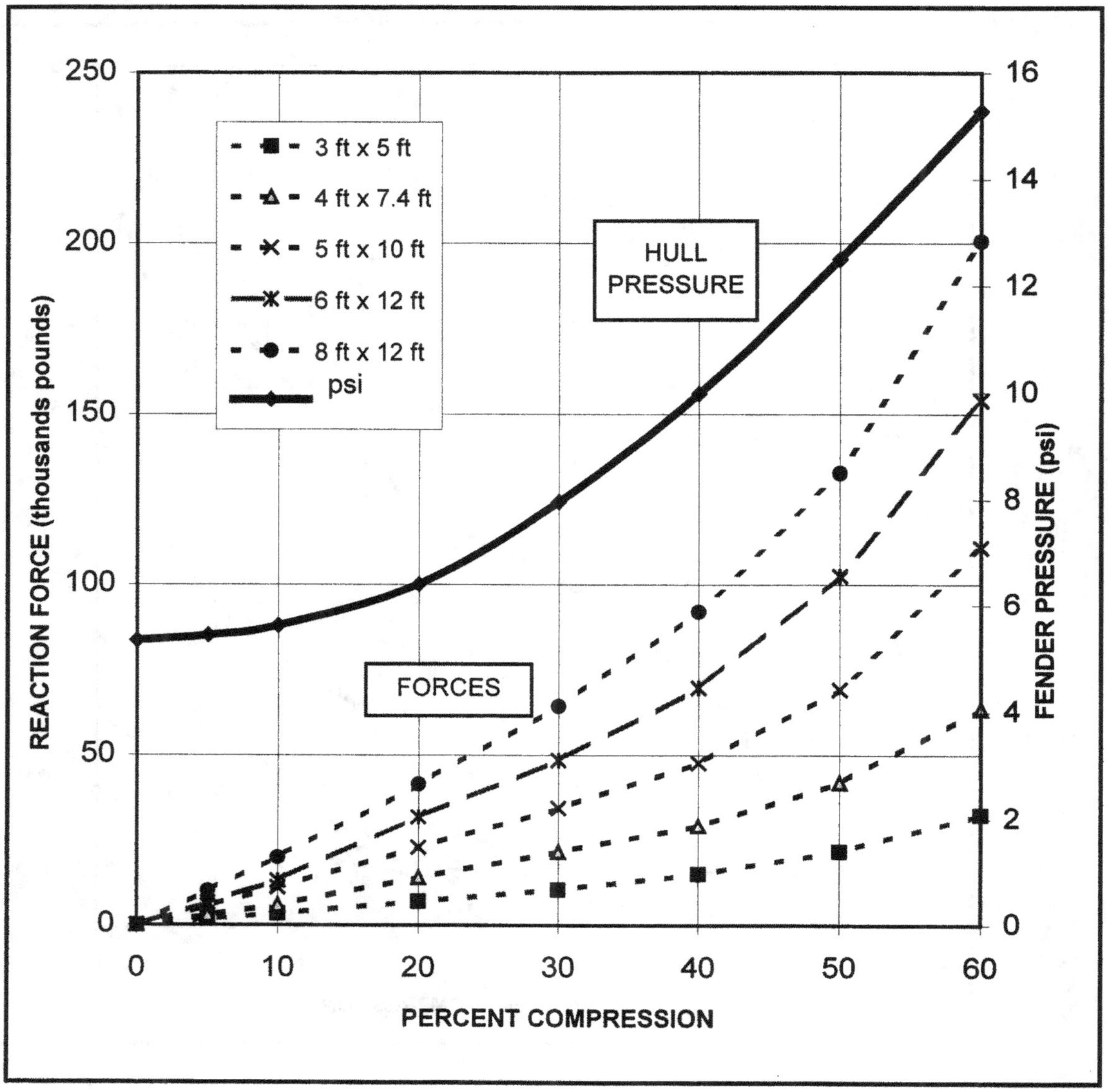

3.9. Pier Fittings

Standard pier and wharf mooring fittings, as shown in Figure 3-37, include:

- Bollards (Figure 3-38)
- Bitts (Figure 3-39)
- Cleats

Cleats are not recommended for ships, unless absolutely necessary, because they are low capacity.

Some of the fittings commonly used on U.S. Navy piers are summarized in Table 3-13. Guidance for placing pier fittings in pier/wharf design is given in MIL-HDBK-1025/1.

Table 3-13 Commonly Used U.S. Navy Pier Mooring Fittings

DESCRIPTION	SIZE	BOLTS	WORKING CAPACITY (kips)
SPECIAL MOORING BOLLARD "A"	Height=48 in. Base 48x48 in.	12 x 2.75-in. dia.	Horz. = 660 @45° = 430 Nom. = 450
SPECIAL MOORING BOLLARD "B"	Height=44.5 in. Base 39x39 in.	8 x 2.25-in. dia.	Horz. = 270 @45° = 216 Nom. = 200
LARGE BOLLARD WITH HORN	Height=44.5 Base 39x39 in.	4 x 1.75-in. dia.	Horz. = 104 @45° = 66 Nom. = 70
LARGE DOUBLE BITT WITH LIP	Height=26 in. Base 73.5x28 in.	10 x 1.75-in. dia.	Nom. = 75[9]
LOW DOUBLE BITT WITH LIP	Height=18 in. Base 57.5x21.5 in.	10 x 1.625-in. dia.	Nom. = 60[9]
42" CLEAT	Height=13 in. Base 26x14.25 in.	6 x 1.125-in. dia.	Nom. = 40
30" CLEAT	Height=13 in. Base 16x16 in.	4 x 1.125-in. dia.	Nom. = 20

---

[9] Working capacity per barrel; after NAVFAC Drawing No. 1404464

Figure 3-37 Pier and Wharf Mooring Fittings Shown in Profile and Plan Views

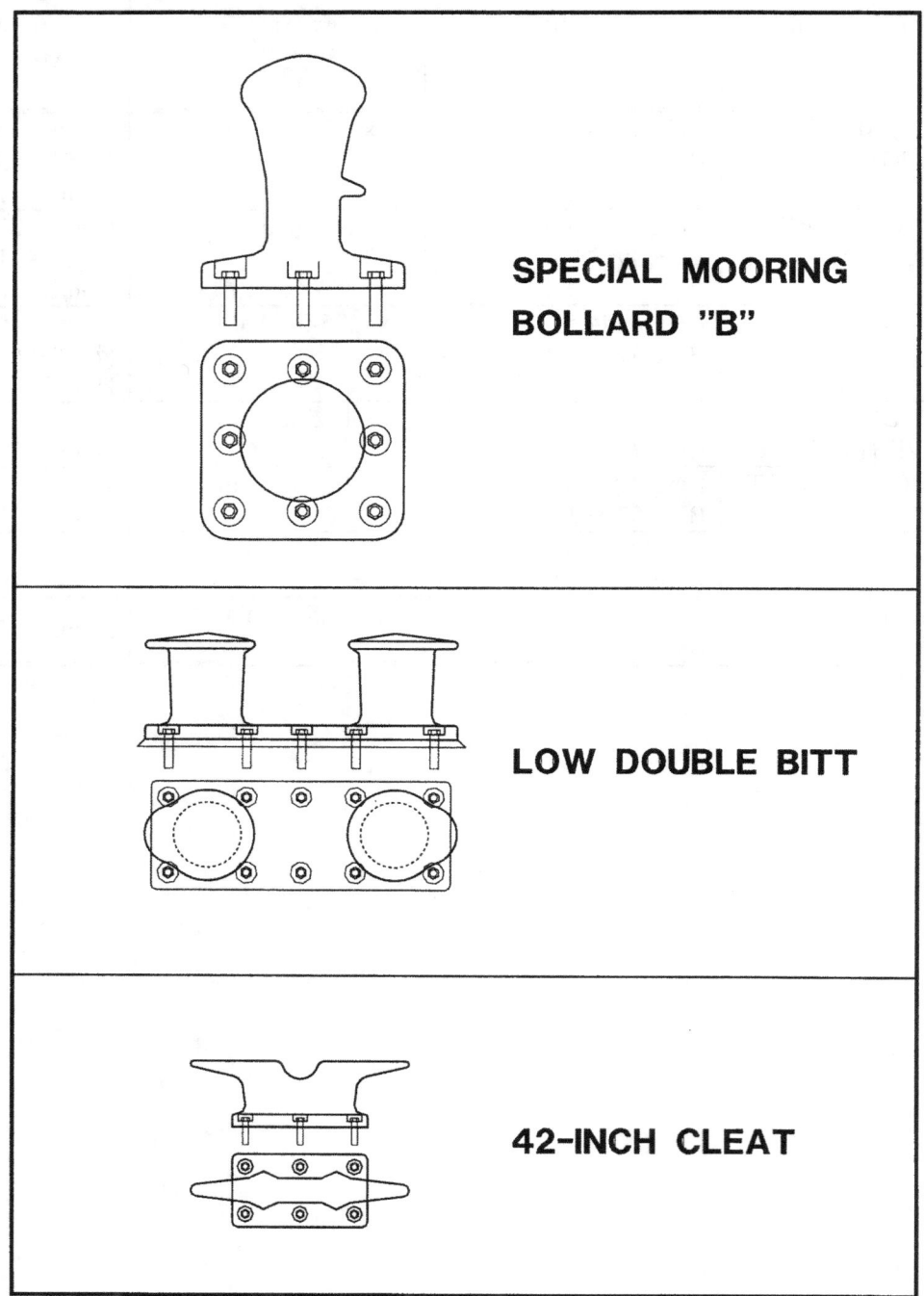

Figure 3-38 Bollard

Figure 3-39 Vessel Secured to Bitts

3.10. Vessel Mooring Equipment

3.10.1. Introduction

A vessel must be provided with adequate mooring equipment to serve its missions. This equipment enables the ship to anchor in a typical soil under design environmental conditions. In addition, the ship can moor to various piers, wharfs, fleet moorings, and other facilities. Equipment on board the ship must be designed for Mooring Service Types I, II and III. Additional mooring hardware, such as specialized padeyes, mooring chains, wire ropes, and lines, can be added for Mooring Service Type IV situations.

3.10.2. Types Of Mooring Equipment

Basic shipboard mooring equipment is summarized in Table 3-14. Additional information is provided in:

- NAVSEA NSTM Chapters 581, 582, 611 and 613; from Naval Sea Systems Command drawings and publications
- Cordage Institute, *Cordage Institute Technical Manual; Guidelines for Deepwater Port Single Point Mooring Design*, Flory et al. (1977)
- *The Choice Between Nylon and Polyester for Large Marine Ropes*, Flory et al. (1988)
- *A Method of Predicting Rope Life and Residual Strength*, Flory et al. (1989)
- *Fiber Ropes for Ocean Engineering in the 21st Century*, Flory et al, (1992a)
- *Failure Probability Analysis Techniques for Long Mooring Lines*, Flory et al. (1992b)

- *Modeling the Long-Term Fatigue Performance of Fibre Ropes*, Hearle et al. (1993)
- Oil Companies International Marine Forum (OCIMF), *Mooring Equipment Guidelines* (1992)
- OCIMF *Recommendations for Equipment Employed in the Mooring of Ships at Single Point Moorings* (1993)
- OCIMF *Prediction of Wind and Current Loads on VLCCs* (1994)
- OCIMF *Single Point Mooring Maintenance and Operations Guide* (1995)
- *Fatigue of SPM Mooring Hawsers*, Parsey (1982).

Table 3-14 Types of Ship Based Mooring Equipment[10]

EQUIPMENT	DESCRIPTION
Drag embedment anchors	One or more anchors required. See Section 7 for anchor information.
Anchor chain	Stud link grade 3 chain (see Section 6.4) is used.
Anchor windlass/wildcat and associated equipment	Equipment for deploying and recovering the anchor(s), including the windlass(s), hawse pipe(s), chain stoppers, chain locker, and other equipment.
Bitts	Bitts for securing mooring lines.
Chocks, mooring rings and fairleads	Fittings through which mooring lines are passed.
Padeyes	Padeyes are provided for specialized mooring requirements and towing.
Mooring lines	Synthetic lines for mooring at piers, wharfs, and other structures. See Section 6.7 for information.
Capstans	Mechanical winches used to aid in handling mooring lines.
Wire ropes	Wire rope is sometimes used for mooring tension members.
Fenders	Marine fenders, as discussed in Section 6.8, are sometimes carried on board.
Winches	Winches of various types can support mooring operations. Some ships use constant tension winches with wire rope automatically paid out/pulled in to adjust to water level changes and varying environmental conditions. Fixed-length synthetic spring lines are used in pier/wharf moorings that employ constant tension winches to keep the ship from 'walking' down the pier.
Other	Various specialized equipment is carried to meet needs (such as submarines).

3.10.3. Equipment Specification

Whenever possible, standard equipment is used on board ships as mooring equipment. The specification, size, number, and location of the equipment is selected to safely moor the ship. Some of the many factors that need to be considered in equipment specification are weight, room required, interaction with other systems, power requirements, reliability, maintenance, inspection, and cost.

---

[10] See NAVSEASYSCOM Naval Ships' Technical Manual for additional information and Section 3.1 for design criteria.

3.10.4. Fixed Bitts

Bitts provide a termination for tension members. Fixed bitts, Figure 3-40, are typically placed in pairs within a short distance forward or aft of a chock location. They are often placed symmetrically on both the port and starboard sides, so that the ship can moor to port or starboard. Capacities of the bitts are based on their nominal diameter. Table 3-15 provides fixed bitt sizes with their associated capacities. An example of fixed bitts in use is shown in Figure 3-41.

Figure 3-40 Fixed and Recessed Shell Bitts

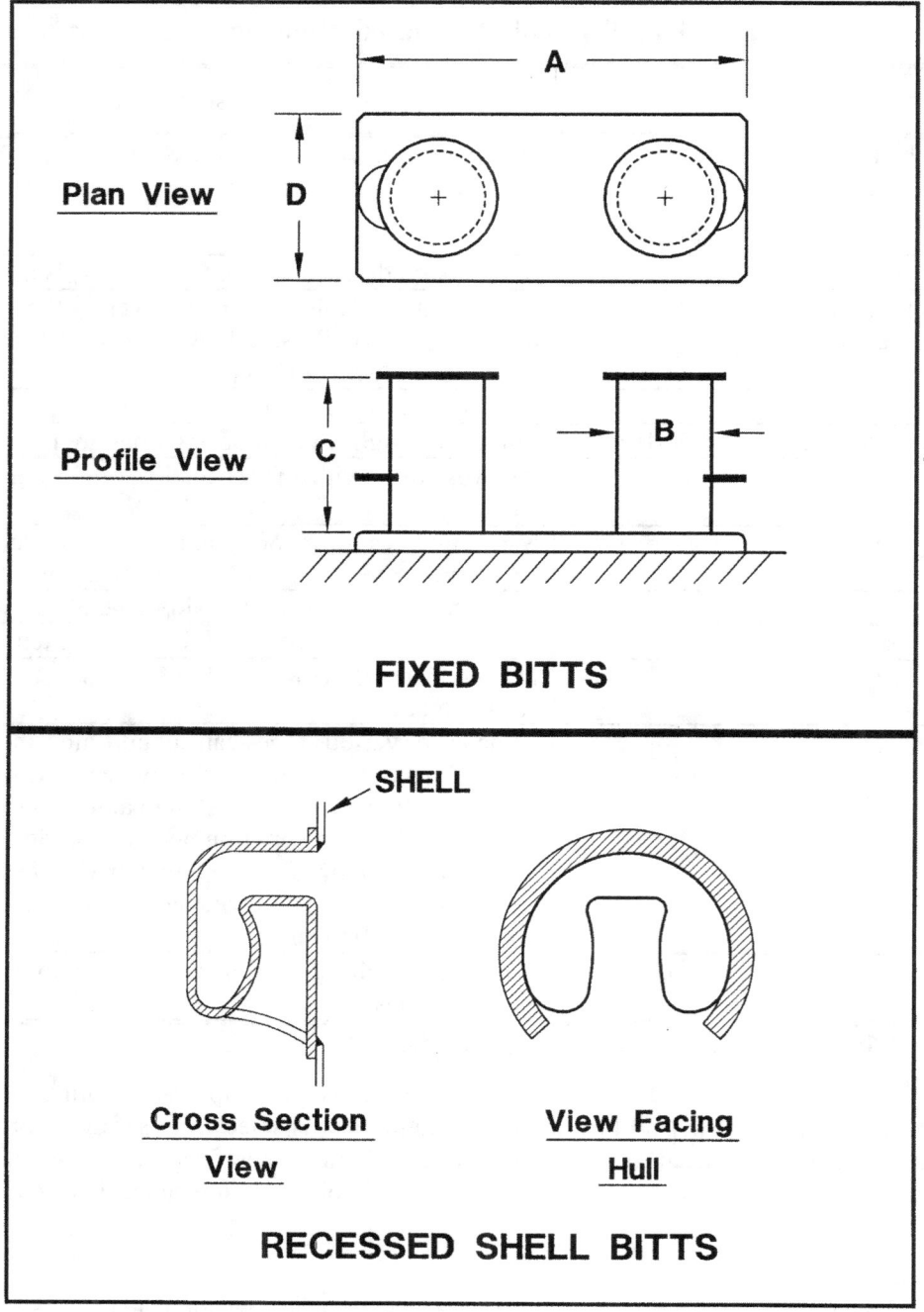

Mooring Systems: Design and Maintenance
Pile Buck International Inc.

Table 3-15 Fixed Ships' Bitts (minimum strength requirements)

NAVSEA FIXED BITTS (after 804-1843362 REV B OF 1987)						
NOMINAL SIZE (inches)	4	8	10	12	14	18
MAX. LINE CIR. (inches)	3	5	6.5	8	10	12
MAX. LINE DIA. (inches)	1.0	1.6	2.1	2.5	3.2	3.8
MAX. MOMENT (lbf-in x 1000)	134	475	1046	1901	3601	6672
MAX. CAPACITY (lbf x 1000)*	26.8	73.08	123.1	181	277	417
A - BASE LENGTH (inches)	16.5	28.63	36.75	44.25	52.5	64
B - BARREL DIA. (inches)	4.5	8.625	10.75	12.75	14	18
C - BARREL HT. (inches)	10	13	17	21	26	32
D - BASE WIDTH (inches)	7.5	13.63	17.25	20.25	22.5	28
MAX. LINE CIR. (mm)	76	127	165	203	254	305
MAX. CAPACITY (newton x 100000)*	1.19	3.25	5.47	8.05	12.32	18.55
A - BASE LENGTH (inches)	419	727	933	1124	1334	1626
B - BARREL DIA. (inches)	114	219	273	324	356	457
C - BARREL HT. (inches)	254	330	432	533	660	813
D - BASE WIDTH (inches)	191	346	438	514	572	711

* force applied at half the barrel height

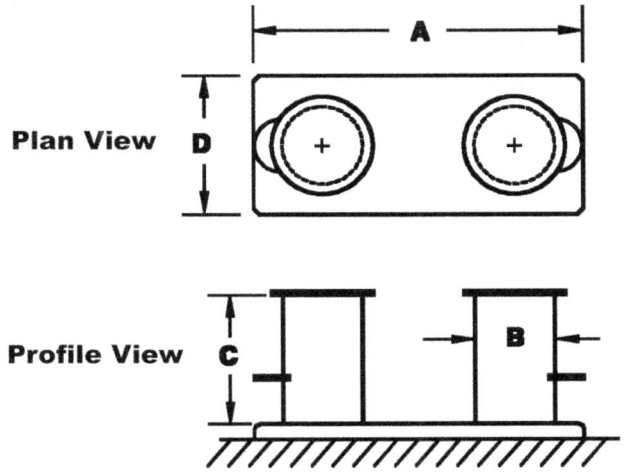

Figure 3-41 Fixed Bitts in Use

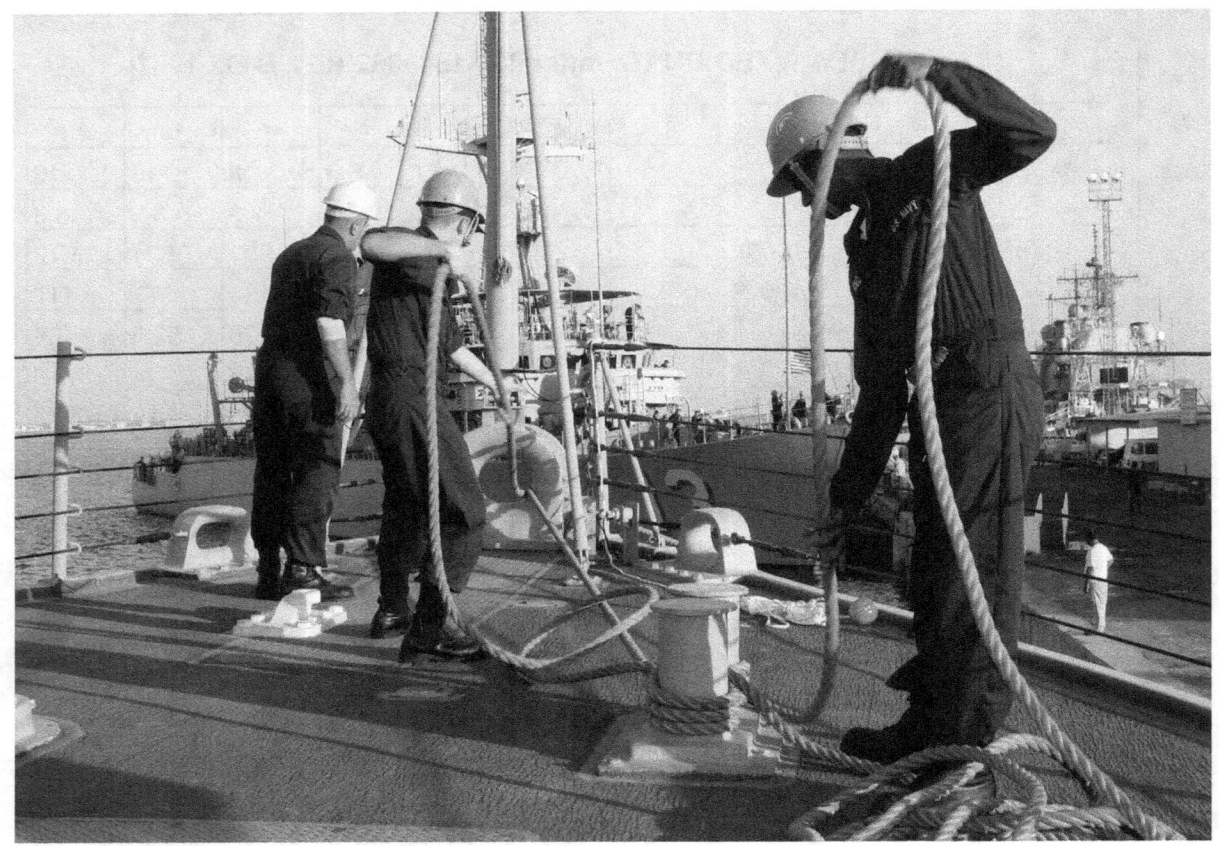

3.10.5. Recessed Shell Bitts

Recessed shell bitts, Figure 3-42, are inset into ships' hulls well above the waterline. These bitts are used to moor lighterage or harbor craft alongside. They also assist in mooring at facilities. The NAVSEA shell bitt has a total working capacity of 92 kips (427 kN) with two lines of 46 kips maximum tension each.

Figure 3-42 Recessed Shell Bitt (minimum strength requirements)

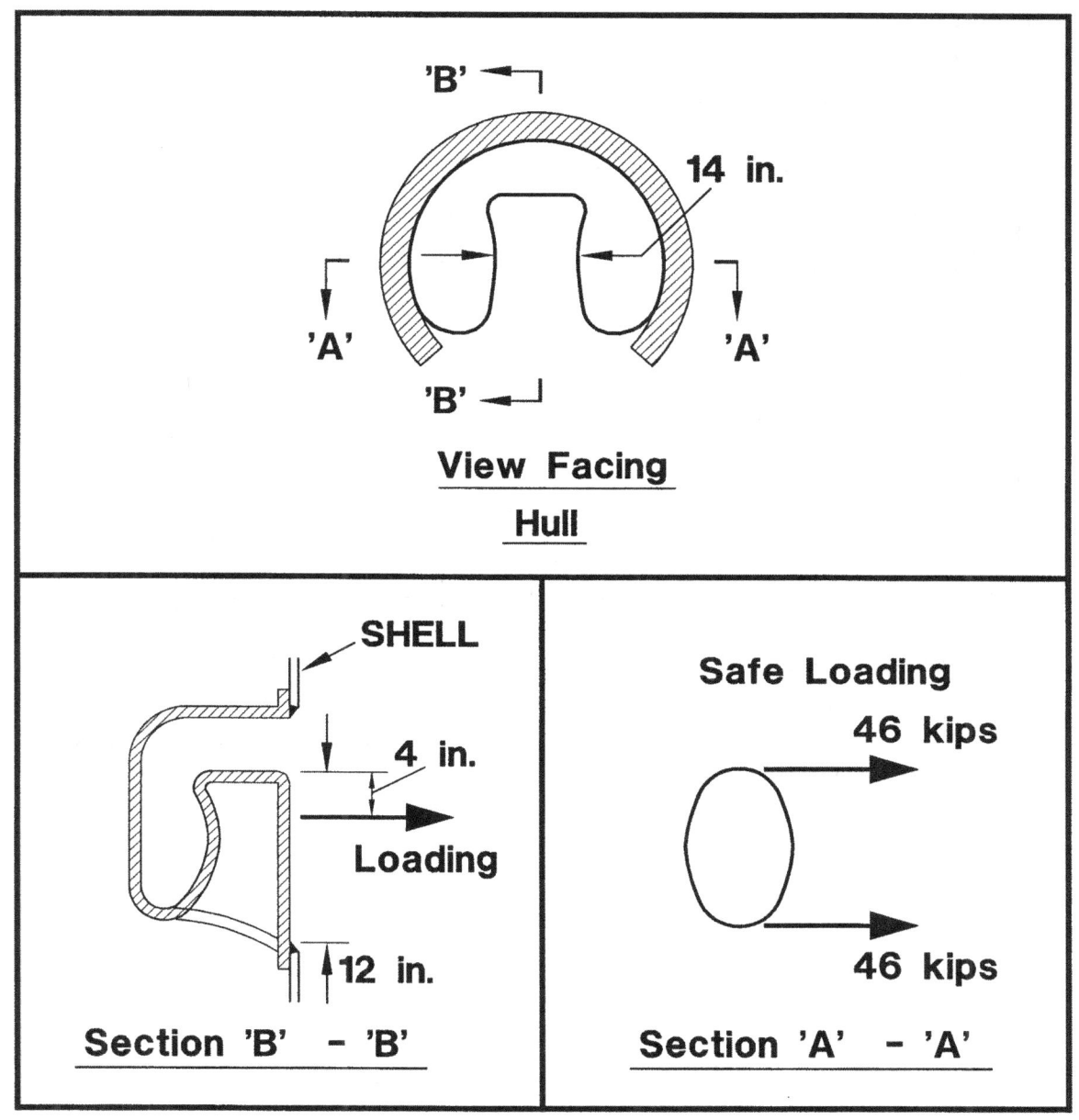

3.10.6. Exterior Shell Bitts

Aircraft carriers have exterior shell bitts, Drawing No. 600-6601101, that are statically proof loaded to 184 kips (820 kN). This proof load is applied 11 inches (280 mm) above the base. This testing is described in the Newport News Shipbuilding testing report for USS HARRY S TRUMAN *Bitts, Chocks and Mooring Rings*.

3.10.7. Chocks

There are many types of chocks, such as closed chocks, Panama chocks, roller chocks, and mooring rings. Closed clocks are often used and characteristics of these fittings are shown in Table 3-16. Figure 3-43 shows a chock in use.

Table 3-16 Closed Chocks (minimum strength requirements)

NAVSEA CLOSED CHOCKS (from Drawing 804-1843363)						
CHOCK SIZE (inches)	6	10	13	16	20	24
MAX. LINE CIR. (inches)	3	5	6.5	8	10	12
LINE BREAK (lbf x 1000)	26.8	73	123	181	277	417
A - HOLE WIDTH (inches)	6	10	13	16	20	24
B - HOLE HEIGHT (inches)	3	5	6.5	8	10	12
C - HEIGHT (inches)	8.5	11.25	13.88	16.75	25.75	25.25
D - BASE THICKNESS (inches)	5.25	6.5	7.5	9	16	13.5
E - LENGTH (inches)	13	19	23	28	38.75	40
MAX. LINE CIR. (mm)	76	127	165	203	254	305
LINE BREAK (newton x 100000)	1.19	3.25	5.47	8.05	12.32	18.55
A - HOLE WIDTH (mm)	152	254	330	406	508	610
B - HOLE HEIGHT (mm)	76	127	165	203	254	305
C - HEIGHT (mm)	216	286	352	425	654	641
D - BASE THICKNESS (mm)	133	165	191	229	406	343
E - LENGTH (mm)	330	483	584	711	984	1016

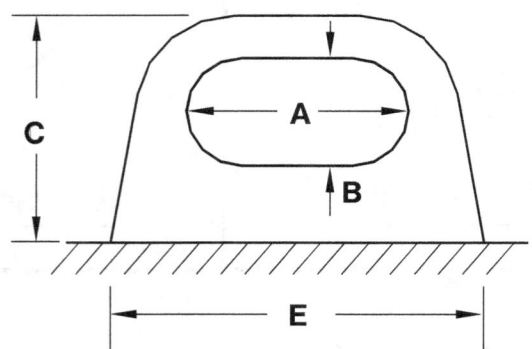

Note: D = thickness at the base

Figure 3-43 Chock in Use

3.10.8. Allowable Hull Pressures

As a ship berths or when it is moored, forces may be exerted by structures, such as fenders, camels, and dolphins, on the ship hull. NFESC TR-6015-OCN, *Foam-Filled Fender Design to Prevent Hull Damage* provides a rational design criterion to prevent yielding of vessel hull plating.

3.11. Sources Of Information

- *Facility Mooring Equipment.* Detailed NAVFAC information, including drawings, specifications, and manuals, is available in the National Institute of Building Sciences, "Construction Criteria Base." Further information can be obtained from the Naval Facilities Engineering Command Criteria Office (Code 15C) and the Naval Facilities Engineering Service Center, Moorings Center of Expertise (Code 551). A list of sources for information on facility mooring equipment is provided in Table 3-17.

- *Ships' Mooring Equipment.* Additional information is available from the Naval Sea Systems Command (NAVSEA 03P), Military Sealift Command (MSC), and the U.S. Coast Guard (USCG). Table 3-18 provides a list of selected referenced materials.

Table 3-17 Sources of Information for Facility Mooring Equipment

ITEM	SOURCE
Standard fittings for waterfront structures	NAVFAC Drawing No. 1404464
Marine fenders	NAVFAC Spec. 02396A "Resilient Foam-Filled Marine

	Fenders"
	NAVFAC Spec. 02397B "Arch-Type Rubber Marine Fenders"
	NAVFAC Spec. 02395 "Prestressed Concrete Fender Piling"
	NAVFAC Spec. Y02882D "Fenders (Yokosuka, JA)"
Camels	MIL-C-28628C(YD) "Camel, Wood, Marine; Single Log Configuration, Untreated"
	"Standard Aircraft Carrier Mooring Camel" NAVFAC Drawings SD-1404045A to 52 and NAVFAC Spec. C39
	"Standard Submarine Mooring Camel" NAVFAC Drawings SD-1404943 to 47 and NAVFAC Spec. C46
	"Standard Attack Submarine Mooring Camel" NAVFAC Drawings SD-1404667 to 70 and NAVFAC Spec. C49
Mooring lines	Cordage Institute Technical Manual
Foam buoys	NFESC purchase descriptions of Mar. 1988, Dec. 1989 and May 1990.
Stud link chain and fittings	NFESC purchase description of Mar. 1995.
NAVMOOR anchors	NFESC purchase description of Nov. 1985 and drawing package of July 1990.
Stud link chain anodes	NFESC purchase description of June 1990.

Table 3-18 Sources of Information for Ships' Mooring Equipment

ITEM	SOURCE
Information on existing U.S. Navy ships, drawings, and mooring hardware	NAVFAC Ships Database
Chocks	NAVSEA Drawing No. 804-1843363 & S1201-921623 (Roller Chock)
Panama chocks	NAVSEA Drawing No. 804-1843363
Fixed bitts	NAVSEA Drawing No. 804-1843362
Recessed shell bitts	NAVSEA Drawing No. 805-1841948
Exterior shell bitts	Newport News Shipbuilding Drawing No. 600-6601101
Cleats	NAVSEA Drawing No. 804-2276338
Capstans/gypsy heads	NAVSEA Drawing No. S260-860303 & MIL-C-17944

Hawser reels	NAVSEA Drawing No. S2604-921841 & 42
Mooring lines	Cordage Institute Technical Manual

Chapter 4. Basic Design Procedure

4.1. Design Approach

Begin the design with specified parameters and use engineering principles to complete the design. Types of parameters associated with mooring projects are summarized in Table 4-1. The basic approach to performing mooring design with the ship known is given in Table 4-2.

Table 4-1 Parameters in a Mooring Project

PARAMETER	EXAMPLES
1. Operational Parameters	Required ship position, amount of motion allowed
2. Ship Configuration	Basic ship parameters, such as length, width, draft, displacement, wind areas, mooring fitting locations, wind/current force, and moment coefficients
3. Facility Configuration	Facility location, water depth, dimensions, locations/type/capacity of mooring fittings/fenders, facility condition, facility overall capacity
4. Environmental Parameters	Wind speed, current speed and direction, water levels, wave conditions and possibility of ice
5. Mooring Configuration	Number/size/type/location of tension members, fenders, camels, etc.
6. Material Properties	Stretch/strain characteristics of the mooring tension and compression members

Table 4-2 Basic Mooring Design Approach With Known Facility for a Specific Site and a Specific Ship

STEP	NOTES
Define customer(s) requirements	Define the ship(s) to be moored, the type of service required, the maximum allowable ship motions, and situations under which the ship will leave.
Determine planning requirements	Define the impact/interaction with other facilities and operations, evaluate explosive arcs, determine permit requirements, establish how the mooring is to be used, review the budget and schedule.
Define site and environmental parameters	Determine the water depth(s), engineering soil parameters, design winds, design currents, design waves, design water levels, and evaluate access.
Ship characteristics	Find the engineering characteristics of the ship(s) including sail areas, drafts, displacements, ship mooring fittings, allowable hull pressures, and other parameters.

Ship forces/moments	Determine the forces, moments, and other key behaviors of the ship(s).
Evaluate mooring alternatives	Evaluate the alternatives in terms of safety, risk, cost, constructability, availability of hardware, impact on the site, watch circle, compatibility, maintenance, inspectability, and other important aspects.
Design Calculations	Perform static and/or dynamic analyses (if required) for mooring performance, anchor design, fender design, etc
Plans/Specs	Prepare plans, specifications, and cost estimates.
Permits	Prepare any required environmental studies and obtain required permits.
Installation planning	Prepare instructions for installation, including safety and environmental protection plans.
Installation monitoring	Perform engineering monitoring of the installation process.
Testing	Perform pull tests of all anchors in mooring facilities to ensure that they hold the required load.
Documentation	Document the design and as-built conditions with drawings and reports.
Instructions	Provide diagrams and instructions to show the customer how to use and inspect the mooring.
Inspection	Perform periodic inspection/testing of the mooring to assure it continues to meet the customer(s) requirements.
Maintenance	Perform maintenance as required and document on as-built drawings.

4.2. General Design Criteria

General design issues shown in Table 4-3 should be addressed during design to help ensure projects meet customers' needs.

Table 4-3 Design Issues

CRITERIA	NOTES
Vessel operating conditions	Under what conditions will the vessel(s) exit? What are the operating mission requirements for the ship? What is the maximum allowable hull pressure?
Allowable motions	How much ship motion in the six degrees-of-freedom will be allowable for the moored ship? This is related to brow positions and use, utilities, ship loading and unloading operations, and other requirements. Note that most ships have a very high buoyancy force and moorings should be designed to allow for water level changes at a site.

User skills	Is the user trained and experienced in using the proposed system? What is the risk that the mooring would be improperly used? Can a design be formulated for easy and reliable use?
Flexibility	How flexible is the design? Can it provide for new mission requirements not yet envisioned? Can it be used with existing facilities/ships?
Constructability	Does the design specify readily available commercial products and is it able to be installed and/or constructed using standard techniques, tolerances, etc.?
Cost	Are initial and life cycle costs minimized?
Inspection	Can the mooring system be readily inspected to ensure continued good working condition?
Maintenance	Can the system be maintained in a cost-effective manner?
Special requirements	What special requirements does the customer have? Are there any portions of the ship that cannot come in contact with mooring elements (e.g., submarine hulls)?

4.2.1. Mooring Service Types

There are several types of standard services that moorings provide for DOD vessels in harbors. Therefore, the facilities and ship's mooring hardware should accommodate the types of services shown in Table 4-4. An example of mooring under adverse conditions is shown in Figure 4-1.

Table 4-4 Mooring Service Types

MOORING SERVICE TYPE	DESCRIPTION
TYPE I	This category covers moorings that are used in winds of less than 34 knots and currents less than 2 knots. Moorings include ammunition facilities, fueling facilities, deperming facilities, and ports of call. Use of these moorings is normally selected concomitant with forecasted weather.
TYPE II	This category covers moorings that for general purpose berthing by a vessel that **will leave** prior to an approaching tropical hurricane, typhoon, or flood.
TYPE III	This category covers moorings that are used for **up to 2 years** by a vessel that **will not leave** prior to an approaching tropical hurricane or typhoon. Moorings include fitting-out, repair, drydocking, and overhaul berthing facilities. Ships experience this service approximately every 5 years. Facilities providing this service are nearly always occupied.
TYPE IV	This category covers moorings that are used for **2 years or more** by a vessel that **will not leave** in case of a hurricane, typhoon, or flood. Moorings include inactive, drydock, ship museum, and training

	berthing facilities.

Figure 4-1 Mooring in Adverse Conditions

4.2.2. Facility Design Criteria For Mooring Service Types

Mooring facilities should be designed using the site-specific criteria given in Table 4-5. This table gives design criteria in terms of environmental design return intervals, R, and in terms of probability of exceedence, P, for 1 year of service life, N = 1.

Table 4-5 Facility Design Criteria for Mooring Service Types

MOORING SERVICE TYPE	WIND[11]	CURRENT[12]	WATER LEVEL	WAVES
TYPE I	< 34 knots	≤ 2 knots	mean lower low to mean higher high	P=1 or R=1 yr
TYPE II	P=0.02 (min.) R=50 yr (min.)	P=0.02 R=50	extreme lower low to mean higher high	P=1 or

---

[11] Use exposure D (American Society of Civil Engineers (ASCE) 7-95, *Minimum Design Loads for Buildings and Other Structures*; flat, unobstructed area exposed to wind flowing over open water for a distance of at least 1 mile or 1.61 km) for determining design wind speeds. Note that min. = minimum return interval or probability of exceedence used for design; max. = maximum wind speed used for design.

[12] To define the design water depth, use T/d = 0.9 for flat keeled ships; for ships with non-flat hulls, that have sonar domes or other projections, take the ship draft, T, as the mean depth of the keel and determine the water depth, d, by adding 0.61 meter (2 feet) to the maximum navigation draft of the ship.

	V_w=64 knots (max.)	yr		R=1 yr
TYPE III	P=0.02 or R=50 yr	P=0.02 or R=50 yr	extreme lower low to high	P=0.02 or R=50 yr
TYPE IV	P=0.01 or R=100 yr	P=0.01 or R=100 yr	extreme water levels	P=0.01 or R=100 yr

4.2.3. Ship Hardware Design Criteria For Mooring Service Types

Ship mooring hardware needs to be designed to accommodate various modes of ship operation. During Type II operation, a ship may be moored in relatively high broadside current and get caught by a sudden storm, such as a thunderstorm. Type III mooring during repair may provide the greatest potential of risk, because the ship is moored for a significant time and cannot get underway. During Type IV mooring, the ship should be aligned with the current, extra padeyes can be welded to the ship hull for mooring, etc., so special provisions can be made for long-term storage. There are several U.S. shipyards where DOD ships can undergo major repairs. The area near Norfolk/Portsmouth, Virginia has the most extreme design criteria, so use conditions derived from that site for the ship's hardware design. Bremerton, Washington, and Pearl Harbor, Hawaii have major U.S. Navy repair shipyards with lower design winds and currents at those sites. Ship mooring hardware environmental design criteria are given in Table 4-6.

Table 4-6 Ship Mooring Hardware Design Criteria[13]

a. Ship Anchor Systems

MAXIMUM WATER DEPTH	MINIMUM WIND SPEED	MINIMUM CURRENT SPEED	CHAIN FACTOR OF SAFETY	ANCHOR HOLDING FACTOR OF SAFETY
240 ft (73 m)	70 knots (36.0 m/s)	4 knots (2.06 m/s)	4.0	1.0

b. Submarine Anchor Systems

MAXIMUM WATER DEPTH	MINIMUM WIND SPEED	MINIMUM CURRENT SPEED	CHAIN FACTOR OF SAFETY	ANCHOR HOLDING FACTOR OF SAFETY
120 ft (36.6 m)	70 knots (36.0 m/s)	4 knots (2.06 m/s)	4.0	1.0

---

[13] Quasi-static design assuming wind and current are co-linear for ship and submarine anchor systems (after NAVSEA DDS-581). For ship mooring systems, Quasi-static design assuming current is broadside and wind can approach from any direction (after NAVSEA DDS-582-1).

c. Ship Mooring Systems

CONDITION	MINIMUM WIND SPEED	MINIMUM CURRENT SPEED	MOORING LINE FACTOR OF SAFETY
Normal weather condition	25 knots (12.9 m/s)	1 knot (0.51 m/s)	9.0
Heavy weather condition	50 knots (25.7 m/s)	3 knots (1.54 m/s)	3.0

4.2.4. Strength

Moorings should be designed and constructed to safely resist the nominal loads in load combinations defined herein without exceeding the appropriate allowable stresses for the mooring components. Normal wear of materials and inspection methods and frequency need to be considered. Due to the probable chance of simultaneous maximum occurrences of variable loads, no reduction factors should be used.

4.2.5. Serviceability

Moorings should be designed to have adequate stiffness to limit deflections, vibration, or any other deformations that adversely affect the intended use and performance of the mooring. At the same time moorings need to be flexible enough to provide for load sharing and allow for events, such as tidal changes.

4.2.6. General Mooring Integrity

For multiple-member moorings, such as for a ship secured to a pier by a number of lines, the mooring system strongly relies on load sharing among several members. If one member is lost, the ship should remain moored. Therefore, design multiple member mooring to ensure that remaining members maintain a factor of safety at least 75% of the intact mooring factors of safety shown in Table 4-7 with any one member missing.

Table 4-7 Minimum Quasi-Static Factors of Safety

COMPONENT	MINIMUM FACTOR OF SAFETY	NOTES
Stockless & balanced fluke anchors	1.5	For ultimate anchoring system holding capacity; use 1.0 for ship's anchoring.[14]
High efficiency drag anchors	2.0	For ultimate anchoring system holding capacity; use 1.0 for ship's anchoring.[14]
Fixed anchors (piles & plates)	3.0	For ultimate anchoring system holding capacity.[14]
Deadweight anchors	-	Use carefully (see Naval Civil Engineering Laboratory (NCEL) *Handbook for Marine*

---

[14] It is recommended that anchors be pull tested.

		Geotechnical Engineering, 1985, available from Pile Buck)
Chain	3.0	For relatively straight lengths.
	4.0	For chain around bends.
		These factors of safety are for the new chain break strength.
Wire rope	3.0	For the new wire rope break strength.
Synthetic line[15]	3.0	For new line break strength.
Ship bitts	[16]	Use American Institute of Steel Construction (AISC) code.
Pier bollards	[16]	Use AISC & other applicable codes.

4.2.7. Quasi-Static Safety Factors

Table 4-7 gives recommended minimum factors of safety for "quasi-static" design based on material reliability.

4.2.8. Allowable Ship Motions

Table 4-8 gives recommended operational ship motion criteria for moored vessels.

- Table 4-8(a) gives maximum wave conditions for manned and moored small craft (Permanent International Association of Navigation Congresses (PIANC), *Criteria for Movements of Moored Ships in Harbors; A Practical Guide*, 1995). These criteria are based on comfort of personnel on board a small boat, and are given as a function of boat length and locally generated.
- Table 4-8(b) gives recommended motion criteria for safe working conditions for various types of vessels (PIANC, 1995).
- Table 4-8(c) gives recommended velocity criteria.
- Table 4-8(d) and Table 4-8(e) give special criteria.

Table 4-8 Recommended Practical Motion Criteria for Moored Vessels

(a) Safe Wave Height Limits for Moored Manned Small Craft
(after PIANC, 1995)

	Beam/Quartering Seas		Head Seas	
Vessel Length (m)	Wave Period (sec)	Maximum Sign Wave	Wave Period (sec)	Maximum Sign Wave

---

[15] Reduce effective strength of wet nylon line by 15%.

[16] For mooring fittings take 3 parts of the largest size of line used on the fitting; apply a load of: 3.0*(minimum line break strength)*1.3 to determine actual stresses, $\sigma_{act.}$; design fittings so ($\sigma_{act.}/\sigma_{allow.}$)<1.0, where $\sigma_{allow.}$ is the allowable stress from AISC and other applicable codes.

		Height, H_s (m)		Height, H_s (m)
4 to 10	<2.0	0.20	<2.5	0.20
	2.0-4.0	0.10	2.5-4.0	0.15
	>4.0	0.15	>4.0	0.20
10-16	<3.0	0.25	<3.5	0.30
	3.0-5.0	0.15	3.5-5.5	0.20
	>5.0	0.20	>5.5	0.30
20	<4.0	0.30	<4.5	0.30
	4.0-6.0	0.15	4.5-7.0	0.25
	>6.0	0.25	>7.0	0.30

(b) Recommended Motion Criteria for Safe Working Conditions[17] (after PIANC, 1995)

Vessel Type	Cargo Handling Equipment	Surge (m)	Sway (m)	Heave (m)	Yaw (°)	Pitch (°)	Roll (°)
Fishing vessels 10-3000 GRT[18]	Elevator crane	0.15	0.15	-	-	-	-
	Lift-on/off	1.0	1.0	0.4	3	3	3
	Suction pump	2.0	1.0	-	-	-	-
Freighters & coasters <10000 DWT[19]	Ship's gear	1.0	1.2	0.6	1	1	2
	Quarry cranes	1.0	1.2	0.8	2	1	3
Ferries, Roll-On/Roll-Off (RO/RO)	Side ramp[20]	0.6	0.6	0.6	1	1	2
	Dew/storm ramp	0.8	0.6	0.8	1	1	4
	Linkspan	0.4	0.6	0.8	3	2	4
	Rail ramp	0.1	0.1	0.4	-	1	1
General cargo 5000-10000 DWT	-	2.0	1.5	1.0	3	2	5
Container vessels	100% efficient	1.0	0.6	0.8	1	1	3
	50% efficient	2.0	1.2	1.2	1.5	2	6
Bulk carriers 30000-150000 DWT	Cranes	2.0	1.0	1.0	2	2	6
	Elevator/bucket-wheel	1.0	0.5	1.0	2	2	2
	Conveyor belt	5.0	2.5	-	3	-	-

---

[17] Motions refer to peak-to-peak values (except for sway, which is zero-to-peak)
[18] GRT = Gross Registered Tons expressed as internal volume of ship in units of 100 ft^3 (2.83 m^3)
[19] DWT = Dead Weight Tons, which is the total weight of the vessel and cargo expressed in long tons (1016 kg) or metric tons (1000 kg)
[20] Ramps equipped with rollers.

| Oil tankers | Loading arms | 3.0[21] | 3.0 | - | - | - | - |
| Gas tankers | Loading arms | 2.0 | 2.0 | - | 2 | 2 | 2 |

(c) Recommended Velocity Criteria for Safe Mooring Conditions for Fishing Vessels, Coasters, Freighters, Ferries and Ro/Ro Vessels (after PIANC, 1995)

Ship Size(DWT)	Surge (m/s)	Sway (m/s)	Heave (m/s)	Yaw (°/s)	Pitch (°/s)	Roll (°/s)
1000	0.6	0.6	-	2.0	-	2.0
2000	0.4	0.4	-	1.5	-	1.5
8000	0.3	0.3	-	1.0	-	1.0

(d) Special Criteria for Walkways and Rail Ramps (after PIANC, 1995)

Parameter	Maximum Value
Vertical velocity	0.2 m/s
Vertical acceleration	0.5 m/s^2

(e) Special Criteria

CONDITION	MAXIMUM VALUES	NOTES
Heave	-	Ships will move vertically with any long period water level change (tide, storm surge, flood, etc.). The resulting buoyancy forces may be high, so the mooring must be designed to provide for these motions due to long period water level changes.
Loading/unloading preposition ships	0.6 m (2 feet)	Maximum ramp motion during loading/unloading moving wheeled vehicles.
Weapons loading/unloading	0.6 m (2 feet)	Maximum motion between the crane and the object being loaded/unloaded.

4.3. Design Methods

4.3.1. Quasi-Static Design

Practical experience has shown that in many situations such as for Mooring Service Types I and II, static analysis tools can be used to reliably determine mooring designs in harbors. Winds are a key

---

[21] For exposed locations, loading arms usually allow for 5.0-meter motion.

forcing factor in mooring harbors. Winds can be highly dynamic in heavy weather conditions. However, practical experience has shown that for typical DOD ships, a wind speed with a duration of 30 seconds can be used, together with static tools, to develop safe mooring designs. The use of the 30-second duration wind speed with static tools and the approach shown in Table 4-9 is called "quasi-static" design.

Table 4-9 Quasi-Static Design Notes

CRITERIA	NOTES
Wind speed	Determine for the selected return interval, R. For typical ships use the wind that has a duration of 30 seconds at an elevation of 10 m.
Wind direction	Assume the wind can come from any direction except in cases where wind data show extreme winds occur in a window of directions.
Current speed	Use conditions for the site (speed and direction).
Water levels	Use the range for the site.
Waves	Neglected. If waves are believed to be important, then dynamic analyses are recommended.
Factors of safety	Perform the design using quasi-static forces and moments, minimum factors of safety in Table 4-7, and design to assure that all criteria are met.

4.3.2. Dynamic Mooring Analysis

Conditions during Mooring Service Types III and IV, and during extreme events can be highly dynamic. Unfortunately, the dynamic behavior of a moored ship in shallow water can be highly complex, so dynamics cannot be fully documented in this book. Information on dynamics is found in:

- *Dynamic Analysis of Moored Floating Drydocks*, Headland et. al. (1989)
- *Advanced Dynamics of Marine Structures*, Hooft (1982)
- *Hydrodynamic Analysis and Computer Simulation Applied to Ship Interaction During Maneuvering in Shallow Channels*, Kizakkevariath (1989)
- David Taylor Research Center (DTRC), SPD-0936-01, *User's Manual for the Standard Ship Motion Program, SMP81*
- *Low Frequency Second Order Wave Exciting Forces on Floating Structures*, Pinkster (1982)
- *Mooring Dynamics Due to Wind Gust Fronts*, Seelig and Headland (1998)
- *A Simulation Model for a Single Point Moored Tanker*, Wichers (1988).

Some conditions when mooring dynamics may be important to design or when specialized considerations need to be made are given in Table 4-10.

Table 4-10 Conditions Requiring Special Analysis

FACTOR	SPECIAL ANALYSIS REQUIRED
Wind	> 45 mph for small craft
	> 75 mph for larger vessels
Wind waves	> 1.5 ft for small craft
	> 4 ft for larger vessels
Wind gust fronts	Yes for SPMs
Current	> 3 knots
Ship waves and passing ship effects	Yes for special cases (see Kizakkevariath, 1989; Occasion, 1996; Weggel and Sorensen, 1984 & 1986)
Long waves (seiches and tidal waves or tsunamis)	Yes
Berthing and using mooring as a break	Yes (see MIL-HDBK-1025/1)
Parting tension member	May be static or dynamic
Ship impact or other sudden force on the ship	Yes (if directed)
Earthquakes (spud moored or stiff systems)	Yes
Explosion, landslide, impact	Yes (if directed)
Tornado (reference NUREG 1974)	Yes
Flood, sudden water level rise	Yes (if directed)
Ice forcing	Yes (if a factor)
Ship/mooring system dynamically unstable (e.g., SPM = single point mooring)	Yes (dynamic behavior of ships at SPMs can be especially complex)
Forcing period near a natural period of the mooring system	Yes; if the forcing period is from 80% to 120% of a system natural period

4.4. Risk

Risk is a concept that is often used to design facilities, because the probability of occurrence of extreme events (currents, waves, tides, storm surge, earthquakes, etc.) is strongly site dependent. Risk is used to ensure that systems are reliable, practical, and economical.

A common way to describe risk is the concept of 'return interval', which is the mean length of time between events. For example, if the wind speed with a return interval of $R = 100$ years is given for a site, this wind speed would be expected to occur, on the average, once every 100 years. However, since wind speeds are probabilistic, the specified 100-year wind speed might not occur at all in any 100-year period. Or, in any 100-year period the wind speed may be equal to or exceed the specified wind speed multiple times.

The probability or risk that an event will be equaled or exceeded one or more times during any given interval is determined from:

Equation 4-1: $P = \left(1-\left(1-\frac{1}{R}\right)^N\right) \times 100\%$

where

- P = probability, in percent, of an event being equaled or exceeded one or more times in a specified interval
- R = return interval (years)
- N = service life (years)

Figure 4-2 shows risk versus years on station for various selected values of return interval. For example, take a ship that is on station at a site for 20 years ($N = 20$). There is a $P = 18.2\%$ probability that an event with a return interval of $R = 100$ years or greater will occur one or more times at a site in a 20-year interval.

Figure 4-2 Risk Diagram

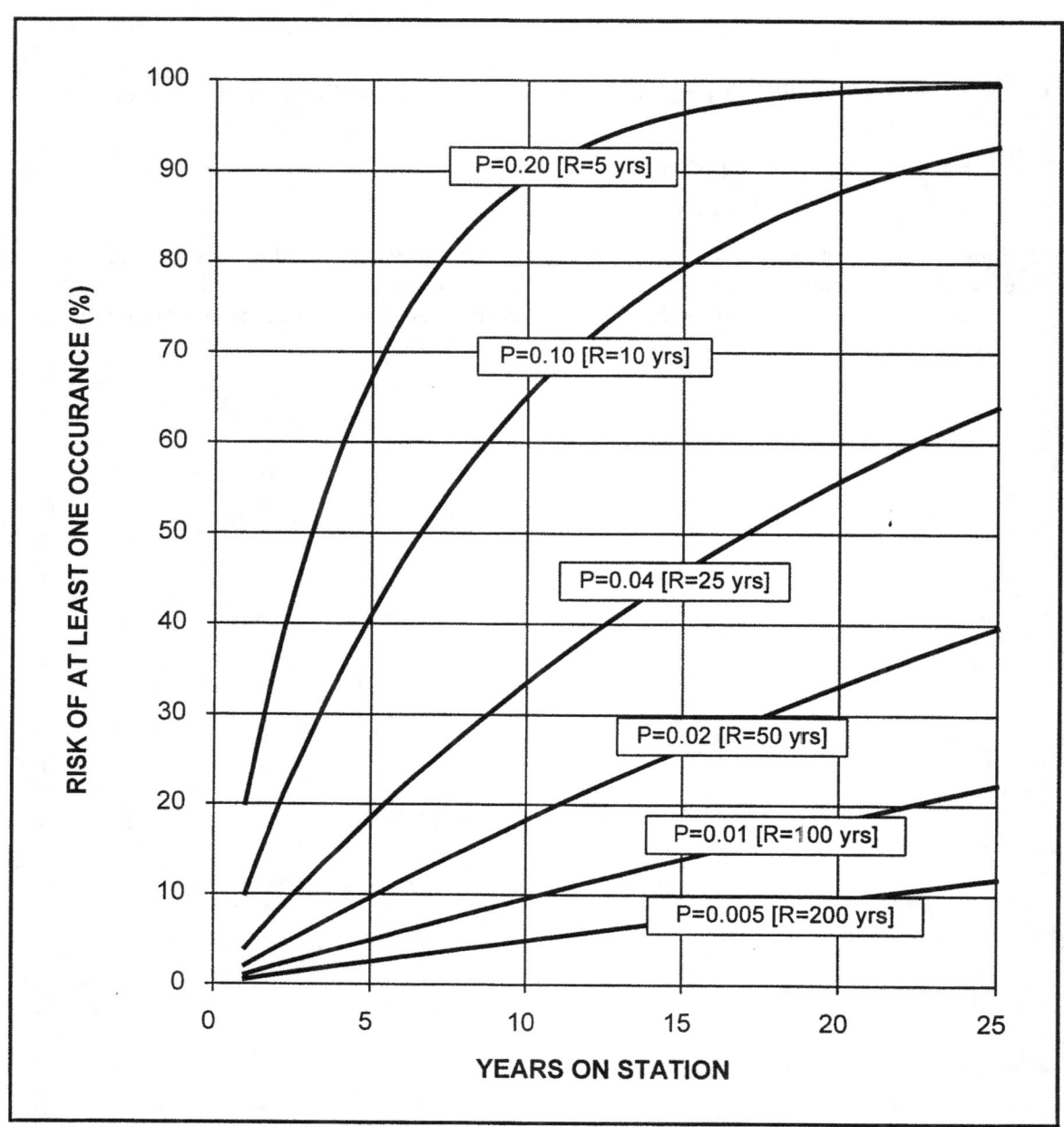

4.5. Coordinate Systems

The various coordinate systems used for ships and mooring design are described below.

4.5.1. Ship Design/Construction Coordinates

A forward perpendicular point (FP), aft perpendicular point (AP), and regular spaced frames along the longitudinal axes of the ship are used to define stations. The bottom of the ship keel is usually used as the reference point or "baseline" for vertical distances. Figure 4-3 illustrates ship design coordinates.

Figure 4-3 Ship Design and Hydrostatic Coordinates

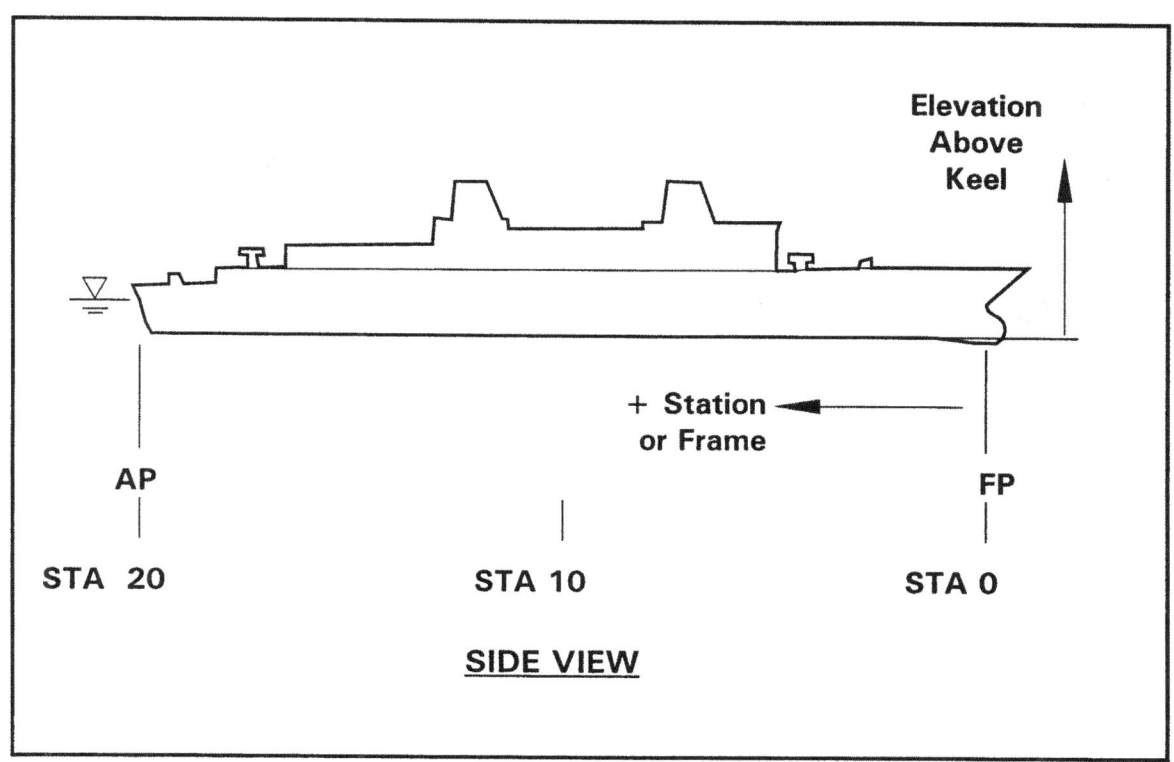

4.5.2. Ship Hydrostatics/Hydrodynamics Coordinates

The forward perpendicular is taken as Station 0, the aft perpendicular is taken as Station 20, and various cross-sections of the ship hull (perpendicular to the longitudinal axis of the ship) are used to describe the shape of the ship hull. Figure 4-3 illustrates ship hydrostatic conventions.

4.5.3. Local Mooring Coordinate System

Environmental forces on ships are a function of angle relative to the vessel's longitudinal centerline. Also, a ship tends to move about its center of gravity. Therefore, the local "right-hand-rule" coordinate system, shown in Figure 4-4, is used in this book. The midship's point is shown as a convenient reference point in Figure 4-4 and Figure 4-5.

Figure 4-4 Local Mooring Coordinate System for a Ship

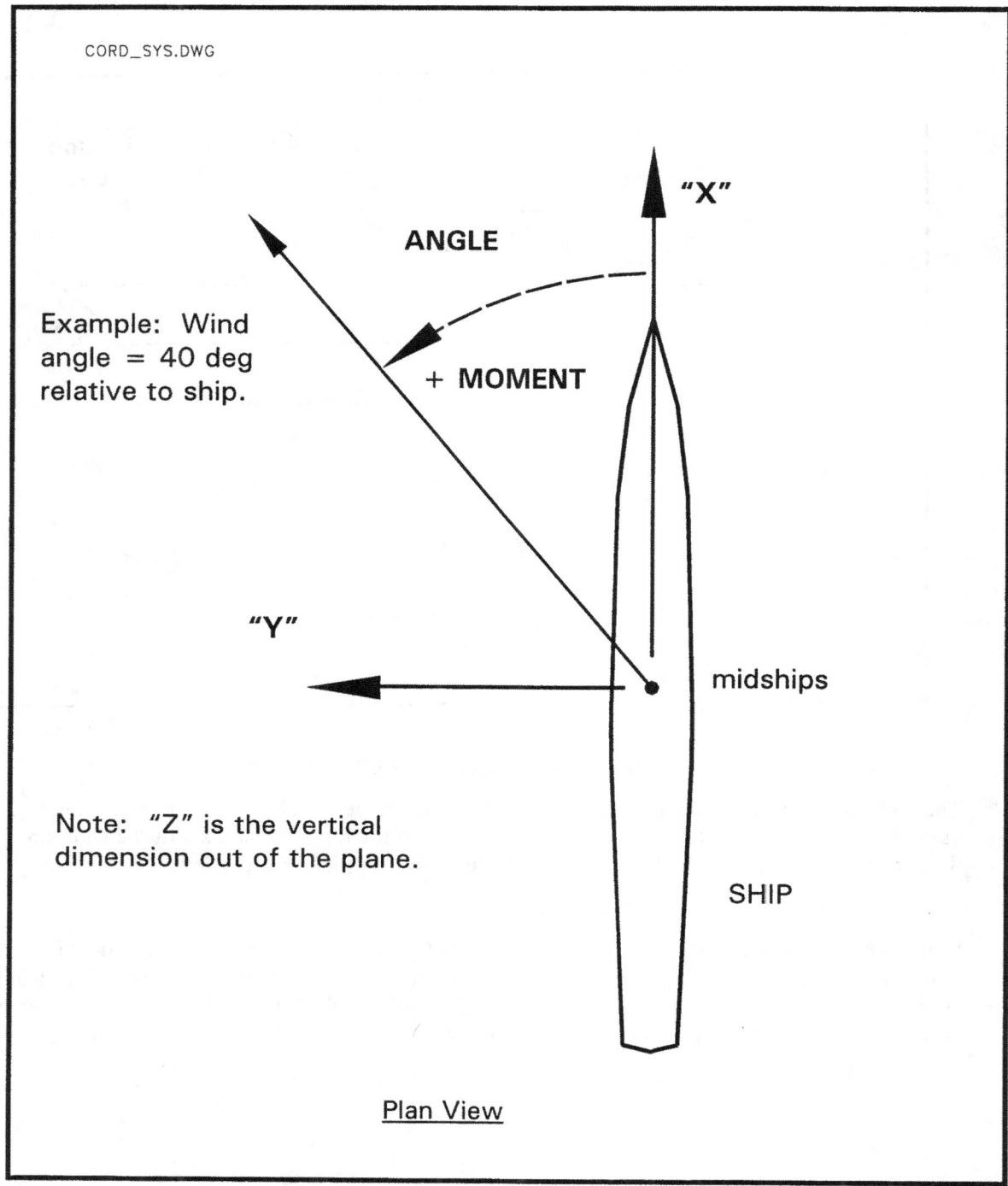

Figure 4-5 Local Mooring Coordinate System for a Ship

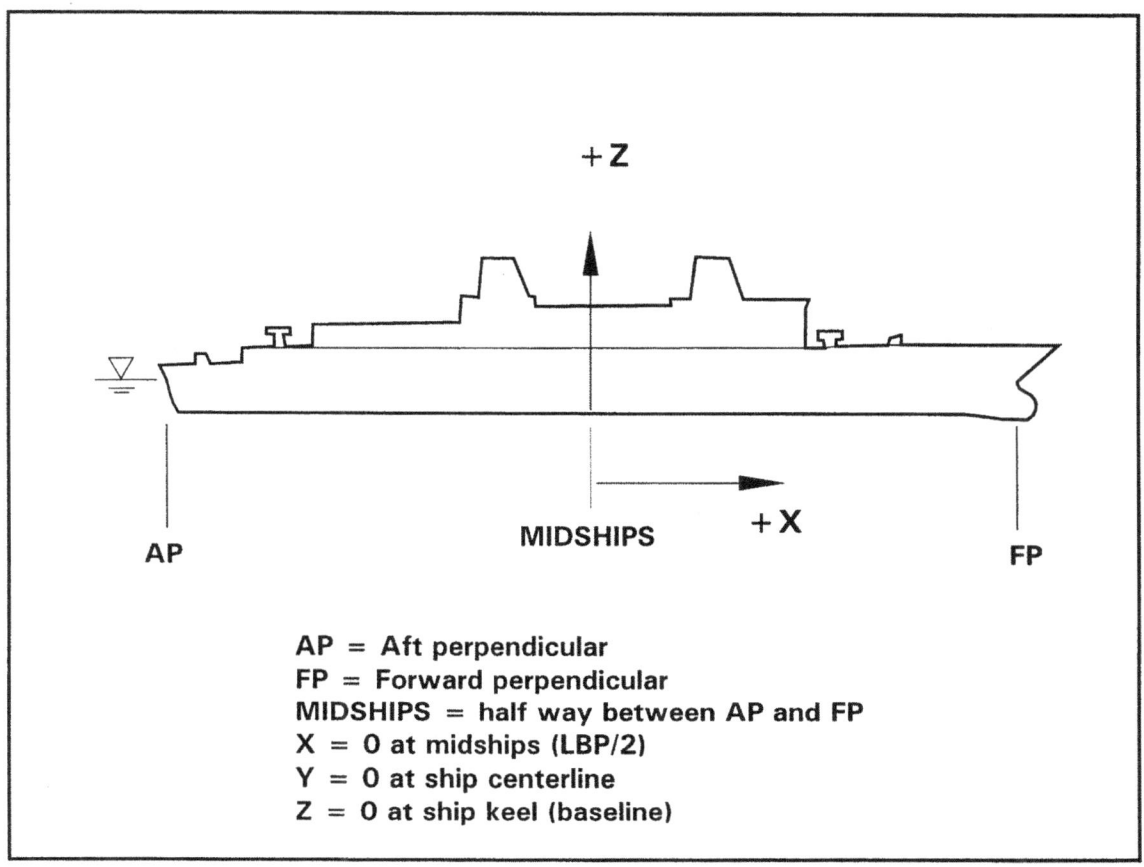

4.5.4. Global Coordinate System

Plane state grids or other systems are often used to describe x and y coordinates. The vertical datum is most often taken as relative to some water level, such as mean lower low water (MLLW).

4.6. Vessel Design Considerations

Some important vessel mooring design considerations are summarized in Table 4-11.

Table 4-11 Design Considerations - Ship

PARAMETER	NOTES
Ship fittings	The type, capacity, location, and number of mooring fittings on the ship are critical in designing moorings.
Ship hardware	The type, capacity, location, and number of other mooring hardware (chain, anchors, winches, etc.) on the ship are critical.
Buoyancy	The ship's buoyancy supports the ship up in the heave, pitch, and roll directions. Therefore, it is usually undesirable to

	have much mooring capacity in these directions. A large ship, for example, may have over a million pounds of buoyancy for a foot of water level rise. If an unusually large water level rise occurs for a mooring with a large component of the mooring force in the vertical direction, this could result in mooring failure.
Hull pressures	Ships are designed so that only a certain allowable pressure can be safely resisted. Allowable hull pressures and fender design are discussed in NFESC TR-6015-OCN, *Foam-Filled Fender Design to Prevent Hull Damage*.
Personnel access	Personnel access must be provided.
Hotel services	Provision must be made for utilities and other hotel services.

4.7. Facility Design Considerations

Some important facility mooring design considerations are summarized in Table 4-12.

Table 4-12 Design Considerations - Facility

PARAMETER	NOTES
Access	Adequate ship access in terms of channels, turning basins, bridge clearance, etc. needs to be provided. Also, tugs and pilots must be available.
Mooring fittings	The number, type, location and capacity of mooring fittings or attachment point have to meet the needs of all vessels using the facility.
Fenders	The number, type, location, and properties of marine fenders must be specified to protect the ship(s) and facility.
Water depth	The water depth at the mooring site must be adequate to meet the customer's needs.
Shoaling	Many harbor sites experience shoaling. The shoaling and possible need for dredging needs to be considered.
Permits	Permits (Federal, state, environmental, historical, etc.) are often required for facilities and they need to be considered.

4.8. Environmental Forcing Design Considerations

Environmental forces acting on a moored ship(s) can be complex. Winds, currents, water levels, and waves are especially important for many designs.

4.8.1. Winds

A change in pressure from one point on the earth to another causes the wind to blow. Turbulence is carried along with the overall wind flow to produce wind gusts. If the mean wind speed and direction do not change very rapidly with time, the winds are referred to as "stationary."

Practical experience has shown that wind gusts with a duration of approximately 30 seconds or longer have a significant influence on typical moored ships with displacements of about 1000 tons or larger. Vessels with shorter natural periods can respond to shorter duration gusts. For the purposes of this book, a 30-second wind duration at a 10-meter (33') elevation is recommended for the design for "stationary" winds. The relationship of the 30-second wind to other wind durations is shown in Figure 4-6.

Figure 4-6 Ratio of Wind Speeds for Various Gusts (after ASCE 7-95)

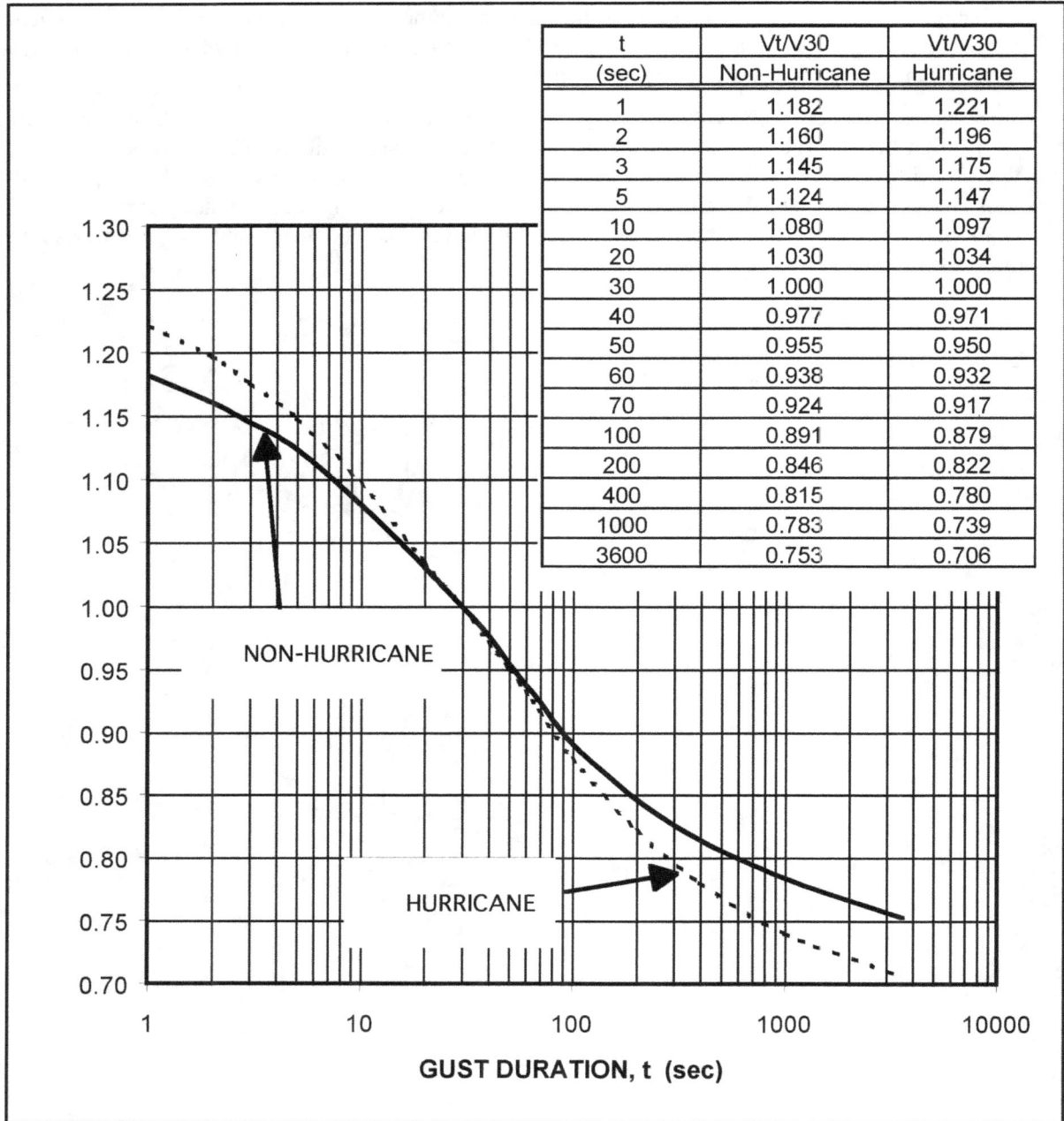

If wind speed and/or direction changes rapidly, such as in a wind gust front, hurricane or tornado, then winds are "non-stationary". Figure 4-7, for example, shows a recording from typhoon OMAR in 1992 at Guam. The eye of this storm went over the recording site. The upper portion of this figure shows the wind speed and the lower portion of the figure is the wind direction. Time on the chart recorder proceeds from right to left. This hurricane had rapid changes in wind speed and direction. As the eye passes there is also a large-scale change in wind speed and direction.

Figure 4-7 Typhoon *OMAR* Wind Chart Recording

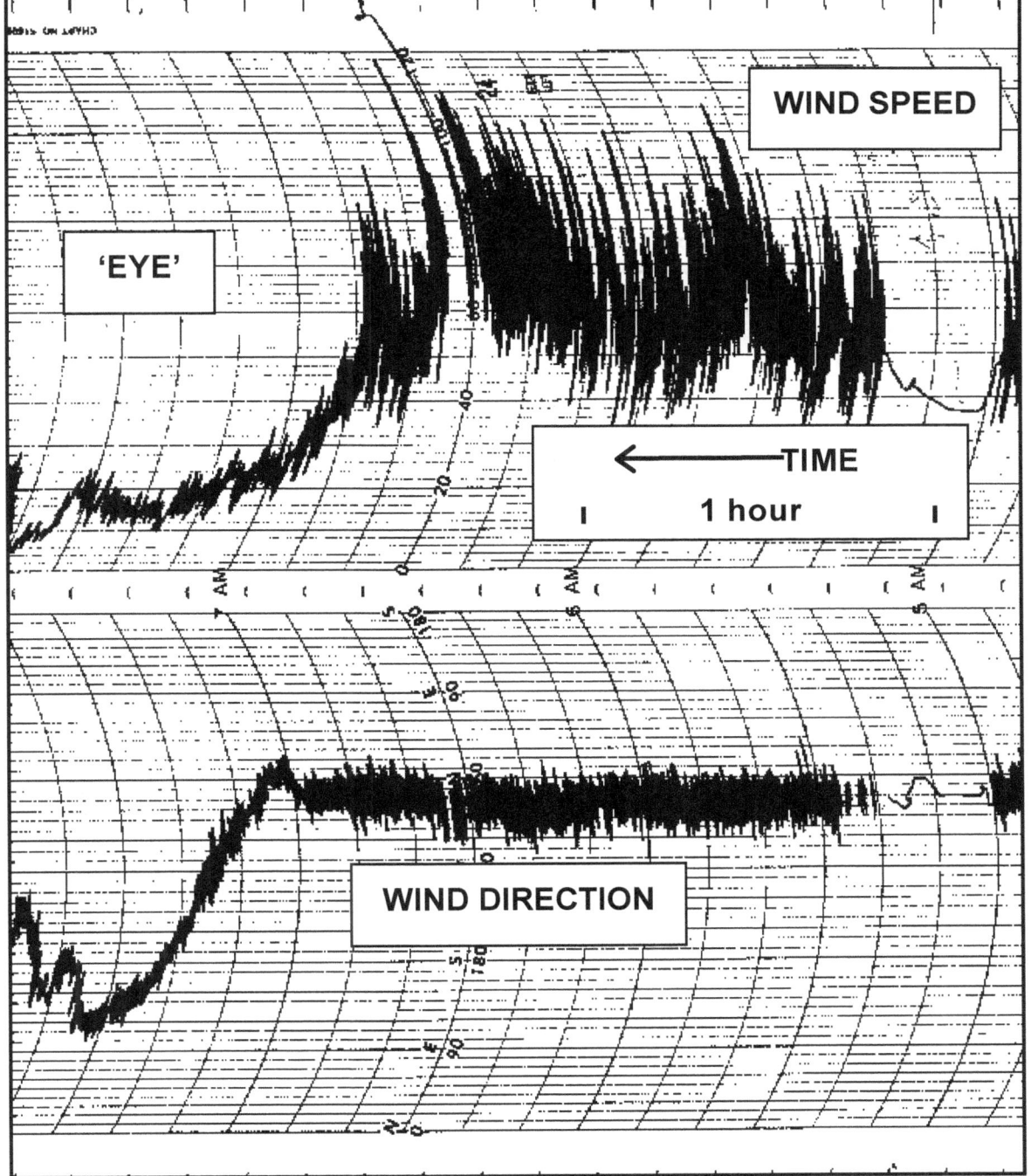

4.8.2. Wind Gust Fronts

A particularly dangerous wind condition that has caused a number of mooring accidents is the wind gust front (*Mooring Dynamics Due to Wind Gust Fronts*, Seelig and Headland, 1998 and CHESNAVFACENGCOM, FPO-1-87(1), *Failure Analysis of Hawsers on BOBO Class MSC Ships at Tinian on 7 December 1986*). This is a sudden change in wind speed that is usually associated with a change in wind direction (*Wind Effects on Structures*, Simiu and Scanlan, 1996). The key problems with this phenomena are: (1) high mooring dynamic loads can be produced in a wind gust front, (2) there is often little warning, (3) little is known about wind gust fronts, and (4) no design criteria for these events have been established.

A study of Guam Agana National Air Station (NAS) wind records was performed to obtain some statistics of wind gust fronts (National Climatic Data Center (NCDC), Letter Report E/CC31:MJC, 1987). The 4.5 years of records analyzed from 1982 through 1986 showed approximately 500 cases of sudden wind speed change, which were associated with a shift in wind direction. These wind shifts predominately occurred in 1 minute or less and never took longer than 2 minutes to reach maximum wind speed. Figure 4-8 shows sudden changes in wind speed and direction that occurred over a 2-1/2 day period in October 1982. These wind gust fronts seemed to be associated with a nearby typhoon.

Figure 4-8 Sample Wind Gust Fronts on Guam, 2-4 October 1982

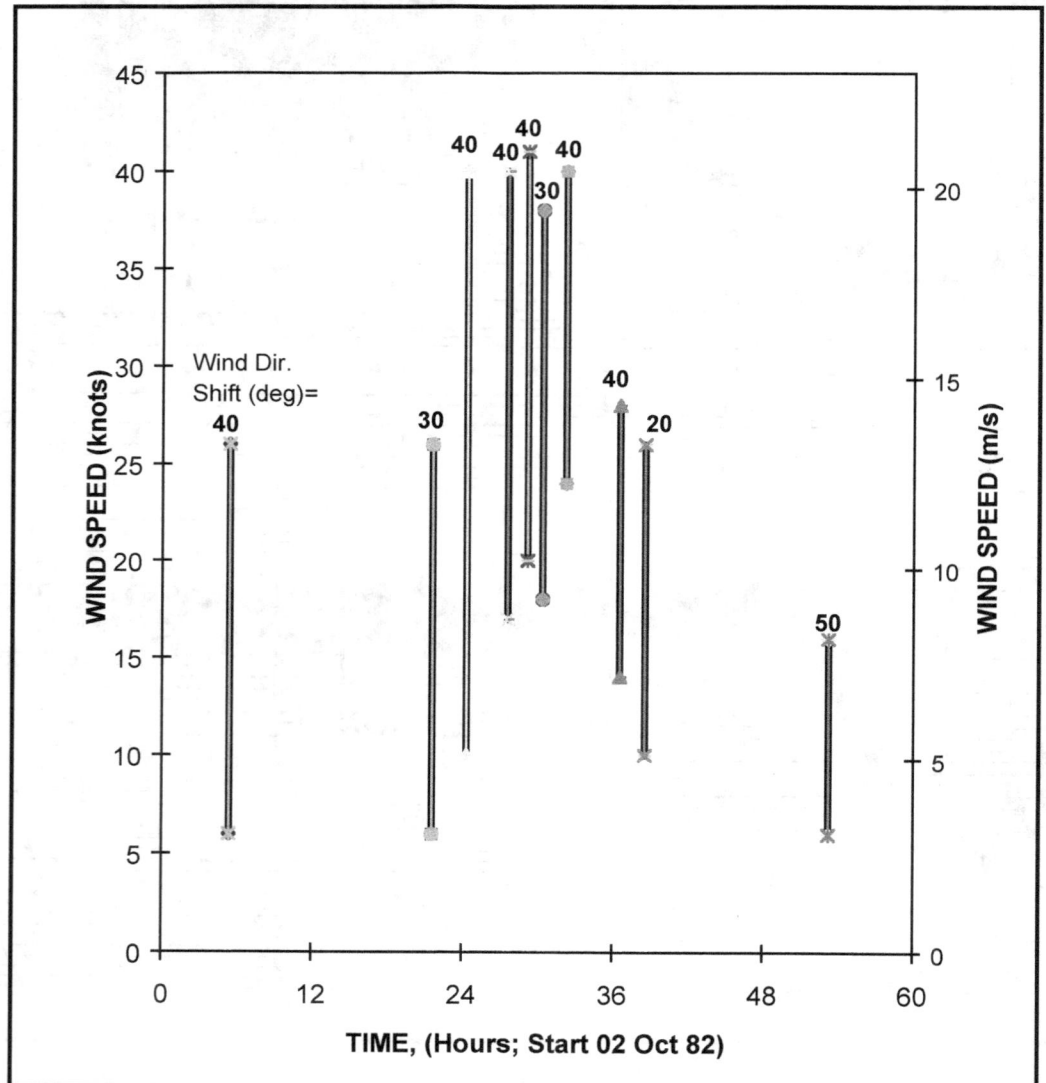

Table 4-13 gives the joint distribution of wind shifts in terms of the amount the increase in wind speed and the wind direction change. Approximately 60% of the wind gust fronts from 1982 through 1986 had wind direction changes in the 30° range, as shown in Figure 4-9.

Table 4-13 Sample Distribution of Wind Gust Fronts on Guam (Agana NAS) from 1982 to 1986

WIND SPEED CHANGE				NUMBER OF OBSERVATIONS WIND DIRECTION CHANGE							
(knots)		(m/s)									
MIN.	MAX.	MIN.	MAX.	20 deg	30 deg	40 deg	50 deg	60 deg	70 deg	80 deg	90 deg
6	10	3.1	5.1	28	241	66	30	4		2	
11	15	5.7	7.7	8	42	18	13	5	3	1	1
16	20	8.2	10.3	6	7	3	2	2			
21	25	10.8	12.9		3	2		1			
26	30	13.4	15.4			1					

Figure 4-9 Distribution of Guam Wind Gust Front Wind Angle Changes

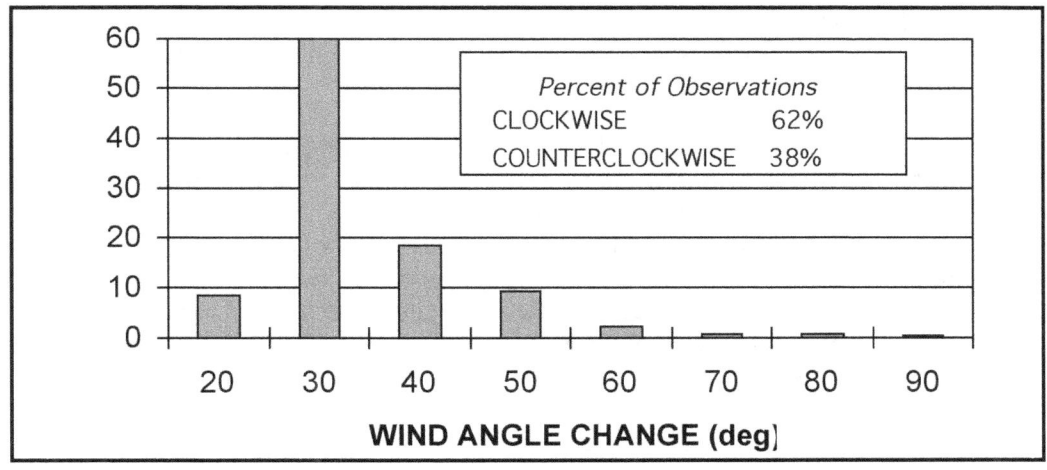

Based on the Guam observations, the initial wind speed in a wind gust front ranges from 0 to 75% of the maximum wind speed, as shown in Figure 4-10. On the average, the initial wind speed was 48% of the maximum in the 4.5-year sample from Guam (NCDC, 1987).

Figure 4-10 Initial Versus Maximum Wind Speeds for Wind Gust Fronts

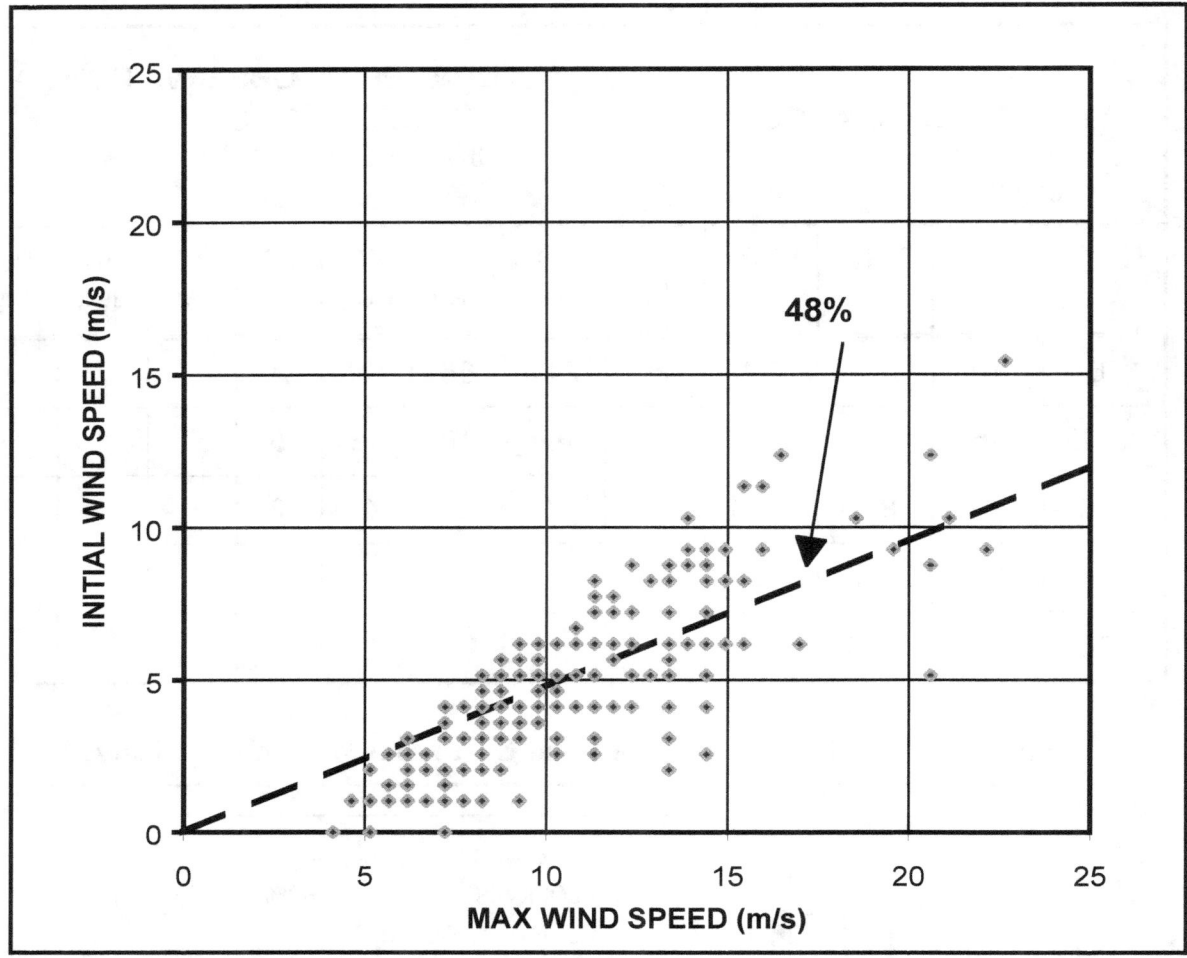

Simiu and Scanlan (1996) report wind gust front increases in wind speed ranging from 3 m/sec to 30 m/sec (i.e., 6 to 60 knots). Figure 4-11 shows the distribution of gust front winds from the 4.5-year sample from 1982 through 1986 on Guam. This figure shows the probability of exceedence on the x-axis in a logarithmic format. The square of the wind gust front speed maximums was plotted on the y-axis, since wind force is proportional to wind speed squared. Figure 4-11 provides a sample of the maximum wind gust front distribution for a relatively short period at one site. Those wind gust fronts that occurred when a typhoon was nearby are identified with an "H." It can be seen that the majority of the higher gust front maximums were associated with typhoons. Also, the typhoon gust front wind speed maxima seem to follow a different distribution that the gust front maxima associated with rain and thunderstorms (see Figure 4-11).

Figure 4-11 Wind Gust Front Maxima on Guam 1982-1986

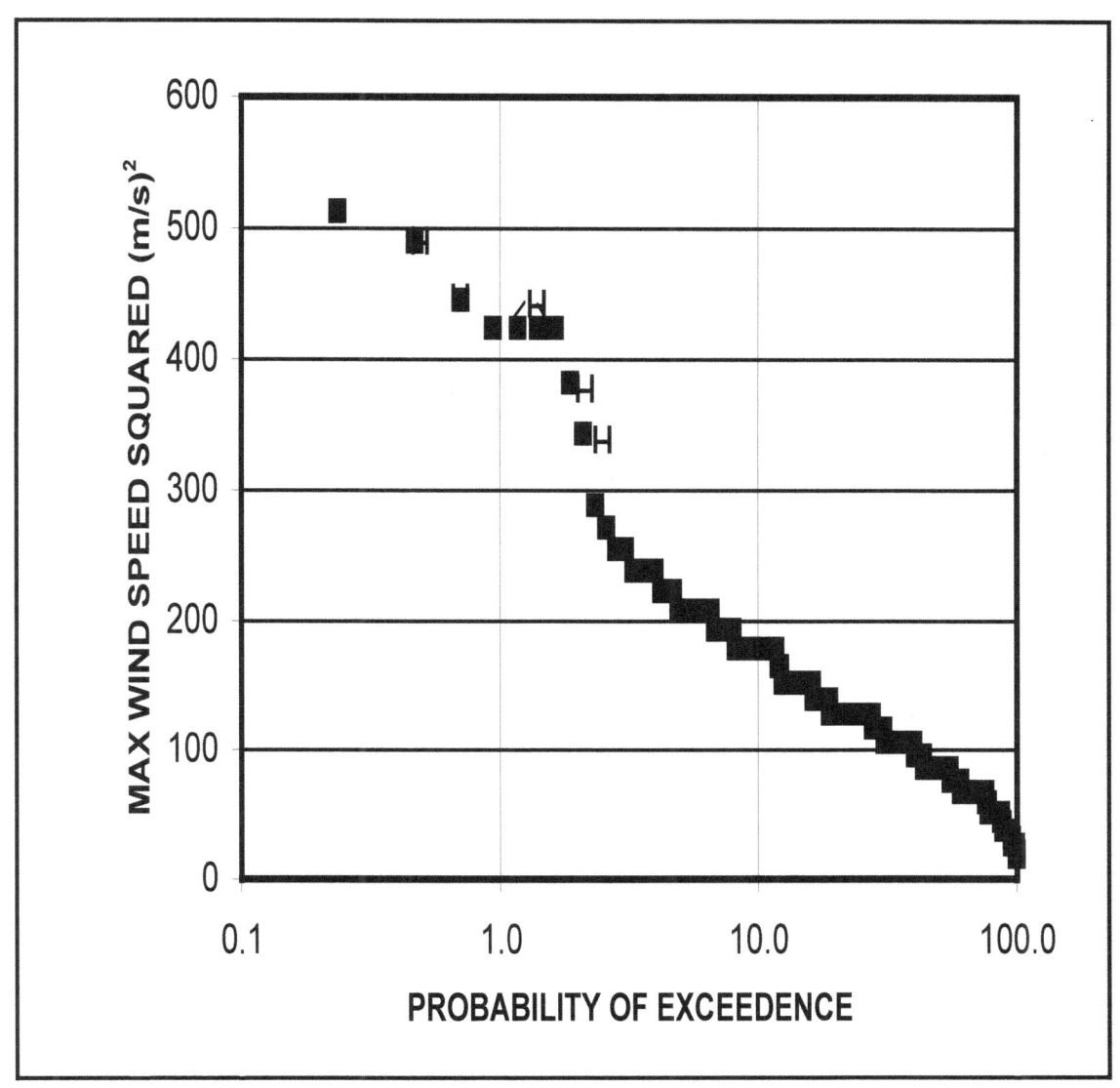

4.8.3. Storms

Table 4-14 gives environmental parameters for standard storms.

Table 4-14 Storm Parameters

(a) Tropical Storms

STORM	LOWER WIND SPEED			UPPER WIND SPEED		
	(m/s)	(mph)	(knts)	(m/s)	(mph)	(knts)
TROPICAL DEPRESSION	10.3	23	20	17	38	33
TROPICAL STORM	18.0	40	35	32.4	74	63

HURRICANE	33.1	74	64	-	-	-

(b) Saffier-Simpson Hurricane Scale

CATE-GORY	WIND SPEED RANGE				OPEN COAST STORM SURGE RANGE			
	LOWER		UPPER		LOWER		UPPER	
	(m/s)	(mph)	(m/s)	(mph)	(m)	(ft)	(m)	(ft)
1	33.1	74	42.5	95	1.22	4	1.52	5
2	42.9	96	49.2	110	1.83	6	2.44	8
3	49.6	111	58.1	130	2.74	9	3.66	12
4	58.6	131	69.3	155	3.96	13	5.49	18
5	69.3	155	-	-	5.49	18	-	-

(c) Beaufort Wind Force[22]

BEAUFORT WIND FORCE/ DESCRIPTION	LOWER WIND SPEED			UPPER WIND SPEED		
	(m/s)	(mph)	(knts)	(m/s)	(mph)	(knts)
0 CALM	0.0	0	0	0.5	1	1
1 LIGHT AIRS	0.5	1	1	1.5	4	3
2 LIGHT BREEZE	2.1	5	4	3.1	7	6
3 GENTLE GREEZE	3.6	8	7	5.1	12	10
4 MODERATE BREEZE	5.7	13	11	8.2	18	16
5 FRESH BREEZE	8.8	20	17	10.8	24	21
6 STRONG BREEZE	11.3	25	22	13.9	31	27
7 MODERATE GALE	14.4	32	28	17.0	38	33
8 FRESH GALE	17.5	39	34	20.6	46	40
9 STRONG GALE	21.1	47	41	24.2	54	47

---

[22] After *Handbook of Ocean and Underwater Engineers*, Myers et al. (1969).

10	WHOLE GALE	24.7	55	48	28.3	63	55
11	STORM	28.8	65	56	32.4	73	63
12	HURRICANE	32.9	74	64	36.6	82	71

(d) World Meteorological Organization Sea State Scale

SEA STATE		Sign. Wave Height (ft) [m]	Sustained Wind Speed (knts) [m/s]	Modal Wave Period Range (sec)
0	CALM/GLASSY	NONE	NONE	-
1	RIPPLED	0-0.3 [0-0.1]	0-6 [0-3]	-
2	SMOOTH	0.3-1.6 [0.1-0.5]	7-10 [3.6-5.1]	3-15
3	SLIGHT	1.6-4.1 [0.5-1.2]	11-16 [5.7-8.2]	3-15.5
4	MODERATE	4.1-8.2 [1.2-2.5]	17-21 [8.7-10.8]	6-16
5	ROUGH	8.2-13.1 [2.5-4.0]	22-27 [11.3-13.9]	7-16.5
6	VERY ROUGH	13.1-19.7 [4.0-6.0]	28-47 [14.4-24.2]	9-17
7	HIGH	19.7-29.5 [6.0-9.0]	48-55 [24.7-28.3]	10-18
8	VERY HIGH	29.5-45.5 [9.0-13.9]	56-63 [28.8-32.4]	13-19
9	PHENOMENAL	>45.5 [>13.9]	>63 [>32.4]	18-24

4.8.4. Currents

The magnitude and direction of currents in harbors and nearshore areas are in most cases a function of location and time. Astronomical tides, river discharges, wind-driven currents, and other factors can influence currents. For example, wind-driven currents are surface currents that result from the stress exerted by the wind on the sea surface. Wind-driven currents generally attain a mean velocity of about 3 to 5% of the mean wind speed at 10 meters (33 feet) above the sea surface. The magnitude of this current strongly decreases with depth.

Currents can be very site specific, so it is recommended that currents be measured at the design site and combined with other information available to define the design current conditions.

4.8.5. Water Levels

At most sites some standard datum, such as mean low water (MLW) or mean lower low water (MLLW), is established by formal methods. Water levels are then referenced to this datum. The water level in most harbors is then a function of time. Factors influencing water levels include astronomical tides, storm surges, river discharges, winds, seiches, and other factors.

The design range in water levels at the site must be considered in the design process.

4.8.6. Waves

Most DOD moorings are wisely located in harbors to help minimize wave effects. However, waves can be important to mooring designs in some cases. The two primary wave categories of interest are:

a) Wind waves. Wind waves can be locally generated or can be wind waves or swell entering the harbor entrance(s). Small vessels are especially susceptible to wind waves.

b) Long waves. These can be due to surf beat, harbor seiching, or other effects.

Ship waves may be important in some cases. The response of a moored vessel to wave forcing includes:

a) A steady mean force.

b) First order response, where the vessel responds to each wave, and

c) Second order response, where some natural long period mode of ship/mooring motion, which usually has little damping, is forced by the group or other nature of the waves.

If any of these effects are important to a given mooring design, then a six-degree-of-freedom dynamic of the system generally needs to be considered in design. Some guidance on safe wave limits is given in Table 4-8.

4.8.7. Water Depths

The bathymetry of a site may be complex, depending on the geology and history of dredging. Water depth may also be a function of time, if there is shoaling or scouring. Water depths are highly site specific, so hydrographic surveys of the project site are recommended.

4.8.8. Environmental Design Information

Some sources of environmental design information of interest to mooring designers are summarized in Table 4-15.

Table 4-15 Some Sources of Environmental Design Information

a. Winds

NAVFAC *Climate Database*, 1998
ANSI/ASCE 7-95 (1996)
National Bureau of Standards (NBS), Series 124, *Hurricane Wind Speeds in the United States*, 1980
Nuclear Regulatory Commission (NUREG), NUREG/CR-2639, *Historical Extreme Winds for the United States – Atlantic and Gulf of Mexico Coastlines*, 1982
Hurricane and Typhoon Havens Handbooks, NRL (1996) and NEPRF (1982)
NUREG/CR-4801, *Climatology of Extreme Winds in Southern California*, 1987
NBS Series 118, *Extreme Wind Speeds at 129 Stations in the Contiguous United States*, 1979

b. Currents

NAVFAC *Climate Database*, 1998
National Ocean Survey records
Nautical Software, *Tides and Currents for Windows*, 1995
U.S. Army Corps of Engineers records

c. Water Levels

NAVFAC *Climate Database*, 1998
Federal Emergency Management Agency records
U.S. Army Corps of Engineers, Special Report No. 7, *Tides and Tidal Datums in the United States*, 1981
National Ocean Survey records
Hurricane and Typhoon Havens Handbooks, NRL (1996) and NEPRF (1982)
Nautical Software (1995)
U.S. Army Corps of Engineers records

d. Waves

Hurricane and Typhoon Havens Handbooks, NRL (1996) and NEPRF (1982)
U.S. Army Corps of Engineers, *Shore Protection Manual* (1984) gives prediction methods

e. Bathymetry

From other projects in the area
National Ocean Survey charts and surveys
U.S. Army Corps of Engineers dredging records

4.9. Operational Considerations

Some important operational design considerations are summarized in Table 4-16.

Table 4-16 Mooring Operational Design Considerations

PARAMETER	NOTES
Personnel experience/ training	What is the skill of the people using the mooring?
Failure	What are the consequences of failure? Are there any design features that can be incorporated that can reduce the impact?
Ease of use	How easy is the mooring to use and are there factors that can make it easier to use?

Safety	Can features be incorporated to make the mooring safer for the ship and personnel?
"Act-of-God" events	Extreme events can occur unexpectedly. Can features be incorporated to accommodate them?
Future use	Future customer requirements may vary from present needs. Are there things that can be done to make a mooring facility more universal?

4.10. Inspection

Mooring systems and components should be inspected periodically to ensure they are in good working order and are safe. Table 4-17 gives inspection guidelines.

Table 4-17 Inspection Guidelines

MOORING SYSTEM OR COMPONENT	MAXIMUM INSPECTION INTERVAL	NOTES
Piers and wharves	1 year	Surface inspection
	3 years	Complete inspection - wood structures
	6 years	Complete inspection - concrete and steel structures
		See NAVFAC MO-104.2, *Specialized Underwater Waterfront Facilities Inspections*; If the actual capacity/condition of mooring fittings on a pier/wharf is unknown, then pull tests are recommended to proof the fittings.
Fleet Moorings	3 years	See CHESNAVFACENGCOM, FPO-1-84(6), *Fleet Mooring Underwater Inspection Guidelines*. Also inspect and replace anodes, if required. More frequent inspection may be required for moorings at exposed sites or for critical facilities.
Synthetic line	6 months	Per manufacturer's recommendations
Ship's chain	36 months	0-3 years of service
	24 months	4-10 years of service
	18 months	>10 years of service
		(American Petroleum Institute (API) RP 2T, *Recommended Practice for Planning, Designing, and Constructing Tension Leg Platforms*)

Wire rope	18 months	0-2 years of service
	12 months	3-5 years of service
	9 months	>5 years of service
		(API RP 2T)

4.11. Maintenance

If excessive wear or damage occurs to a mooring system, then it must be maintained. Fleet mooring chain, for example, is allowed to wear to a diameter of 90% of the original steel bar diameter. As measured diameters approach 90%, then maintenance is scheduled. Moorings with 80 to 90% of the original chain diameter are restricted to limited use. If a chain diameter reaches a bar diameter of 80% of the original diameter, then the mooring is condemned. Figure 4-12 illustrates some idealized models of chain wear.

Figure 4-12 Idealized Models of Chain Wear

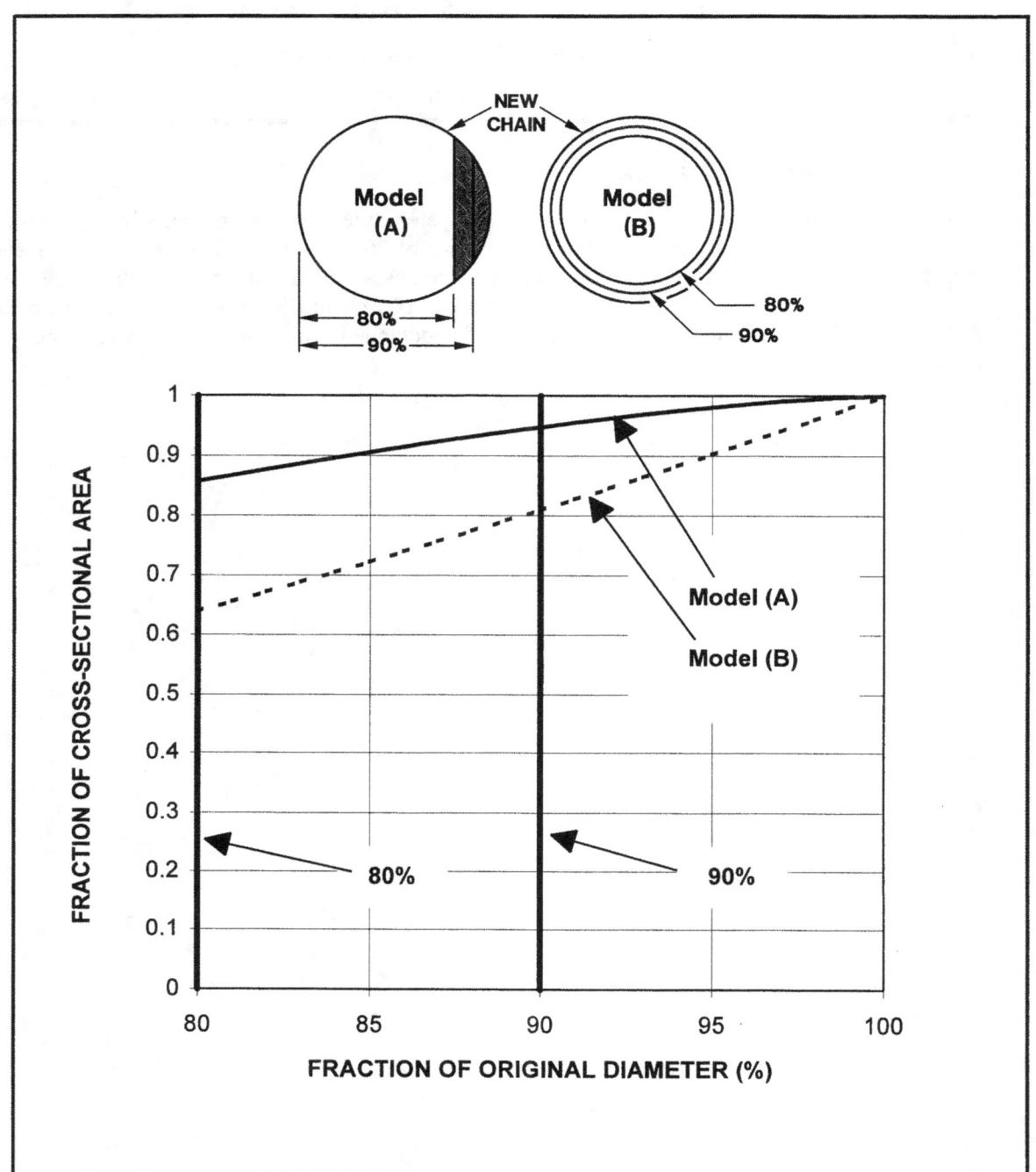

4.12. General Mooring Guidelines

Experience and practical considerations show that the recommendations given in Table 4-18 will help ensure safe mooring. These ideas apply to both ship mooring hardware and mooring facilities.

Table 4-18 Design Recommendations

IDEA	NOTES
Allow ship to move with rising and falling water levels	The weight and buoyancy forces of ships can be very high, so it is most practical to design moorings to allow ships to move in the vertical direction with changing water levels. The design range of water levels for a specific site should be determined in the design process.
Ensure mooring system components have similar strength	A system is only as strong as its weakest segment; a system with components of similar strength can be the most economical. Mooring lines should not have a break strength greater than the capacity of the fittings they use.
Ensure load sharing	In some moorings, such as at a pier, many lines are involved. Ensuring that members will share the load results in the most economical system.
Bridle design	In cases where a ship is moored to a single point mooring buoy with a bridle, ensure that each leg of the bridle can withstand the full mooring load, because one member may take the full load as the vessel swings.
Provide shock absorbing in mooring systems	Wind gusts, waves, passing ships, etc., will produce transient forces on a moored ship. Allowing some motion of the ship will reduce the dynamic loads. 'Shock absorbers' including marine fenders, timber piles, synthetic lines with stretch, chain catenaries, sinkers, and similar systems are recommended to allow a moored ship to move in a controlled manner.
Limit the vertical angles of lines from ship to pier	Designing ships and piers to keep small vertical line angles has the advantages of improving line efficiency and reducing the possibility of lines pulling off pier fittings.
Select drag anchors to have a lower ultimate holding capacity than the breaking strength of chain and fittings	Design mooring system that uses drag anchor, so that the anchor will drag before the chain breaks.
Limit the loading on drag anchors to horizontal tension	Drag anchors work on the principle of 'plowing' into the soils. Keeping the mooring catenary angle small at the seafloor will aid in anchor holding. Have at least one shot of chain on the seafloor to help ensure the anchor will hold.
Pull test anchors whenever possible to the full design load	Pull testing anchors is recommended to ensure that all facilities with anchors provide the required holding capacity.

Chapter 5.   Static Environmental Forces And Moments On Vessels

5.1.   Scope

In this section design methods are presented for calculating static forces and moments on single and multiple moored vessels. Examples show calculation methods.

5.2.   Engineering Properties Of Water And Air

The effects of water and air at the surface of the earth are of primary interest in this section. The engineering properties of both are given in Table 5-1.

Table 5-1 Engineering Properties of Air and Water

(a) Standard Salt Water at Sea Level at 15°C (59°F)

PROPERTY	SI SYSTEM	ENGLISH SYSTEM
Mass density, ρ_w	1026 kg/m^3	1.9905 slug/ft^3
Weight density, γ_w	10060 N/m^3	64.043 lbf/ft^3
Volume per long ton (LT)	0.9904 m^3/LT	34.977 ft^3/LT
Kinematic viscosity, ν	1.191E-6 m^2/sec	1.2817E-5 ft^2/sec

(b) Standard Fresh Water at Sea Level at 15°C (59°F)

PROPERTY	SI SYSTEM	ENGLISH OR INCH-POUND SYSTEM
Mass density, ρ_w	999.0 kg/m^3	1.9384 slug/ft^3
Weight density, γ_w	9797 N/m^3	62.366 lbf/ft^3
Volume per long ton (LT)	1.0171 m^3/LT	35.917 ft^3/LT
Volume per metric ton (ton or 1000 kg or 1 Mg)	1.001 m^3/ton	35.3497 ft^3/ton
Kinematic viscosity, ν	1.141E-6 m^2/sec	1.2285E-5 ft^2/sec

(c) Air at Sea Level at 20°C (68°F)[23]

PROPERTY	SI SYSTEM	ENGLISH OR INCH-POUND SYSTEM
Mass density, ρ_a	1.221 kg/m^3	0.00237 slug/ft^3
Weight density, γ_a	11.978 N/m^3	0.07625 lbf/ft^3

---

[23] Note that humidity and even heavy rain has relatively little effect on the engineering properties of air (personal communication with the National Weather Service, 1996)

| Kinematic viscosity, ν | 1.50E-5 m^2/sec | 1.615E-4 ft^2/sec |

5.3. Principal Coordinate Directions

There are three primary axes for a ship:

- X - Direction parallel with the ship's Longitudinal axis
- Y - Direction perpendicular to a vertical plane through the ship's longitudinal axis
- Z - Direction perpendicular to a plane formed by the "X" and "Y" axes

There are six principal coordinate directions for a ship (Figure 5-1):

- Surge - In the "X"-direction
- Sway - In the "Y"-direction
- Heave - In the "Z"-direction
- Roll - Angular about the "X"-axis
- Pitch - Angular about the "Y"-axis
- Yaw - Angular about the "Z"-axis

Figure 5-1 Principal Directions for a Ship

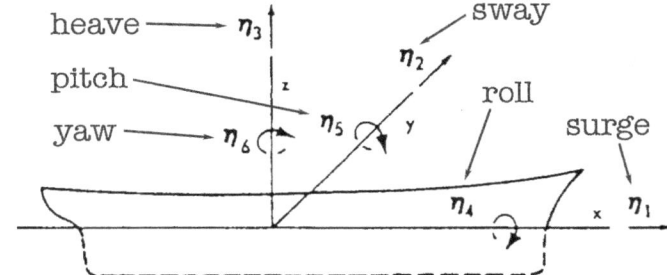

Of primary interest are: (1) forces in the surge and sway directions in the "X-Y" plane, and (2) moment in the yaw direction about the "Z"-axis. Ship motions occur about the center of gravity of the ship.

5.4. Static Wind Forces/Moments

Static wind forces and moments on stationary moored vessels are computed in this section. Figure 5-2 shows the definition of some of the terms used in this section. Figure 5-3 shows the local coordinate system.

Figure 5-2 Definition of Terms

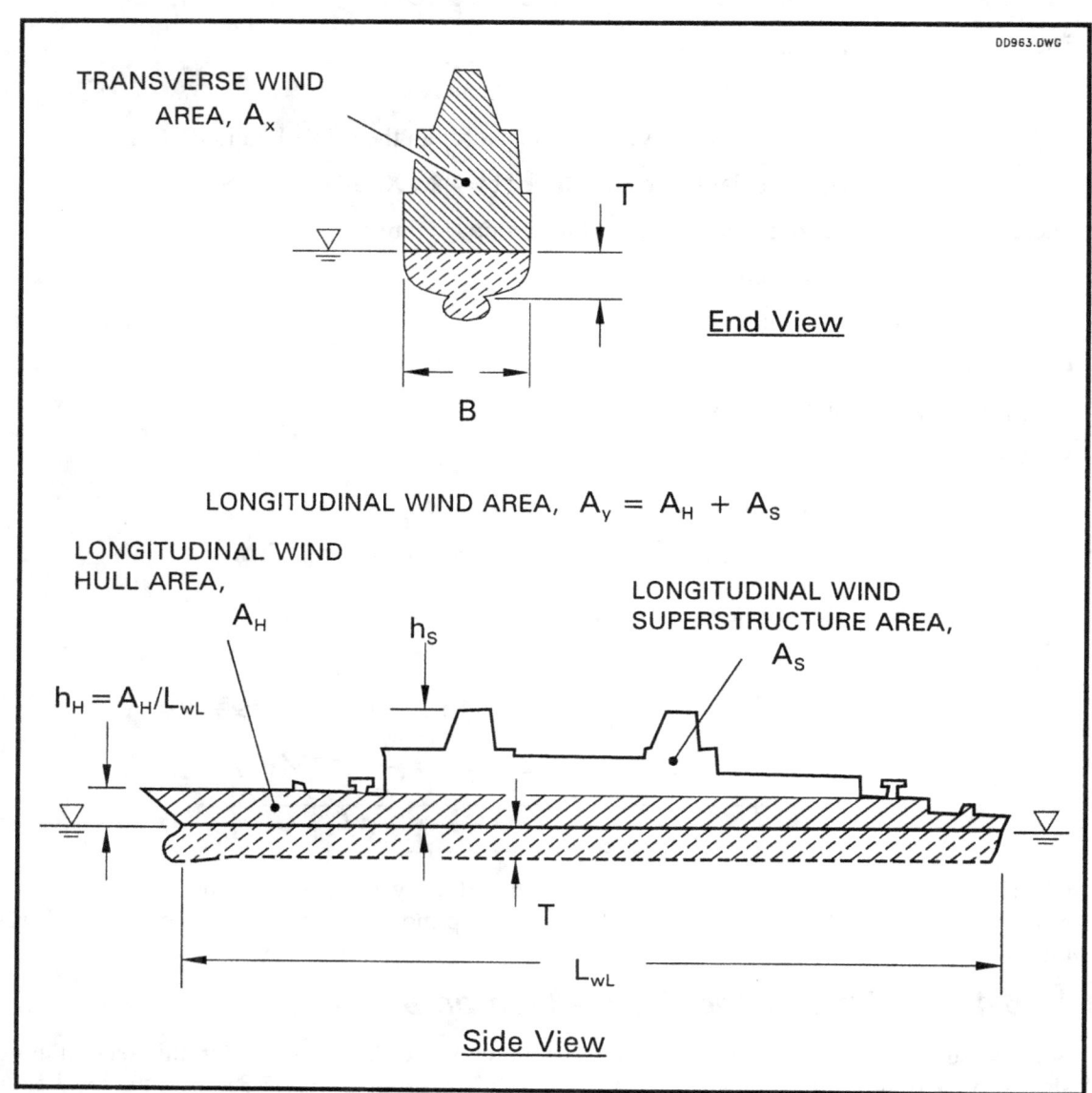

Figure 5-3 Local Coordinate System for a Ship

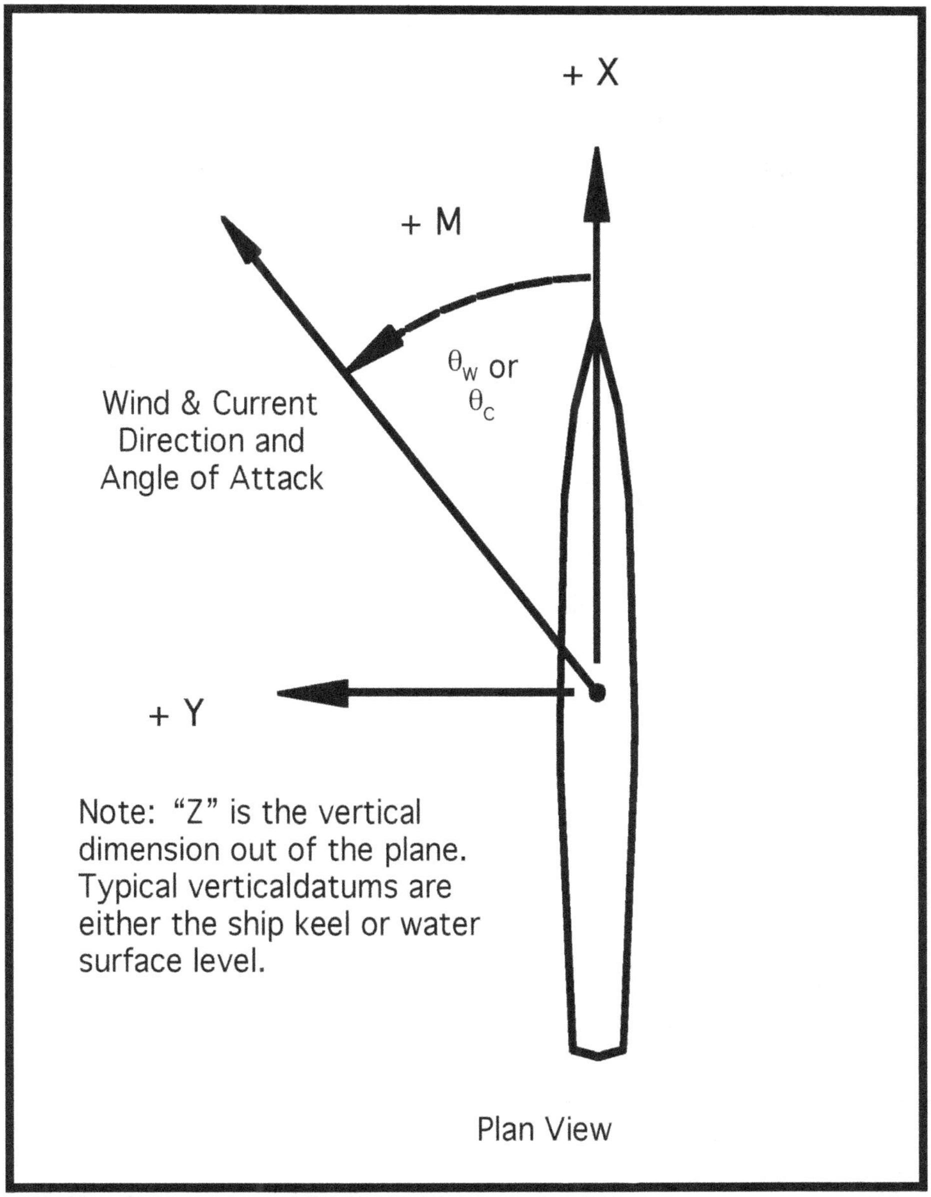

[Figure: Plan view of ship showing local coordinate system with +X axis along ship centerline (forward), +Y axis to port, +M moment direction, and wind & current direction and angle of attack θ_w or θ_c. Note: "Z" is the vertical dimension out of the plane. Typical vertical datums are either the ship keel or water surface level.]

5.4.1. Static Transverse Wind Force

The static transverse wind force is defined as that component of force perpendicular to the vessel centerline. In the local ship coordinate system, this is the force in the "Y" or sway direction. Transverse wind force is determined from the equation:

$$\text{Equation 5-1:} \quad F_{yw} = \rho_a A_y C_{yw} f_{yw}\{\theta_w\} \frac{V_w^2}{2}$$

where

- F_{yw} = transverse wind force (N)
- ρ_a = mass density of air (from Table 5-1)
- V_w = wind speed (m/s)

- A_y = longitudinal projected area of the ship (m^2)
- C_{yw} = transverse wind force drag coefficient
- $f_{yw}\{\theta_w\}$ = shape function for transverse force
- θ_w = wind angle (degrees)

The transverse wind force drag coefficient depends upon the hull and superstructure of the vessel and is calculated using the following equation, adapted from Naval Civil Engineering Laboratory (NCEL), TN-1628, *Wind-Induced Steady Loads on Ships*.

Equation 5-2: $$C_{yw} = C \frac{\left(\dfrac{h_s + h_H}{2h_R}\right)^{\frac{2}{7}} A_s + \left(\dfrac{h_H}{2h_R}\right)^{\frac{2}{7}} A_H}{A_y}$$

where
- C = empirical coefficient, see Table 5-2. Based on scale model wind tunnel tests, C = 0.92 +/-0.1 (NCEL, TN-1628). Figure 5-4 illustrates some ship types.
- h_R = 10 m = reference height (32.8 ft)
- h_H = average height of the hull (m) = A_H/L_{wL}
- A_H = longitudinal wind area of the hull (m^2)
- L_{wL} = ship length at the waterline (m)
- h_s = height of the superstructure above the waterline (m)
- A_s = longitudinal wind area of the superstructure (m^2)

Table 5-2 Sample Wind Coefficients for Ships

SHIP	C	NOTES
Hull dominated	0.82	Aircraft carriers, drydocks
Typical	0.92	ships with moderate superstructure
Extensive superstructure	1.02	Destroyers, cruisers

Figure 5-4 Sample Ship Profiles

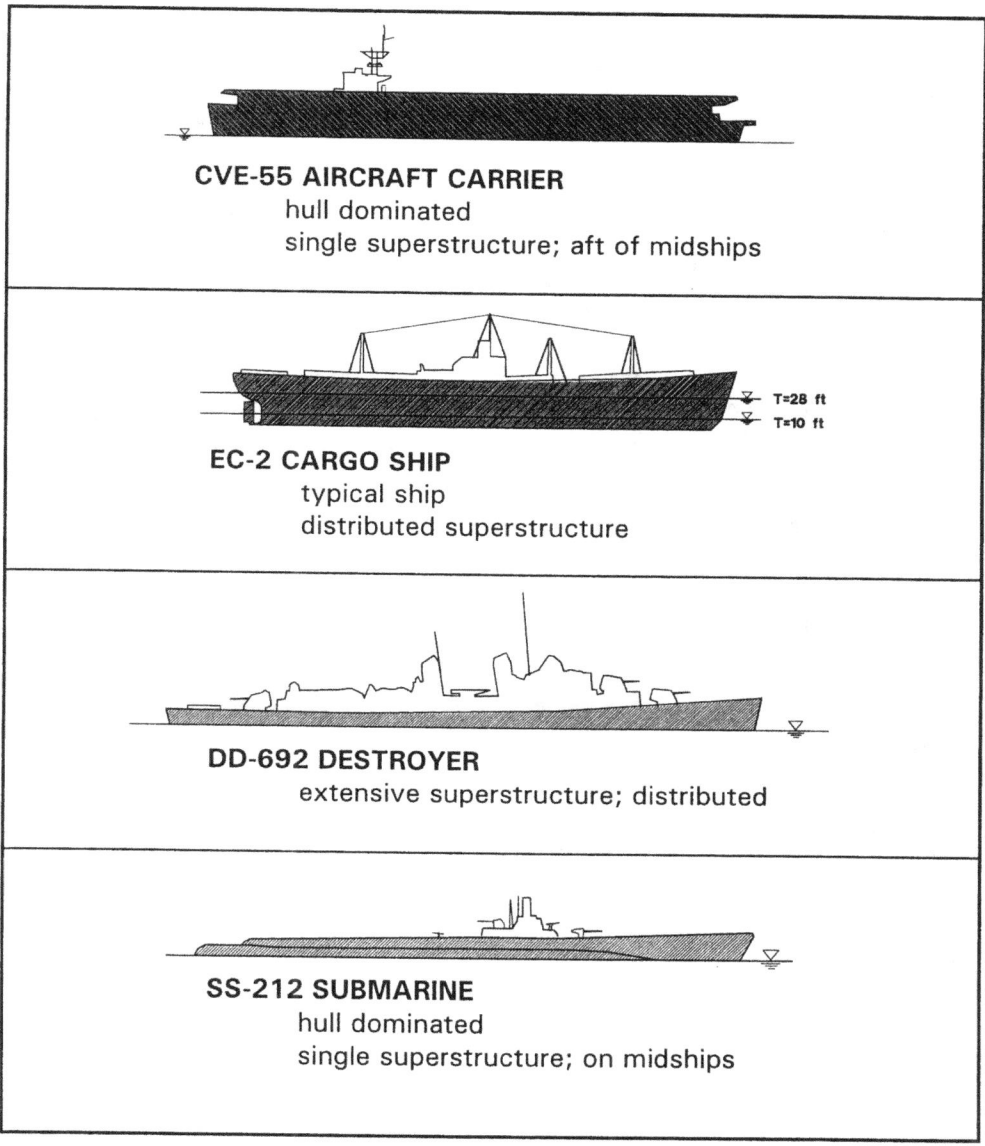

The shape function for the transverse wind force (NCEL, TN-1628) is given by:

$$\textbf{Equation 5-3:} \quad f_{yw}\{\theta_w\} = +\frac{20}{19}\sin\theta_w - \frac{1}{19}\sin\{5\theta_w\}$$

where

- $f_{yw}\{\theta_w\}$ = transverse wind coefficient shape function
- θ_w = wind angle (degrees)

Equation 4 is positive for wind angles $0 < \theta_w < 180°$ and negative for wind angles $180 < \theta_w < 360°$. Figure 5-5 shows the shape and typical values for Equation 5-3.

These two components were derived by integrating wind over the hull and superstructure areas to obtain effective wind speeds (NCEL, TN-1628). The following example illustrates calculations of the transverse wind force drag coefficient.

Figure 5-5 Shape Function for Transverse Wind Force

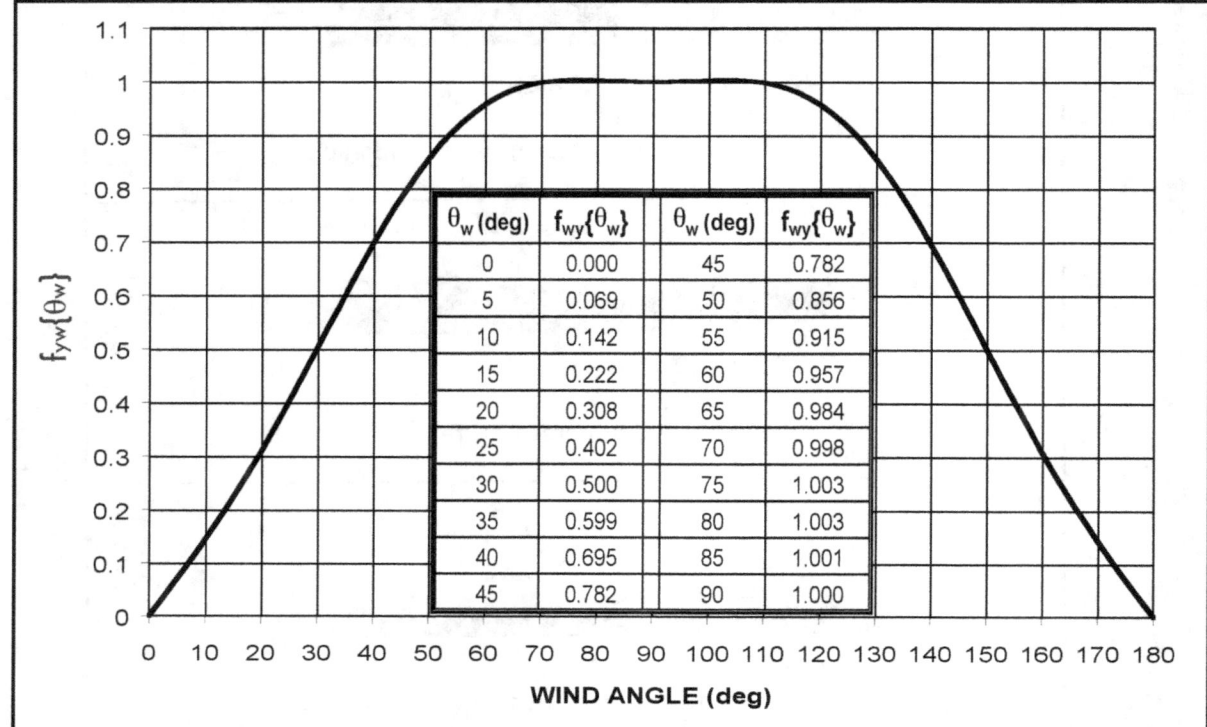

θ_w (deg)	$f_{wy}\{\theta_w\}$	θ_w (deg)	$f_{wy}\{\theta_w\}$
0	0.000	45	0.782
5	0.069	50	0.856
10	0.142	55	0.915
15	0.222	60	0.957
20	0.308	65	0.984
25	0.402	70	0.998
30	0.500	75	1.003
35	0.599	80	1.003
40	0.695	85	1.001
45	0.782	90	1.000

Example 5-1 Transverse Wind Force Drag Coefficient

Find the transverse wind force drag coefficient on the destroyer shown in Figure 5-6.

Figure 5-6 Drag Coefficient Example

PARAMETER	VALUE (SI UNITS)	VALUE (ENGLISH)
L_{wL}	161.23 m	529 ft
A_Y	2239 m^2	24100 ft^2
A_H	1036 m^2	11152 ft^2
A_S	1203 m^2	12948 ft^2
$h_H = A_H/L_{wL}$	6.43 m	21.1 ft
h_S	23.9 m	78.4 ft

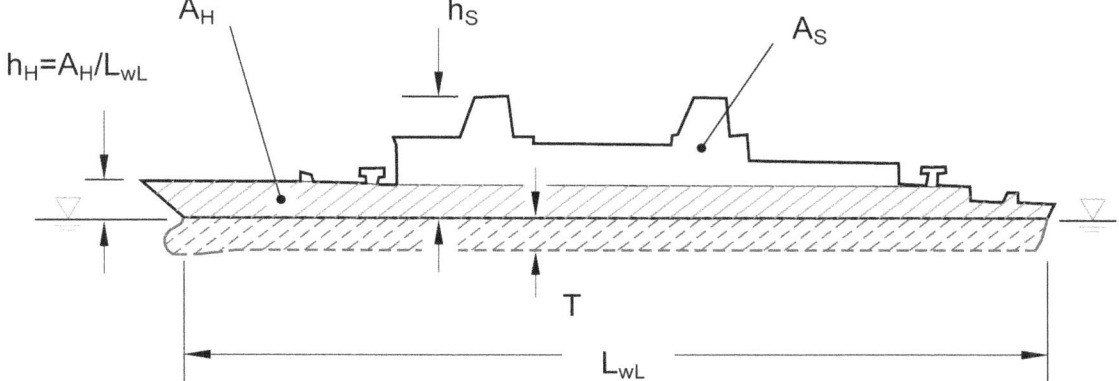

SOLUTION: For this example the transverse wind force drag coefficient from Equation 5-2 is:

$$C_{yw} = \frac{((0.5(23.9m+6.43m))/10m)^{2/7} 1203m^2 + (0.5*6.43m/10m)^{2/7} 1036.1m^2}{2239m^2} C$$

$$C_{yw} = 0.940C$$

Destroyers have extensive superstructure, so a recommended value of C = 1.02 is used to give a transverse wind force drag coefficient of C_{yw} = 0.940*1.02 = 0.958.

Note that for cases where an impermeable structure, such as a wharf, is immediately next to the moored ship, the exposed longitudinal wind area and resulting transverse wind force can be reduced. Figure 5-7 shows an example of a ship next to a wharf. For Case (A), wind from the water, there is no blockage in the transverse wind force and elevations of the hull and superstructure are measured from the water surface. For Case (B), wind from land, the longitudinal wind area of the hull can be reduced by the blocked amount and elevations of hull and superstructure can be measured from the wharf elevation.

Figure 5-7 Ship Adjacent to a Wharf

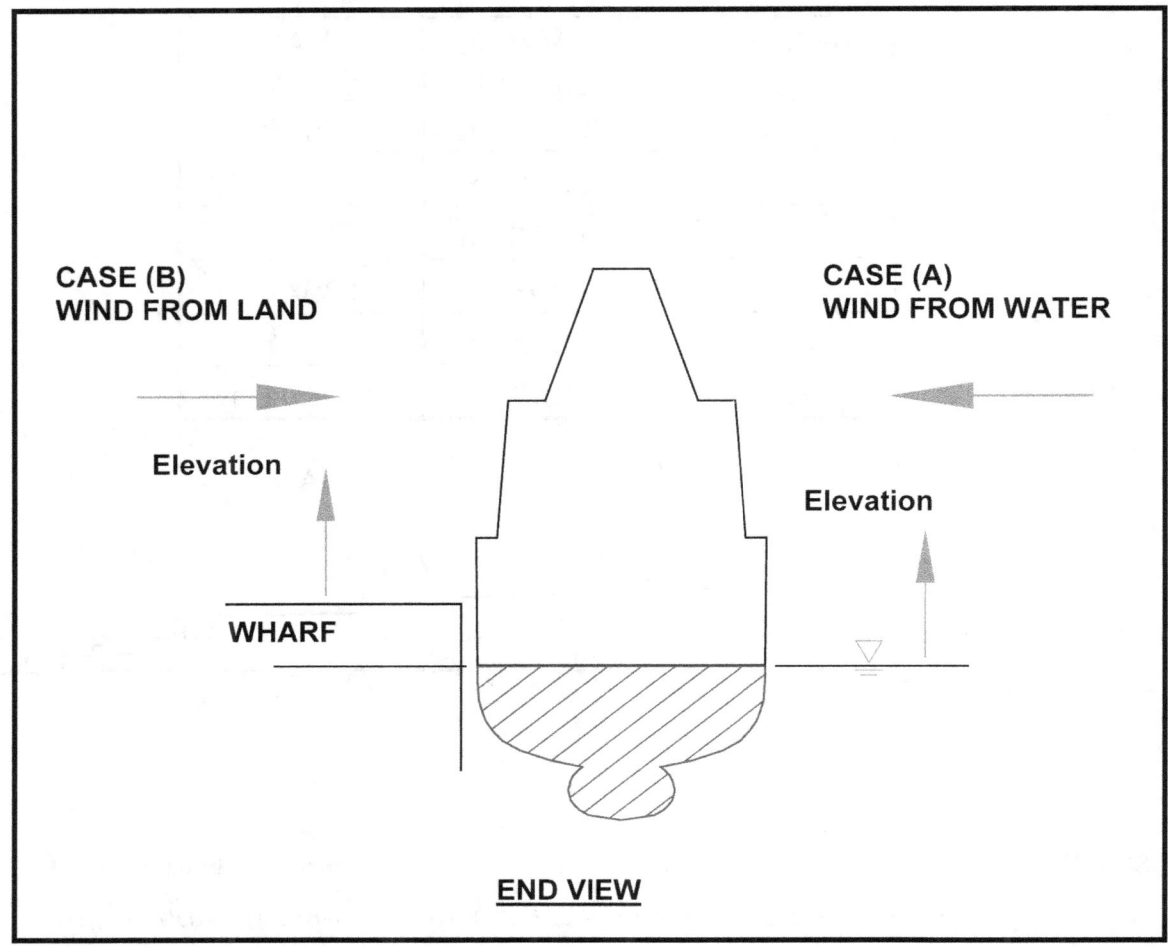

END VIEW

5.4.2. Static Longitudinal Wind Force

The static longitudinal wind force on a vessel is defined as that component of wind force parallel to the centerline of the vessel. This is the force in the "X" or surge direction in Figure 5-3. Figure 5-2 shows the definition of winds areas.

The longitudinal force is determined from:

$$\textbf{Equation 5-4:} \quad F_{xw} = \rho_a A_x C_{xw} f_{xw}(\theta_w) \frac{V_w^2}{2}$$

where

- F_{xw} = longitudinal wind force (N)
- A_x = transverse wind area of the ship (m^2)
- C_{xw} = longitudinal wind force drag coefficient
- $f_{xw}(\theta_w)$ = shape function for longitudinal force

The longitudinal wind force drag coefficient, C_{xw}, depends on specific characteristics of the vessel. Additionally, the wind force drag coefficient varies depending on bow (C_{xwB}) or stern (C_{xwS}) wind loading. Types of vessels are given in three classes: hull dominated, normal, and excessive superstructure. Recommended values of longitudinal wind force drag coefficients are given in Table 5-3.

Table 5-3 Recommended Ship Longitudinal Wind Force Drag Coefficients

VESSEL TYPE	C_{xwB}	C_{xwS}
Hull Dominated (aircraft carriers, submarines, passenger liners)	0.40	0.40
Normal[24]	0.70	0.60
Center-Island Tankers[24]	0.80	0.60
Significant Superstructure (destroyers, cruisers)	0.70	0.80

The longitudinal shape function also varies over bow and stern wind loading regions. As the wind direction varies from headwind to tailwind, there is an angle at which the force changes sign. This is defined as θ_x and is dependent on the location of the superstructure relative to midships. Recommended values of this angle are given in Table 5-4.

Table 5-4 Recommended Values of θ_x

LOCATION OF SUPERSTRUCTURE	θ_x (deg)
Just forward of midships	100
On midships	90
Aft of midships (tankers)	80
Warships	70
Hull dominated	60

Shape functions are given for general vessel categories below:

CASE I SINGLE DISTINCT SUPERSTRUCTURE

The shape function for longitudinal wind load for ships with single, distinct superstructures and hull-dominated ships is given below (examples include aircraft carriers, EC-2, and cargo vessels):

Equation 5-5: $f_{xw}(\theta_w) = \cos(\phi)$

Where θ_x = incident wind angle that produces no net longitudinal force (Table 5-4,) and

---

[24] An adjustment of up to +0.10 to ☐ and ☐ should be made to account for significant cargo or cluttered decks.

Equation 5-6:
$$\phi_- = 90°\left(\frac{\theta_w}{\theta_x}\right) \text{ for } \theta_w < \theta_x$$
$$\phi_+ = 90°\left(1 + \frac{\theta_w - \theta_x}{180° - \theta_x}\right) \text{ for } \theta_w > \theta_x$$

Values of $f_{xw}(\theta_w)$ are symmetrical about the longitudinal axis of the vessel. So when $\theta_w > 180°$, use $360° - \theta_w$ as θ_w in determining the shape function.

CASE II DISTRIBUTED SUPERSTRUCTURE:

Equation 5-7: $f_{xw}(\theta_w) = \frac{10}{9}\sin\gamma - \frac{1}{9}\sin(5\gamma)$

Where

Equation 5-8:
$$\gamma_- = 90°\left(1 + \frac{\theta_w}{\theta_x}\right) \text{ for } \theta_w < \theta_x$$
$$\gamma_+ = 90°\left(2 + \frac{\theta_w - \theta_x}{180° - \theta_x}\right) \text{ for } \theta_w > \theta_x$$

Values of $f_{xw}(\theta_w)$ are symmetrical about the longitudinal axis of the vessel. So when $\theta_w > 180°$, use $360 - \theta_w$ as θ_w in determining the shape function. Note that the maximum longitudinal wind force for these vessels occurs for wind directions slightly off the ship's longitudinal axis.

Example 5-2 Longitudinal Wind Drag Coefficient

Find the longitudinal wind drag coefficient for a wind angle of 40° for the destroyer shown in Figure 5-6.

SOLUTION: For this destroyer, the following values are selected:

- $\theta_x = 70°$ from Table 5-4
- $C_{xwB} = 0.70$ from Table 5-3
- $C_{xwS} = 0.80$ from Table 5-3

This ship has a distributed superstructure (Case II) and the wind angle is less than the crossing value, so Equation 5-8 is used to determine the shape function:

$$\gamma_- = \left(\frac{90°}{70°}\right)40° + 90° = 141.4°$$

$$f_{xw}(\theta_w) = 1.11\sin(141.4°) - 0.11\sin(5 \times 141.4°) = 0.72$$

At the wind angle of 40°, the wind has a longitudinal component on the stern. Therefore, the wind longitudinal drag coefficient for this example is:

$$C_{xw}f_{xw}(\theta_w) = 0.8 * 0.72 = 0.57$$

5.4.3. Static Wind Yaw Moment

The static wind yaw moment is defined as the product of the associated transverse wind force and its distance from the vessel's center of gravity. In the local ship coordinate system, this is the moment about the "Z" axis. Wind yaw moment is determined from the equation:

Equation 5-9: $M_{xyw} = \rho_a A_y L C_{xyw}\{\theta_w\} \dfrac{V_w^2}{2}$

where

- M_{xyw} = wind yaw moment (N*m)
- L = length of ship (m)
- $C_{xyw}\{\theta_w\}$ = normalized yaw moment coefficient = moment arm divided by ship length

The normalized yaw moment coefficient depends upon the vessel type. Equation 5-10 gives equations for computing the value of the yaw moment coefficient and Table 5-5 gives empirical parameter values for selected vessel types. The normalized yaw moment coefficient is found from:

Equation 5-10: $C_{xyw}\{\theta_w\} = -a_1 \sin\left(180° \dfrac{\theta_w}{\theta_z}\right), \; 0 < \theta_w < \theta_z$

Equation 5-11: $C_{xyw}\{\theta_w\} = a_2 \sin\left(180° \dfrac{\theta_w - \theta_z}{180° - \theta_z}\right), \; \theta_z \leq \theta_w < 180°$

where

- a_1 = negative peak value (from Table 5-5)
- a_2 = positive peak value (from Table 5-5)
- θ_z = zero moment angle (degrees) (from Table 5-5)

This coefficient is symmetrical about the longitudinal axis of the vessel.

Table 5-5 Normalized Wind Yaw Moment Variables

SHIP TYPE	Zero Moment Angle (θ_z)	Negative Peak (a_1)	Positive Peak (a_2)	NOTES
Liner	80	0.075	0.14	
Carrier	90	0.068	0.072	
Tanker	95	0.077	0.07	Center island w/ cluttered deck
Tanker	100	0.085	0.04	Center island w/ trim deck
Cruiser	90	0.064	0.05	
Destroyer	68	0.02	0.12	
Others:	130	0.13	0.025	stern superstructure
	102	0.096	0.029	aft midships superstructure
	90	0.1	0.1	midships superstructure
	75	0.03	0.05	forward midships superstructure
	105	0.18	0.12	bow superstructure

A plot of the yaw normalized moment coefficient for the example shown in Figure 5-6 is given as Figure 5-8.

Figure 5-8 Sample Yaw Normalized Moment Coefficient

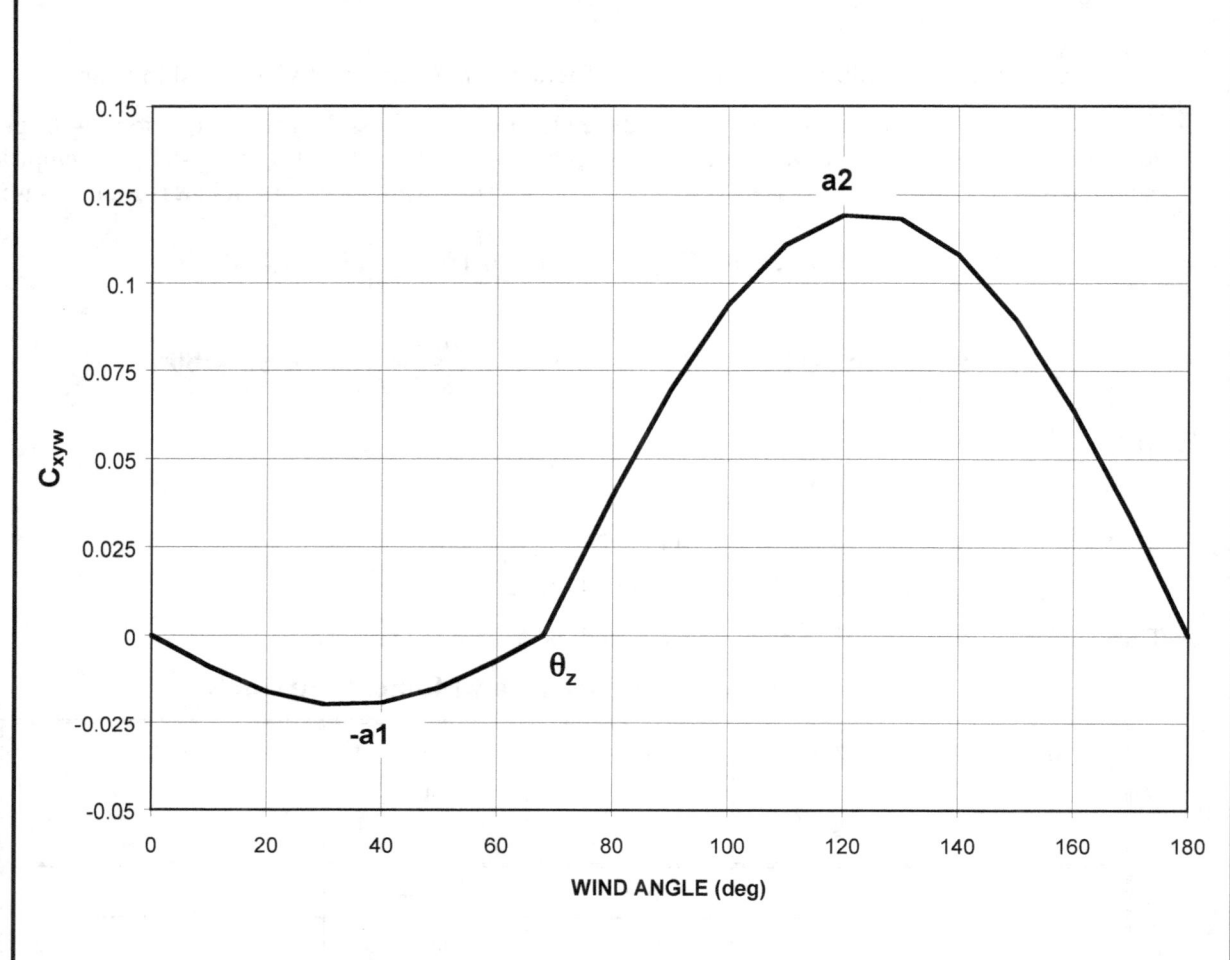

5.5. Static Current Forces/Moments

Methods to determine static current forces and moments on stationary moored vessels in the surge and sway directions and yaw moment are presented in this section. These planar directions are of primary importance in many mooring designs.

5.5.1. Static Transverse Current Force

The transverse current force is defined as that component of force perpendicular to the vessel centerline. If a ship has a large underkeel clearance, then water can freely flow under the keel, as shown in Figure 5-9(a). If the underkeel clearance is small, as shown in Figure 5-9(b), then the ship more effectively blocks current flow, and the transverse current force on the ship increases.

Figure 5-9 Examples of Ratios of Ship Draft (T) to Water Depth (d)

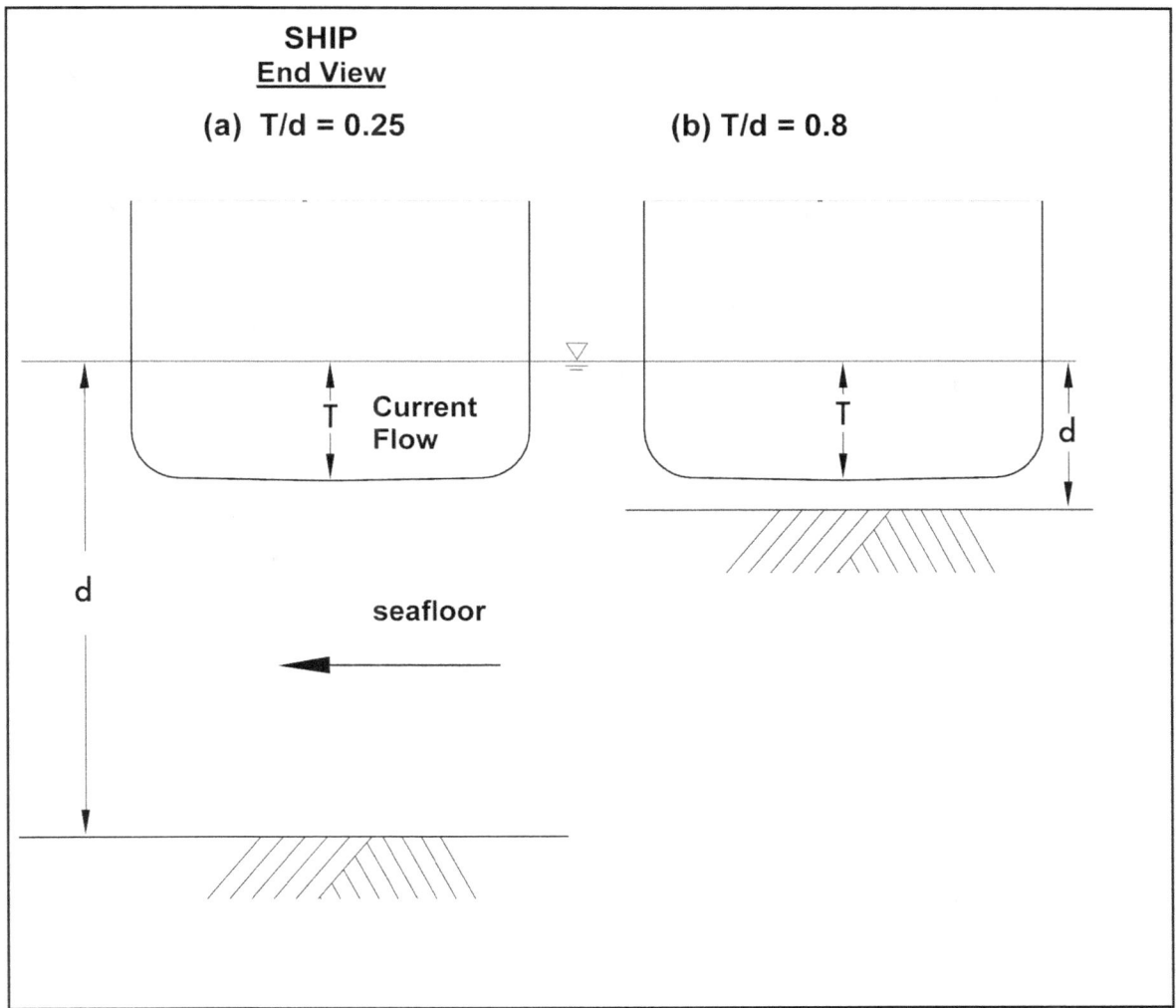

These effects are considered and the transverse current force is determined from the equation:

$$\textbf{Equation 5-12:} \quad F_{yc} = \rho_w L_{wL} \, T \, C_{yc} \, \sin\theta_c \frac{V_c^2}{2}$$

Where
- F_{yc} = transverse current force (N)
- ρ_w = mass density of water (from Table 5-1)
- V_c = current velocity (m/s)
- L_{wL} = vessel waterline length (m)
- T = average vessel draft (m)
- C_{yc} = transverse current force drag coefficient
- θ_c = current angle (degrees)

The transverse current force drag coefficient as formulated in *Broadside Current Forces on Moored Ships*, Seelig et al. (1992) is shown in Figure 5-10.

Figure 5-10 Broadside Current Drag Coefficient

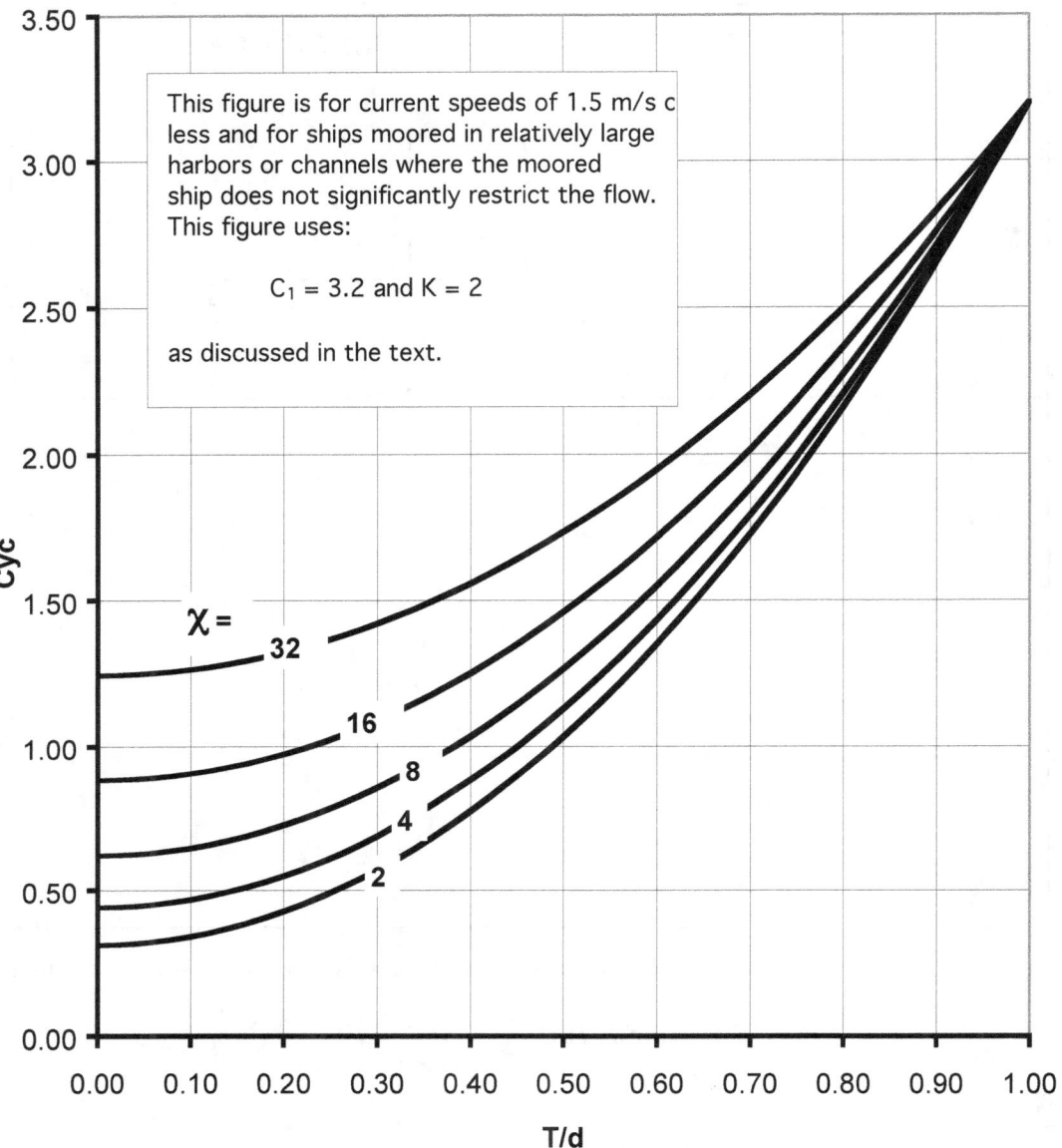

This drag coefficient can be determined from:

$$\text{Equation 5-13:} \quad C_{yc} = C_0 + (C_1 - C_0)\left(\frac{T}{d}\right)^K$$

Where

- C_1 = shallow water current force drag coefficient for the following conditions:
  - T/d = 1.0
  - Currents < 1.5 m/s (3 knots or 5 ft/sec)
- d = water depth (m)

- K = dimensionless exponent; laboratory data from ship models shows:
    - $K = 2$: Wide range of ship and barge tests; most all of the physical model data available can be fit with this coefficient
    - $K = 3$: From a small number of tests on a fixed cargo ship and for a small number of tests on an old aircraft carrier, CVE-55
    - $K = 5$: From a small number of tests on an old submarine hull, SS-212

And

Equation 5-14: $C_0 = 0.22\sqrt{\chi}$, $T/d \approx 0.0$

Where

- C_O = deepwater current force drag coefficient
- χ = dimensionless ship parameter
- $\chi = \dfrac{A_m L_{wL}^2}{BV}$

    - A_m = the immersed cross-sectional area of the ship at midsection (m^2)
    - B = the beam (maximum ship width at the waterline) (m)
    - V = the submerged volume of the ship, which can be found by taking the displacement of the vessel divided by the unit weight of water, given in Table 5-1 (m^3)).

The immersed cross-sectional area of the ship at midships, A_m, can be determined from

Equation 5-15: $A_m = C_m BT$

Values of the midship coefficient, C_m, are provided in the NAVFAC Ship's Database for DOD ships.

The above methods for determining the transverse current force are recommended for normal design conditions with moderate current speeds of 1.5 m/s (3 knots or 5 ft/sec) or less and in relatively wide channels and harbors (see Seelig et al., 1992).

If the vessel is moored broadside in currents greater than 1.5 m/s (3 knots or 5 ft/sec), then scale model laboratory data show that there can be significant vessel heel/roll, which effectively increases the drag force on the vessel. In some model tests in shallow water and at high current speeds this effect was so pronounced that the model ship capsized. Mooring a vessel broadside in a high current should be avoided, if possible.

Scale physical model tests show that a vessel moored broadside in a restricted channel has increased current forces. This is because the vessel decreases the effective flow area of a restricted channel, which causes the current speed and current force to increase.

For specialized cases where:

(1) vessels are moored in current of 1.5 m/s (3 knots or 5 ft/sec) or more, and/or

(2) for vessels moored in restricted channels

then the designer should contact the Moorings Center of Expertise, NFESC ECDET, Washington Navy Yard Bldg. 218, 901 M St. SE, Washington DC 20374-5063.

Example 5-3 Current Force

Find the current force on an FFG-7 vessel produced by a current of $\theta_c = 90°$ to the ship centerline with a speed of 1.5 m/s (2.9 knots or 4.9 ft/sec) in salt water for a given ship draft. At the mooring location, the harbor has a cross-sectional area much larger than the submerged ship longitudinal area, $L_{wL}*T$.

SOLUTION: Dimensions and characteristics of this vessel are summarized in the lower right portion of Figure 5-11. Transverse current drag coefficients predicted using Equation 5-13 are shown on this figure as a solid bold line. Physical scale model data (U.S. Naval Academy (USNA), EW-9-90, *Evaluation of Viscous Damping Models for Single Point Mooring Simulation*) are shown as symbols in the drawing, showing that Equation 5-13 provides a reasonable estimate of drag coefficients. Predicted current forces for this example are given in Table 5-6.

Figure 5-11 Example of Transverse Current Drag Coefficients

Data taken from tests conducted at the US Naval Academy at scales 1/24.75 and 1/80. Some data taken at 5 and 6 knots is not included. (Kreibel, 1992)

$\chi = 14.89$

Model data points

FFG-7
$C_m = 0.78$
$L_{wL} = 124.36$ m
$B = 11.58$ m
$T = 4.389$ m
$D = 3590$ long ton (LT)
$V = 3590$ LT * 0.9904 m³/LT
$\quad = 3555.7$ m³
$A_m = 0.78 *B*T = 39.64$ m²
$\chi = L_{wL}^2 * A_m /(B*V) = 14.89$
$C_0 = 0.8489$
$C_1 = 3.2$
$K = 2$

X-axis: T/d
Y-axis: Cyc

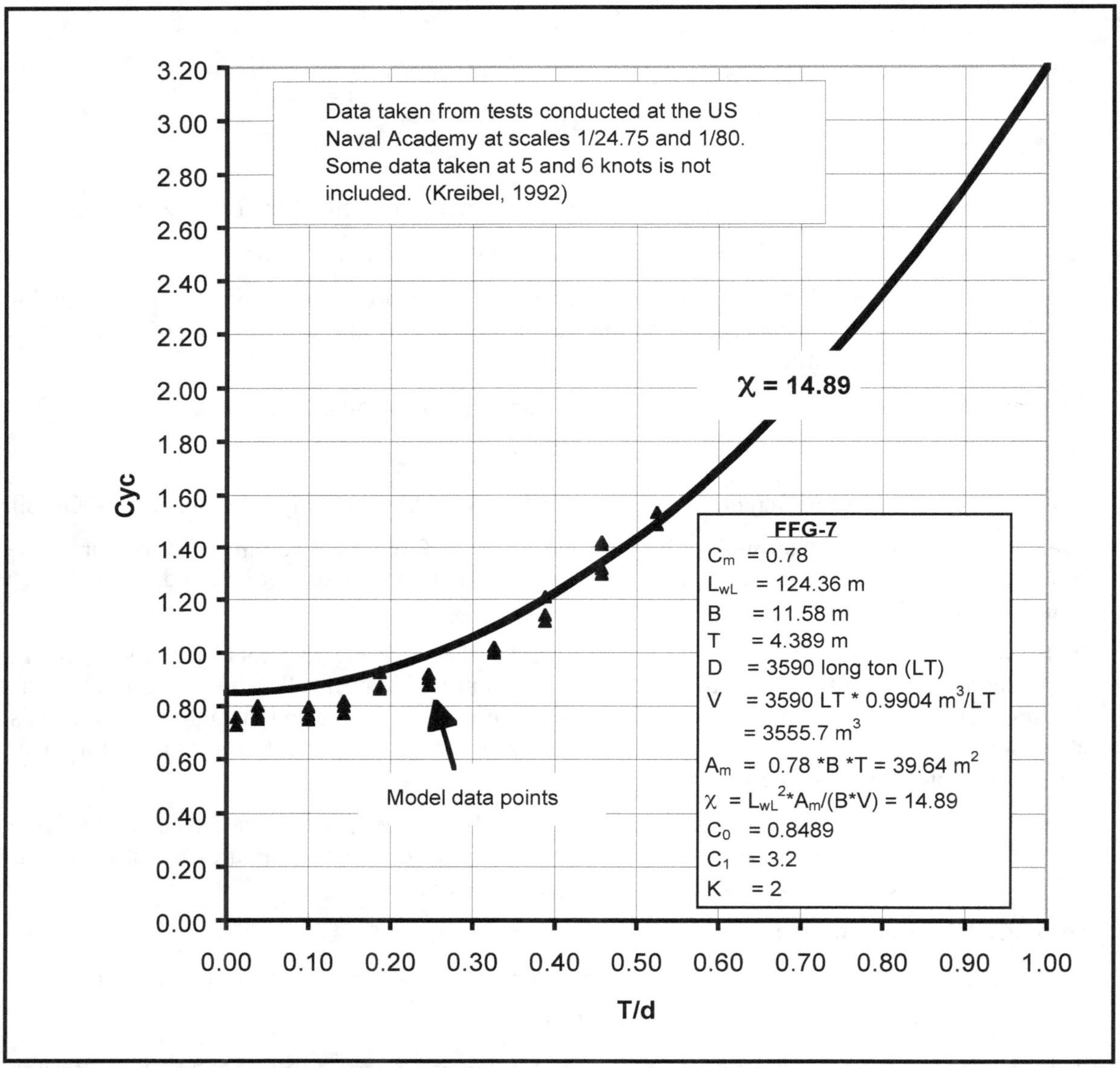

Table 5-6 Predicted Transverse Current Forces on FFG-7 for a Current Speed of 1.5 m/s (2.9 knots)

T/d	d (m)	D (ft)	F_{yc} (MN)	F_{yc} (kips)

0.096	45.7	150	0.55	123
0.288	15.2	50	0.66	148
0.576	7.62	25	1.03	231
0.72	6.096	20	1.30	293
0.96	4.572	15	1.90	427

This example shows that in shallow water the transverse current force can be three times or larger than in deep water for an FFG-7.

5.5.2. Static Longitudinal Current Force

The longitudinal current force is defined as that component of force parallel to the centerline of the vessel. This force is determined from the following equation (Naval Civil Engineering Laboratory (NCEL), TN-1634, *STATMOOR – A Single-Point Mooring Static Analysis Program*):

Equation 5-16: $F_{xc} = F_{x\,FORM} + F_{x\,FRICTION} + F_{x\,PROP}$

where

- F_{xc} = total longitudinal current load (N)
- F_{xFORM} = longitudinal current load due to form drag (N)
- $F_{xFRICTION}$ = longitudinal current load due to skin friction (N)
- F_{xPROP} = longitudinal current load due to propeller drag (N)

The three elements of the general longitudinal current load equation, F_{xFORM}, $F_{xFRICTION}$, and F_{xPROP} are described below:

Equation 5-17: $F_{xFORM} = \rho_w B T C_{xcb} \cos(\theta_c) \dfrac{V_c^2}{2}$

where

- B = maximum vessel width at the waterline (m)
- C_{xcb} = longitudinal current form drag coefficient = 0.1

Equation 5-18: $F_{xFRICTION} = \rho_w S C_{xca} \cos(\theta_c) \dfrac{V_c^2}{2}$

where

- S = wetted surface area (m^2); estimated using

Equation 5-19: $S = 1.7\, T L_{wL} + \left(\dfrac{D}{T \gamma_w}\right)$

- D = ship displacement (N)
- C_{xca} = longitudinal skin friction coefficient, estimated using:

Equation 5-20: $C_{xca} = \dfrac{0.075}{(\log_{10} R_N - 2)^2}$

- R_N = Reynolds Number, from

Equation 5-21: $R_N = \left|\dfrac{V_c L_{wL} \cos(\theta_c)}{\nu}\right|$

- ν = kinematic viscosity of water, from Table 5-1

Equation 5-22: $F_{xPROP} = \rho_w A_p C_{PROP} \cos(\theta_c) \dfrac{V_c^2}{2}$

where

- C_{PROP} = propeller drag coefficient = 1.0
- A_p = propeller expanded blade area (m^2)

Equation 5-23: $A_p = \dfrac{A_{Tpp}}{1.067 - 0.229\left(\dfrac{p}{d}\right)} = \dfrac{A_{Tpp}}{0.838}$

- A_{Tpp} total projected propeller area (m^2) for an assumed propeller pitch ratio of $p/d = 1.0$

Equation 5-24: $A_{Tpp} = \dfrac{L_{wL} B}{A_R}$

A_R is a dimensionless area ratio for propellers. Typical values of this parameter for major vessel groups are given in Table 5-7.

Table 5-7 A_R for Major Vessel Groups

SHIP	AREA RATIO, A_R
Destroyer	100
Cruiser	160
Carrier	125
Cargo	240
Tanker	270
Submarine	125

Note that in these and all other engineering calculations discussed in this book, the user must be careful to keep units consistent.

Example 5-4 Longitudinal Current Force

Find the longitudinal current force with a bow-on current of $\theta_c = 180°$ with a current speed of 1.544 m/sec (3 knots) on a destroyer in salt water with the characteristics shown in Table 5-8.

Table 5-8 Example Destroyer

PARAMETER	SI SYSTEM	ENGLISH OR INCH-POUND SYSTEM
L_{wL}	161.2 m	529 ft

T	6.4 m	21 ft
B	16.76 m	55 ft
D, ship displacement	7,930 Mg	7,810 long tons
C_m; estimated	0.83	0.83
S; est. from Equation 5-19	2963 m^2	31 897 ft^2
A_R; from Table 5-7	100	100
R_N; from Equation 5-21	2.09E8	2.09E8
C_{xca}; est. from Equation 5-20	0.00188	0.00188
A_p; est. from Equation 5-23	32.256 m^2	347.2 ft^2

SOLUTION: Table 5-9 shows the predicted current forces. Note that these forces are negative, since the bow-on current is in a negative "X" direction. For this destroyer, the force on the propeller is approximately two-thirds of the total longitudinal current force. For commercial ships, with relatively smaller propellers, form and friction drag produce a larger percentage of the current force.

Table 5-9 Example Longitudinal Current Forces on a Destroyer

FORCE	SI SYSTEM	ENGLISH OR INCH-POUND SYSTEM	PERCENT OF TOTAL FORCE
F_{xFORM}; Equation 5-17	-13.1 kN	-2.95 kip	22%
$F_{xFRICTION}$; Equation 5-18	-6.8 kN	-1.53 kip	12%
F_{xPROP}; Equation 5-22	-39.4 kN	-8.87 kip	66%
Total F_{xc} =	-59.4 kN	-13.4 kip	100%

5.5.3. Static Current Yaw Moment

The current yaw moment is defined as that component of moment acting about the vessel's vertical "Z"-axis. This moment is determined from the equation:

Equation 5-25: $$M_{xyc} = F_{yc}\left(\frac{e_c}{L_{wL}}\right)L_{wL}$$

Where

- M_{xyc} = current yaw moment (N*m)
- F_{yc} = transverse current force (N)
- e_c/L_{wL} = ratio of eccentricity to vessel waterline length
- e_c = eccentricity of F_{yc} (m)
- L_{wL} = vessel waterline length (m)

The dimensionless moment arm e_c/L_{wL} is calculated by choosing the slope and y-intercept variables from Table 5-10 which are a function of the vessel hull. The dimensionless moment arm is dependent upon the current angle to the vessel,

$$\textbf{Equation 5-26:} \quad \frac{e}{L_{wL}} = a + b\theta_c, \quad \theta_c = 0° \text{ to } 180°$$

$$\textbf{Equation 5-27:} \quad \frac{e}{L_{wL}} = -a - b(360° - \theta_c), \quad \theta_c = 180° \text{ to } 360°$$

Where

- a = y-intercept (refer to Table 5-10) (dimensionless)
- b = slope per degree (refer to Table 5-10)

The above methods for determining the eccentricity ratio are recommended for normal design conditions with moderate current speeds of less than 1.5 m/s (3 knots or 5 ft/sec). Values provided in Table 5-10 are based upon least squares fit of scale model data taken for the case of ships with level keels. Data are not adequately available for evaluating the effect of trim on the current moment.

Table 5-10 Current Moment Eccentricity Ratio Variables

SHIP	a Y-INTERCEPT	b SLOPE PER DEGREE	NOTES
SERIES 60	-0.291	0.00353	Full hull form typical of cargo ships
FFG	-0.201	0.00221	"Rounded" hull typical of surface warships
CVE-55	-0.168	0.00189	Old attack aircraft carrier
SS-212	-0.244	0.00255	Old submarine

5.6. Wind And Current Forces And Moments On Multiple Ships

If ships are moored in close proximity to one another then the nearby ship(s) can influence the forces/moments on a given ship. The best information available on the effects of nearby ships are results from physical model tests, because the physical processes involved are highly complex.

Ships are often moored close to one another to make optimum use of valuable harbor space. Another benefit of nearby ships is to take advantage of "sheltering" effects of one ship on another. For example, the transverse wind force for two identical ships across a pier will be less than for the two ships moored at separate piers.

Examination of laboratory scale-model wind tunnel and flume tests taken at the U.S. Navy David Taylor Model Basin for from 1 to 6 aircraft carriers, destroyers, cargo ships and submarines shows that this data provides much valuable design information. However, the effects of some of the parameters on the transverse force and moments are sometimes complex.

The results are therefore provided in graphical forms for design engineer use. The intent is for these materials to be reviewed and applied with sound engineering judgment. Additional information, background discussion, tabular and graphical data are provided in NFESC TR-6003-OCN, *Wind and Current Forces/Moments on Multiple Ships*, Seelig 1997.

Figure 5-12 shows the ships tested, Figure 5-13 illustrates the coordinate system used and Figure 5-14 shows definition of some terms. Figure Figure 5-15, Figure 5-16 and Figure 5-17 illustrate some of the many ship arrangements tested.

Figure 5-12 Plan and Profile Views of Ships Tested

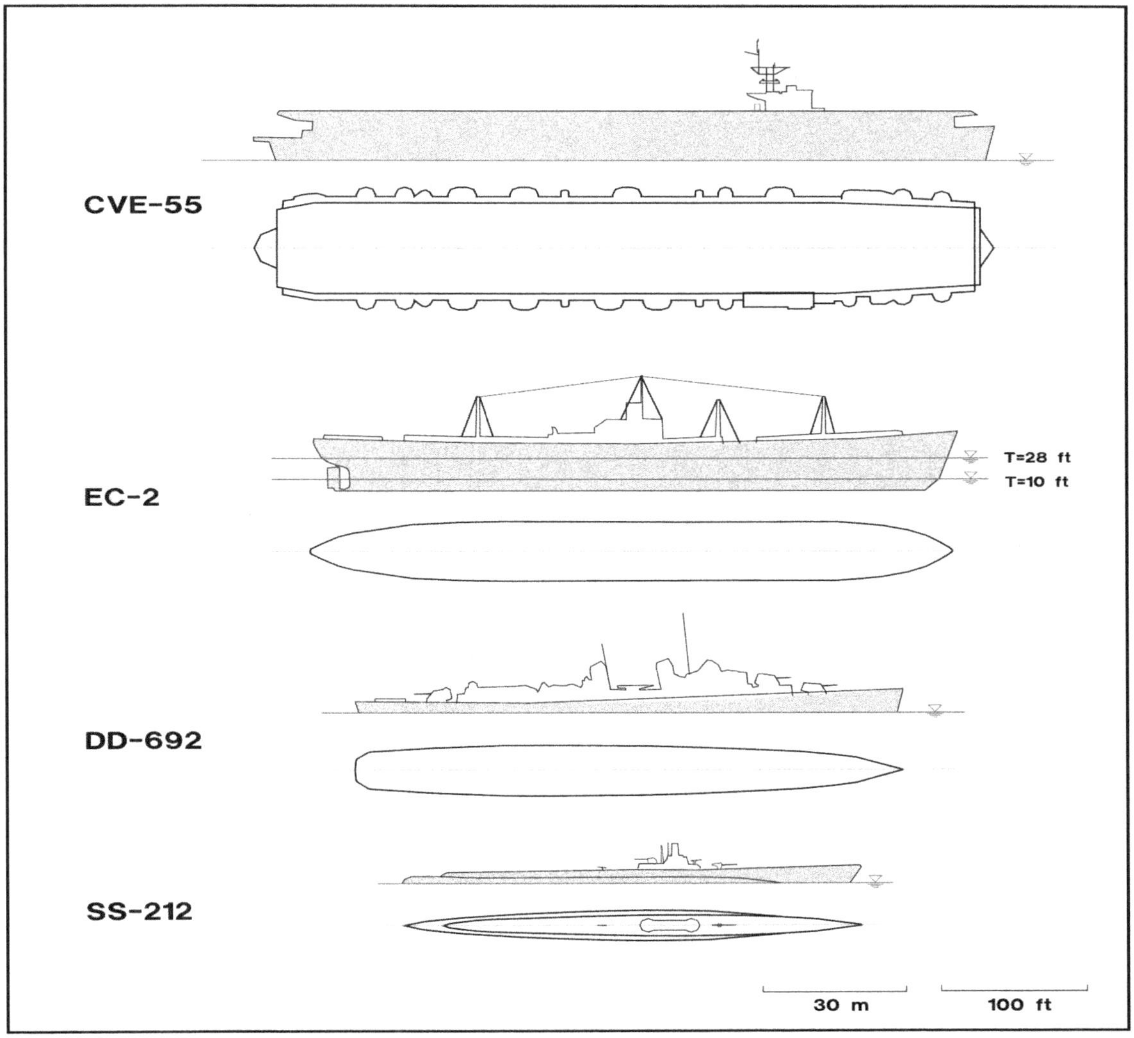

Figure 5-13 Coordinate System

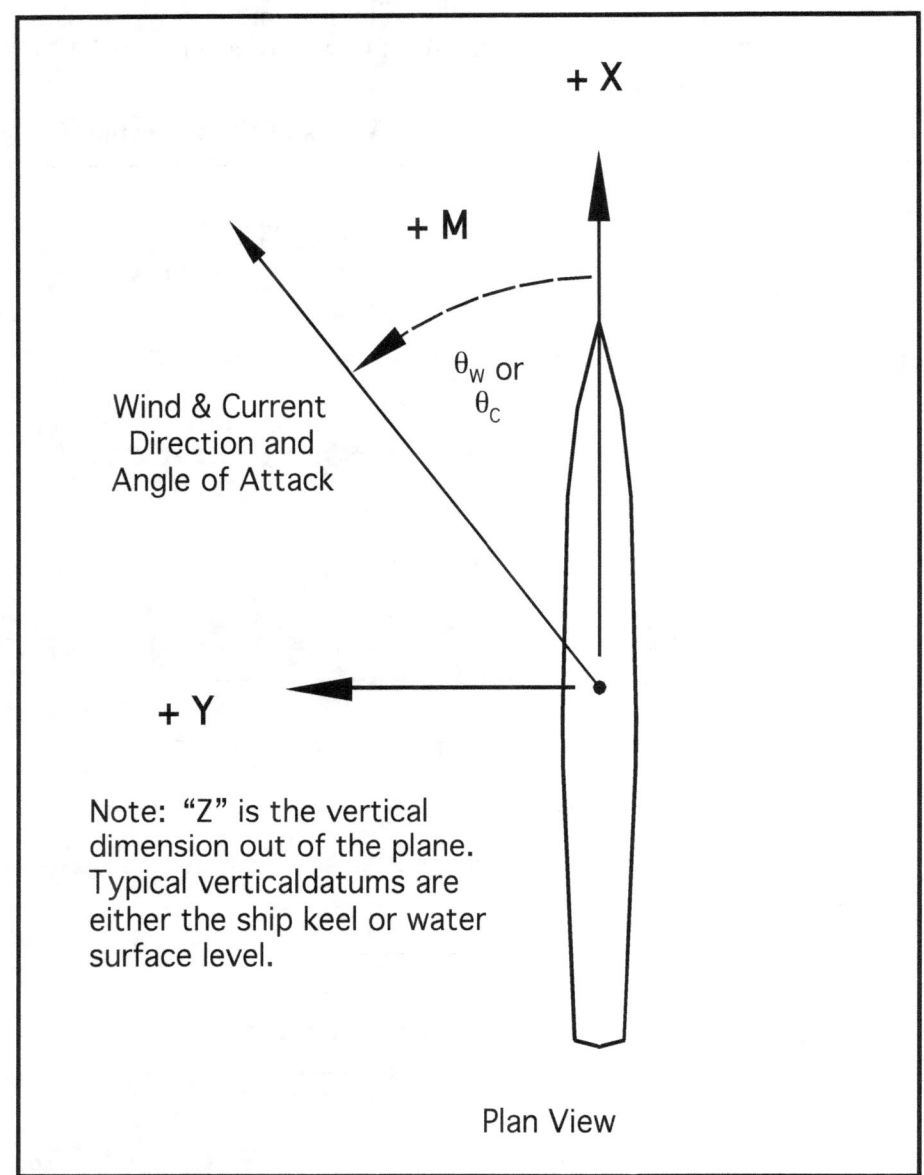

Figure 5-14 Definition of Some Terms

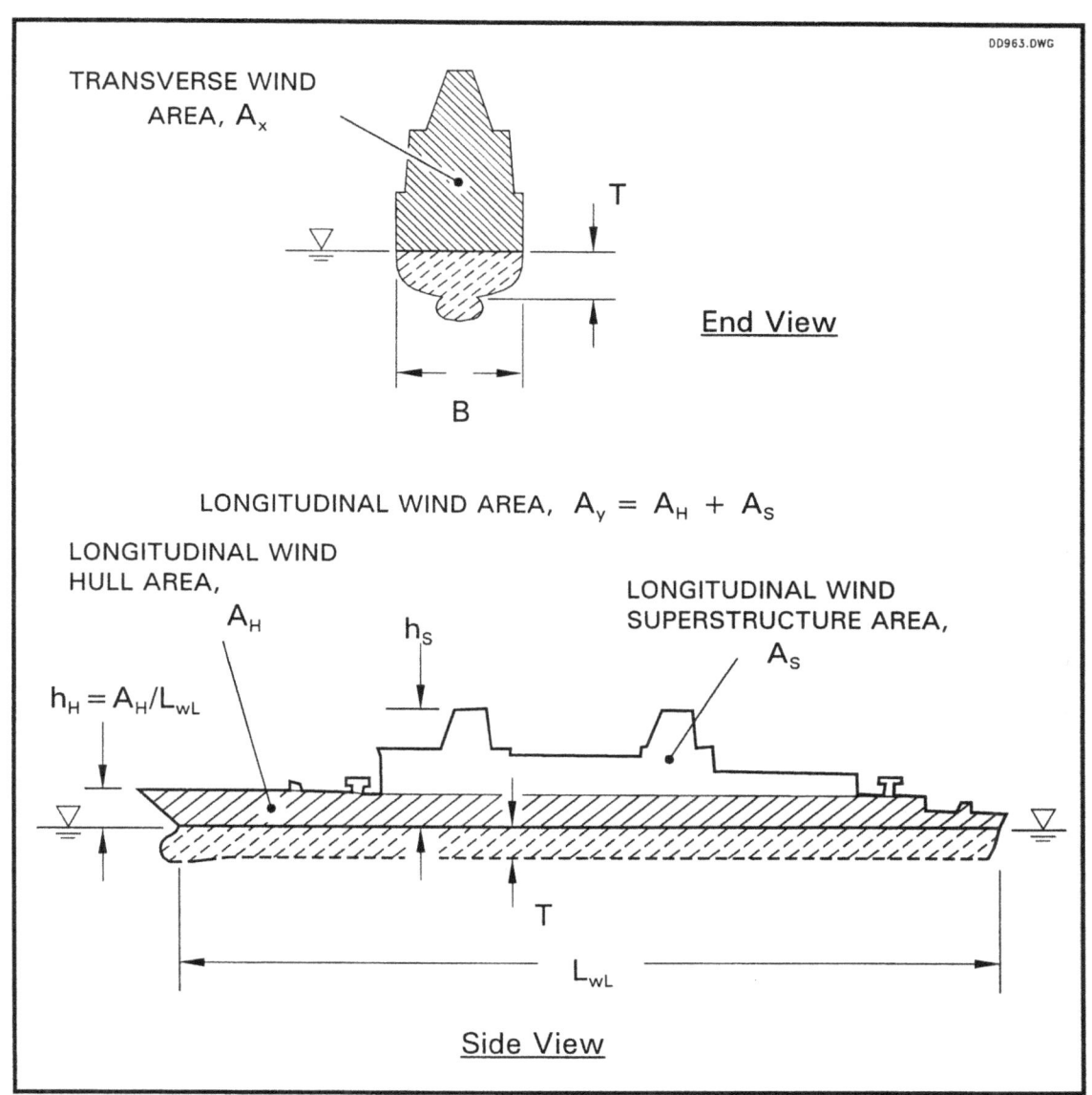

Figure 5-15 Sample Condition

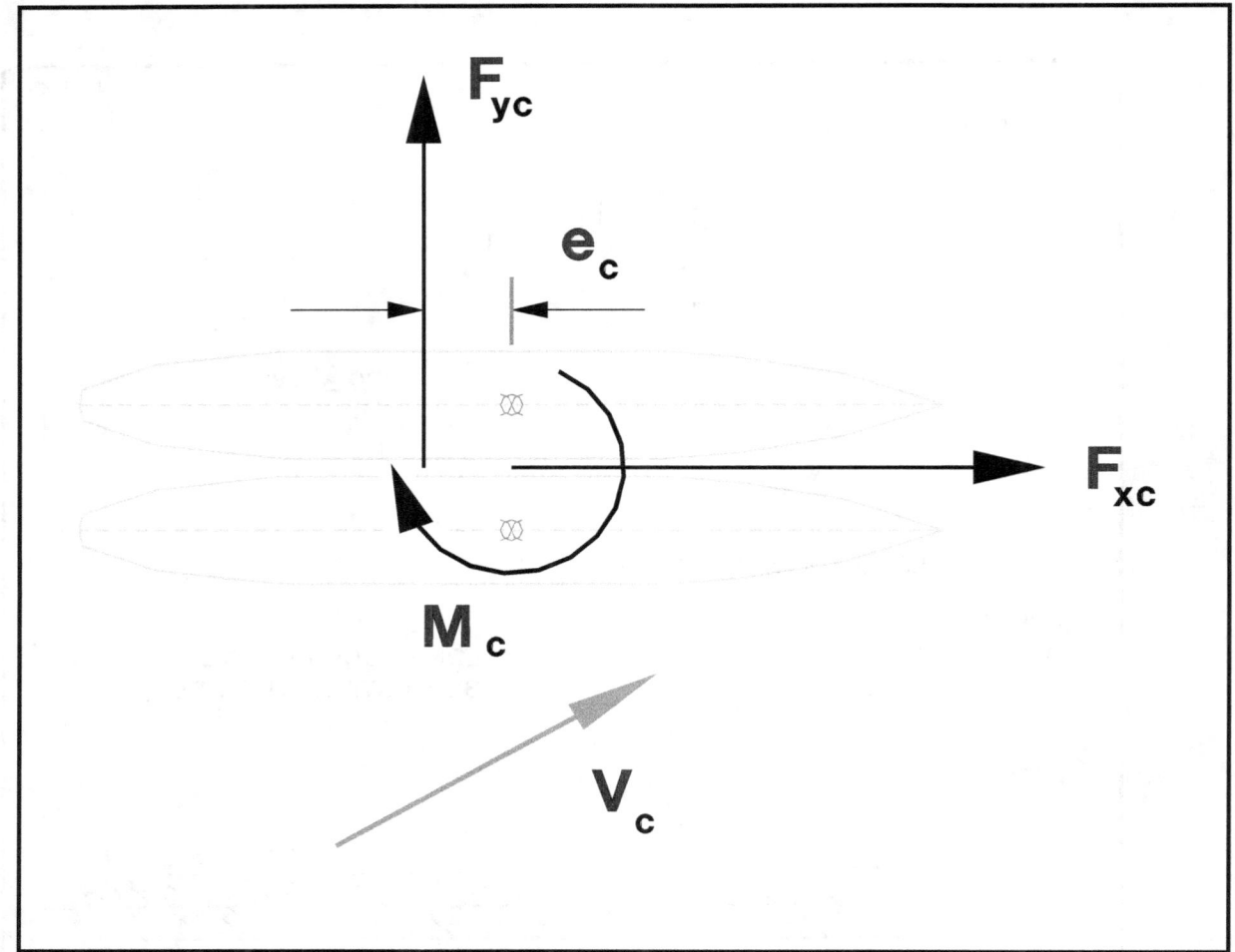

Figure 5-16 Example of Four Ships Moored to Each Other

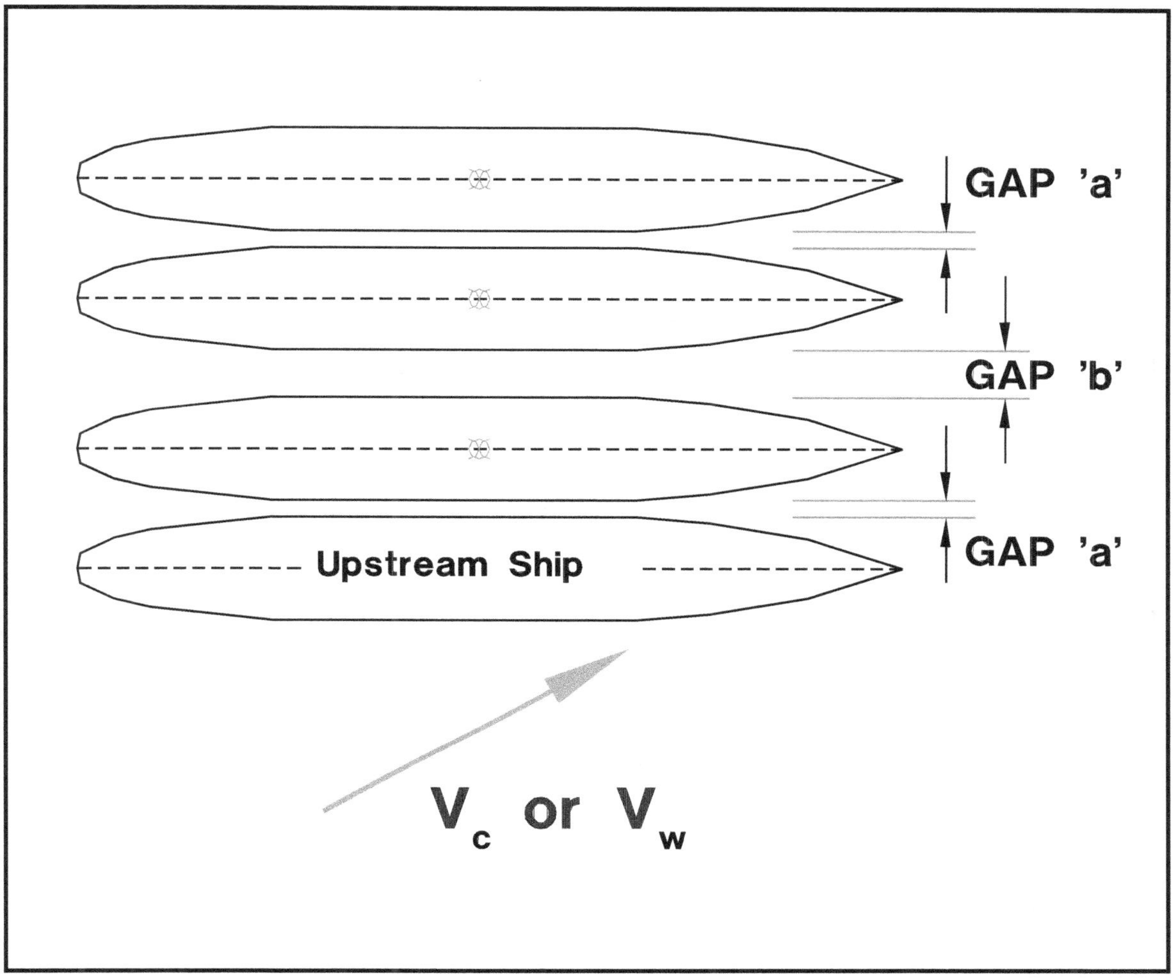

Figure 5-17 Example of Two Ships Moored to Each Other

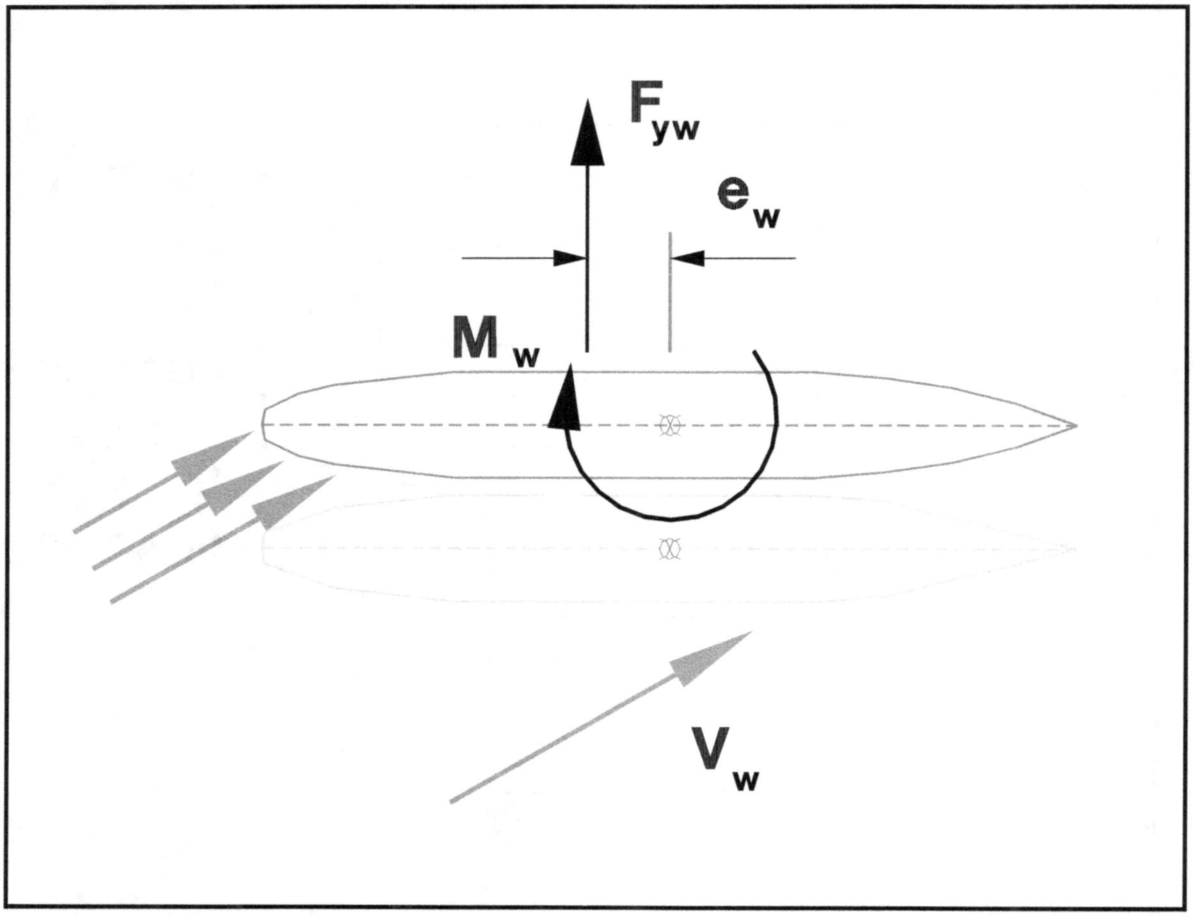

Table 5-11 shows the ship classes tested and location of results.

Table 5-11 Multiple Ship Testing

SHIP CLASS	FIGURES	NOTES
CVE-55 attack aircraft carrier	Figure 5-18	Test arrangement
CVE-55 Wind	Figure 5-19, Figure 5-20, Figure 5-21, Figure 5-22	Results
CVE-55 Current	Figure 5-23, Figure 5-24, Figure 5-25, Figure 5-26	Results for T/d = 0.55
DD-692 destroyer	Figure 5-27	Test arrangement
DD-692 Wind	Figure 5-28, Figure 5-29, Figure 5-30, Figure 5-31	Results
DD-692 Current	Figure 5-32, Figure 5-33, Figure 5-34, Figure 5-35	Results for T/d = 0.425
EC-2 cargo liberty ship	Figure 5-36	Test Arrangement
EC-2 Wind	Figure 5-37, Figure 5-38, Figure 5-39, Figure 5-40, Figure 5-41, Figure 5-42, Figure 5-43	Results
EC-2 Current	Figure 5-44, Figure 5-45, Figure 5-46, Figure 5-47, Figure 5-48, Figure 5-49	Results for T/d = 0.4
SS-212 submarine	Figure 5-50	Test arrangement
SS-212 Current	Figure 5-51, Figure 5-52, Figure 5-53, Figure 5-54, Figure 5-55	Results for T/d = 0.648
SS-212 Wind	Figure 5-56, Figure 5-57, Figure 5-58, Figure 5-59	Results

Figure 5-18 CVE-55 Ship Nests Tested

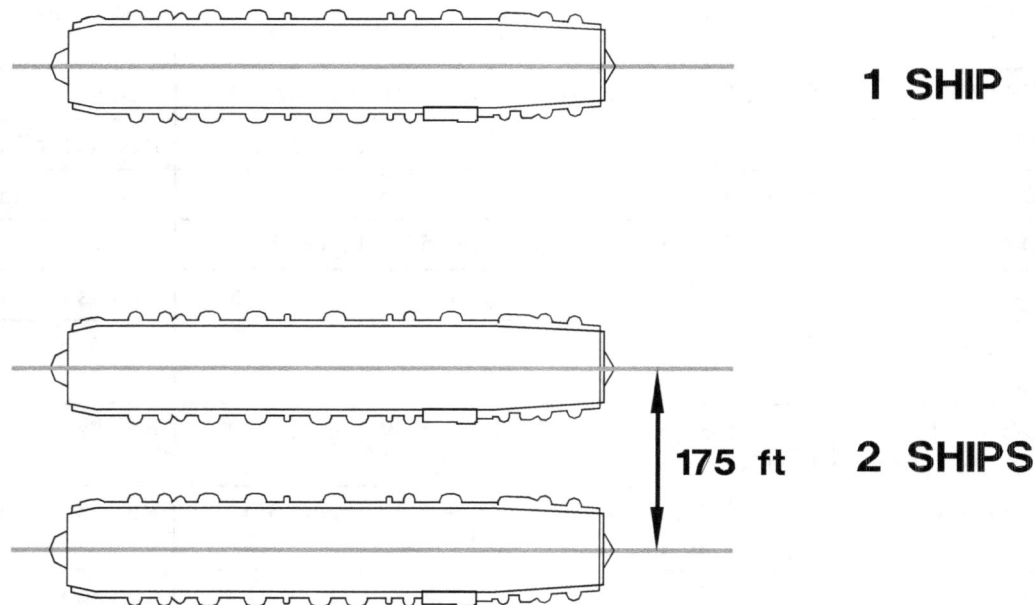

Figure 5-19 CVE-55 Lateral Wind Forces

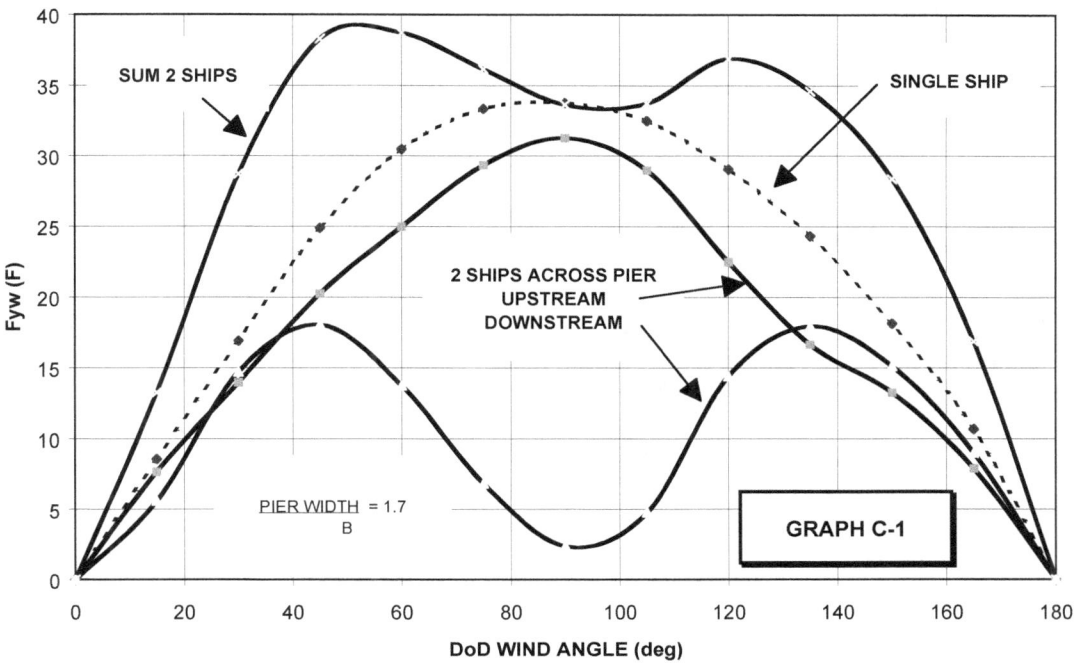

Figure 5-20 CVE-55 Lateral Wind Forces Divided by the Force on One Ship

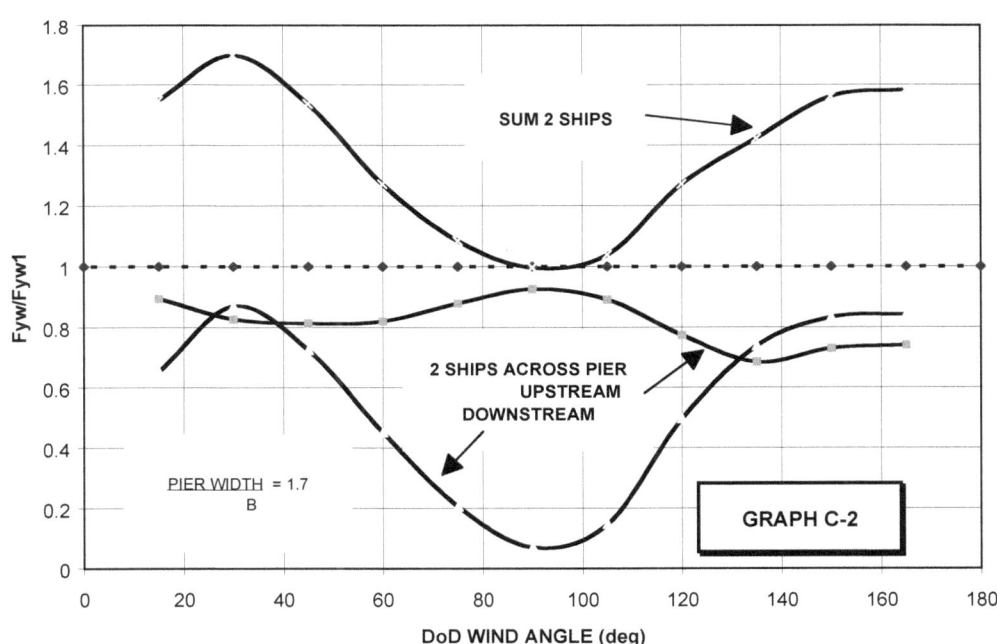

Figure 5-21 CVE-55 Wind Moments on One and Two Ships

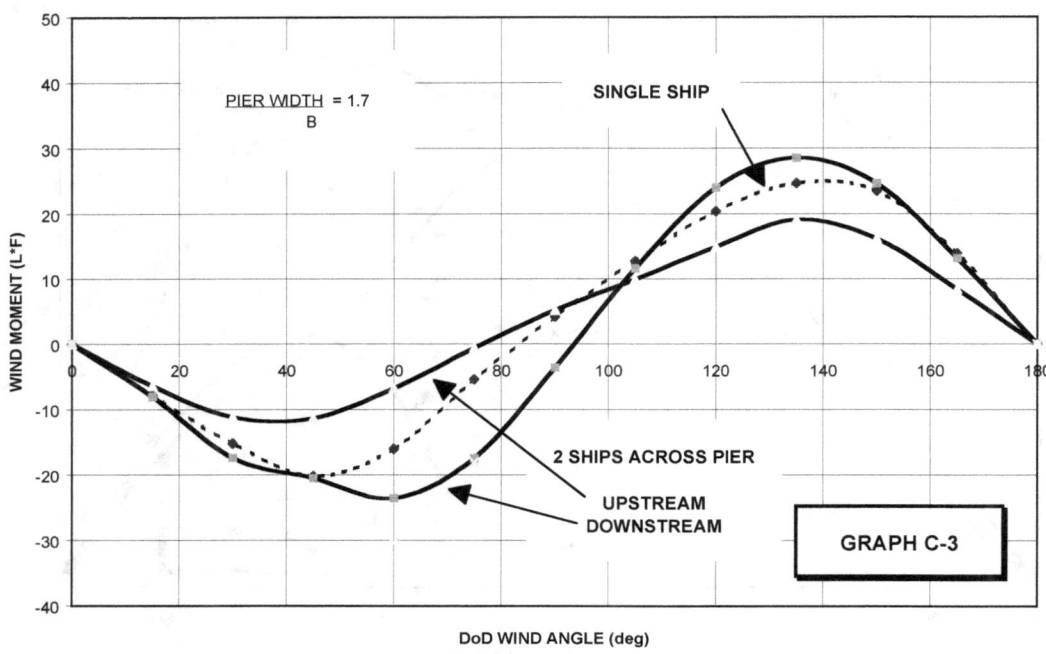

Figure 5-22 CVE-55 Wind Moment Arm for One and Two Ships

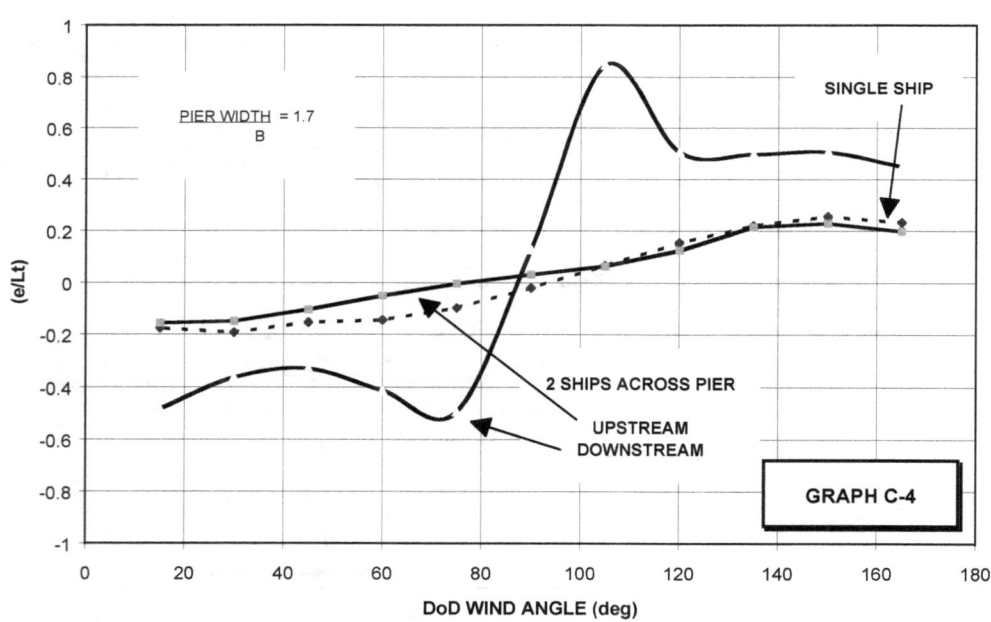

Figure 5-23 CVE-55 Lateral Current Force

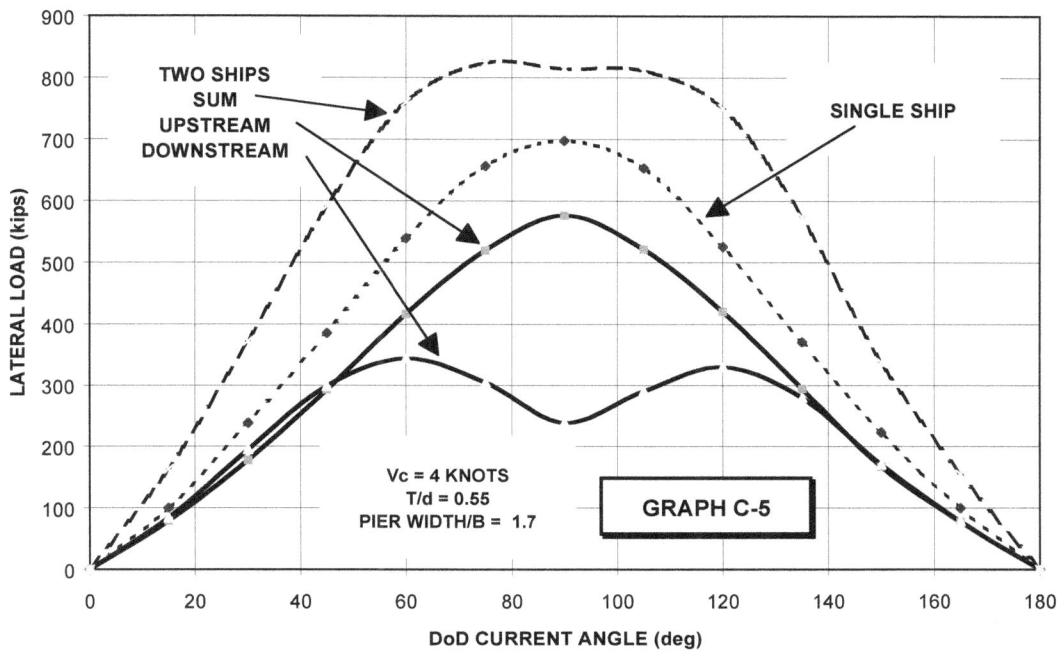

Figure 5-24 CVE-55 Current Force Divided by Force for One Ship

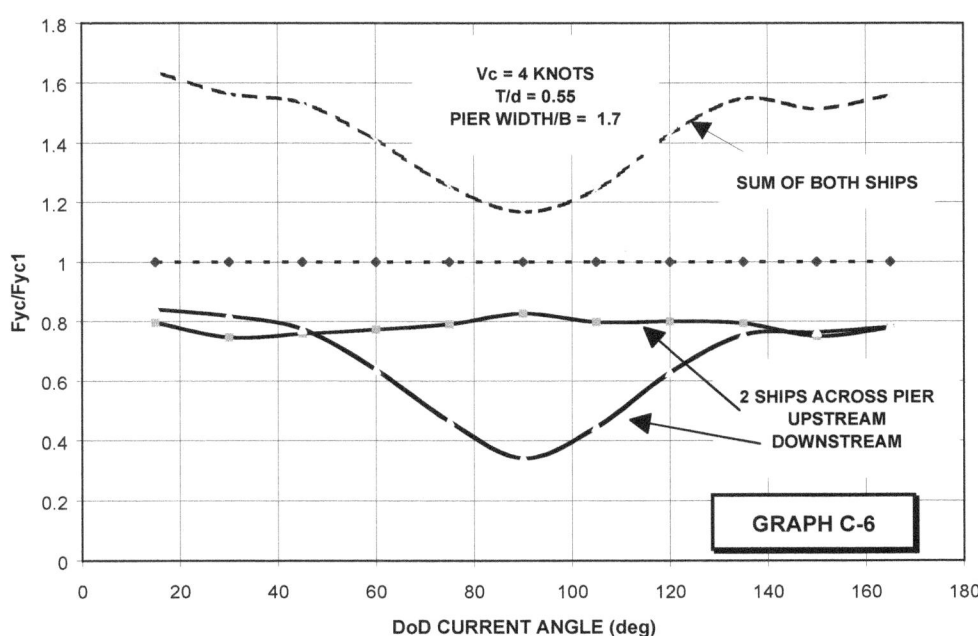

Figure 5-25 CVE-55 Current Moment

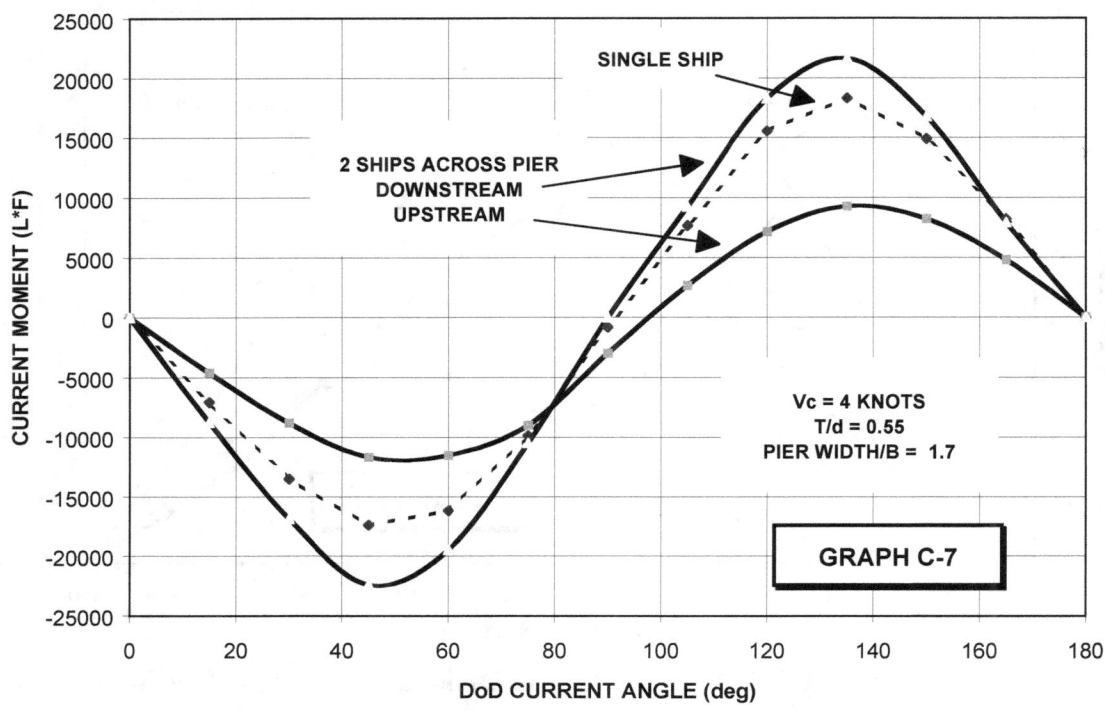

Figure 5-26 CVE-55 Current Moment Arm for One and Two Ships

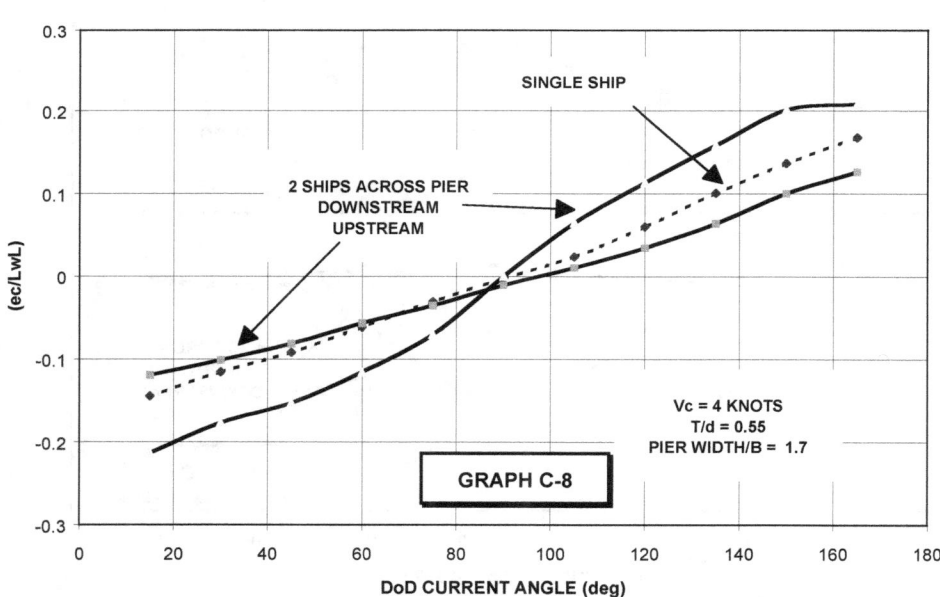

Figure 5-27 DD-692 Ship Nests Tested

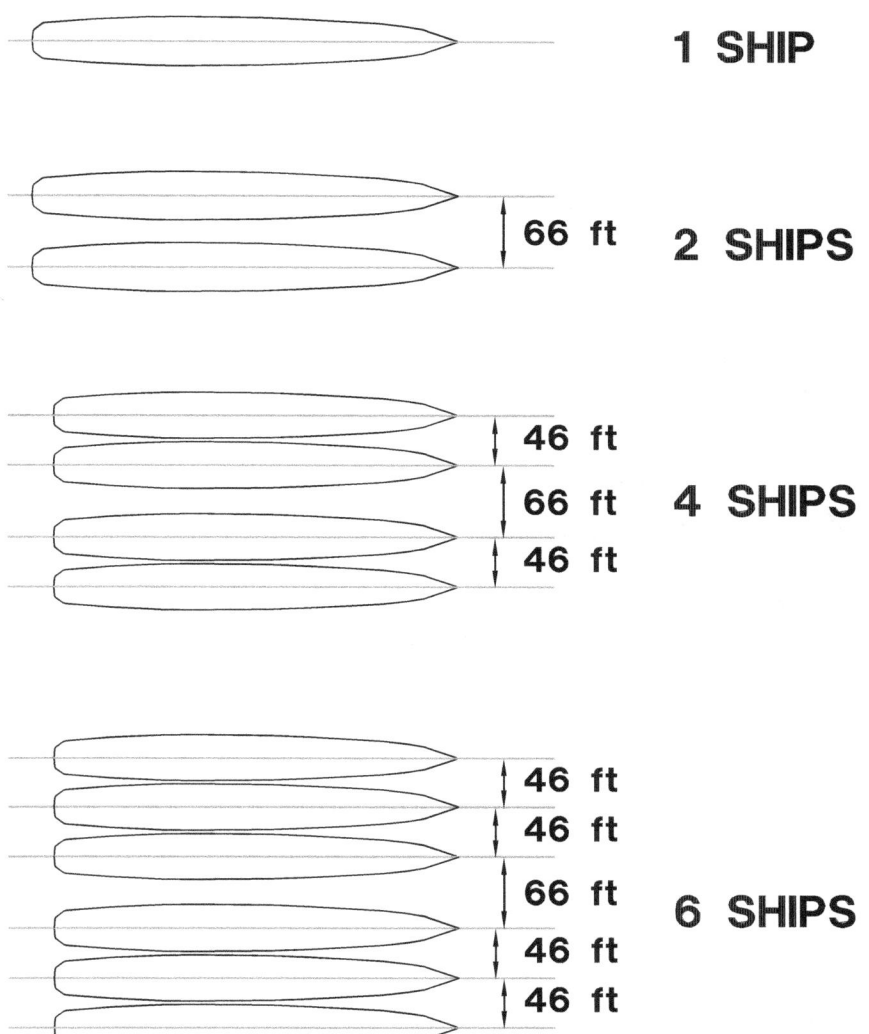

Figure 5-28 DD-692 Lateral Wind Forces

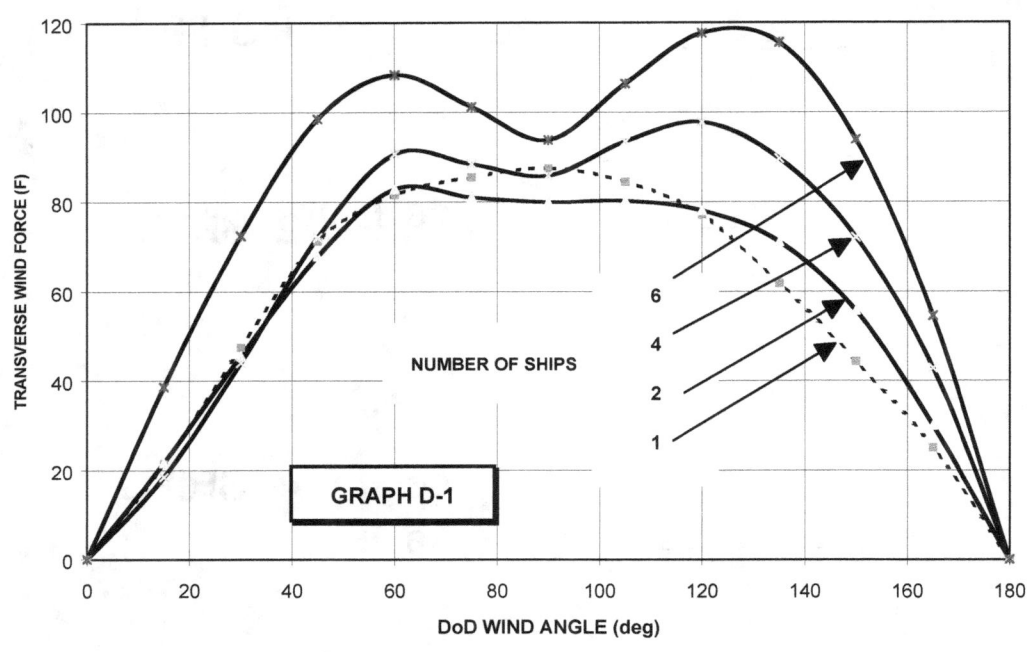

Figure 5-29 DD-692 Lateral Wind Force Divided by Force on One Ship

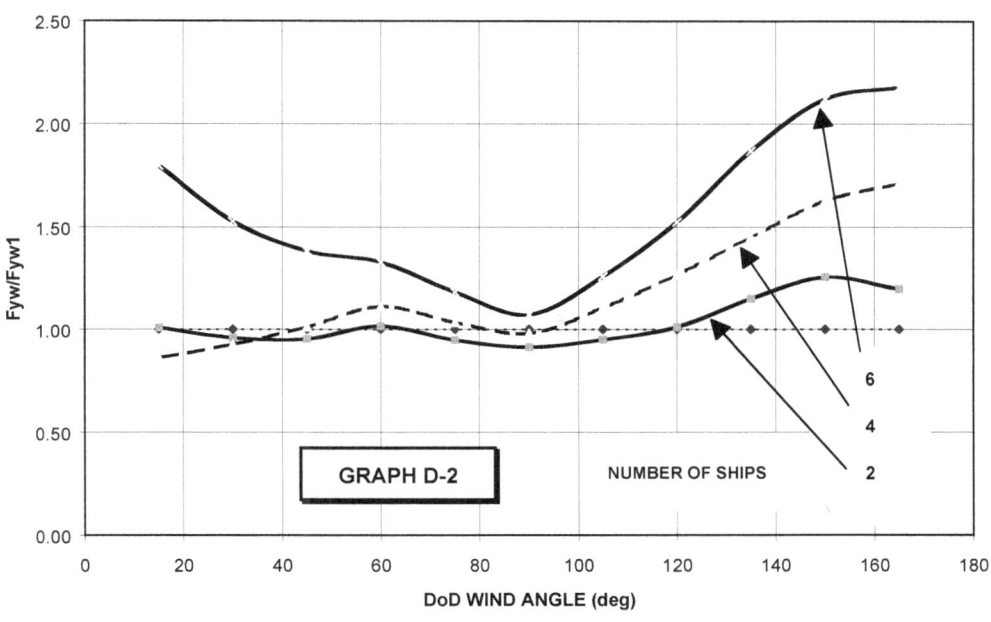

Figure 5-30 DD-692 Lateral Wind Moments

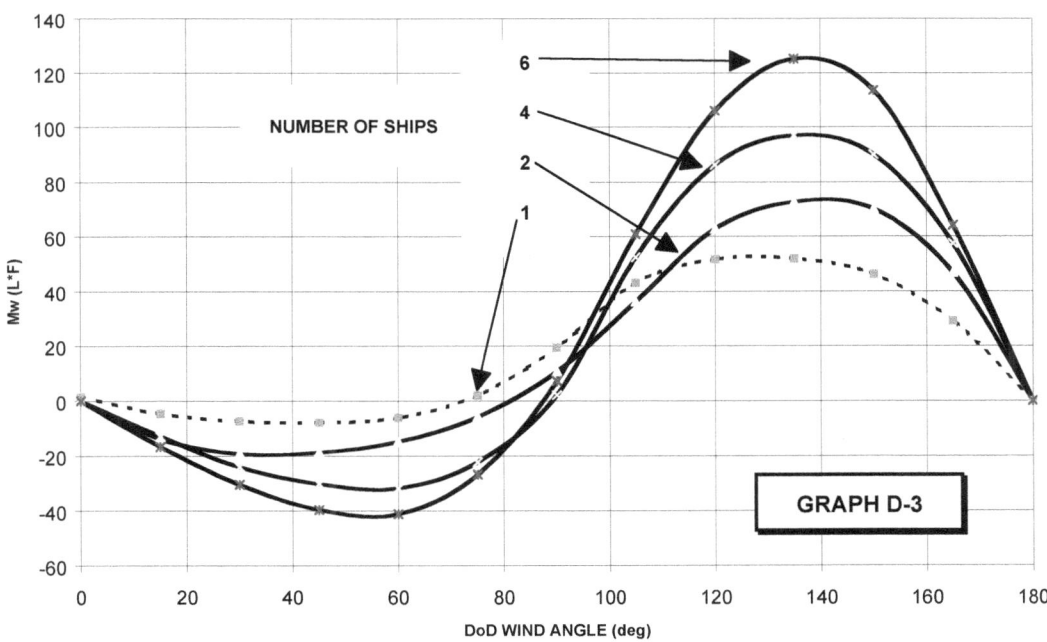

Figure 5-31 DD-692 Wind Moment Arm

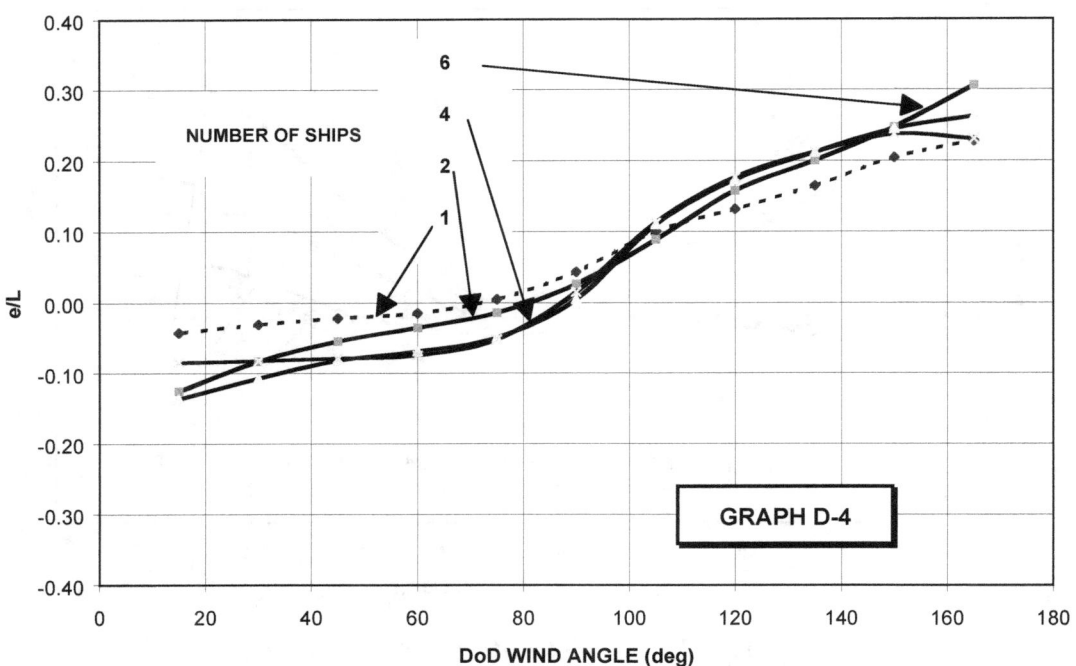

Figure 5-32 DD-692 Lateral Current Force

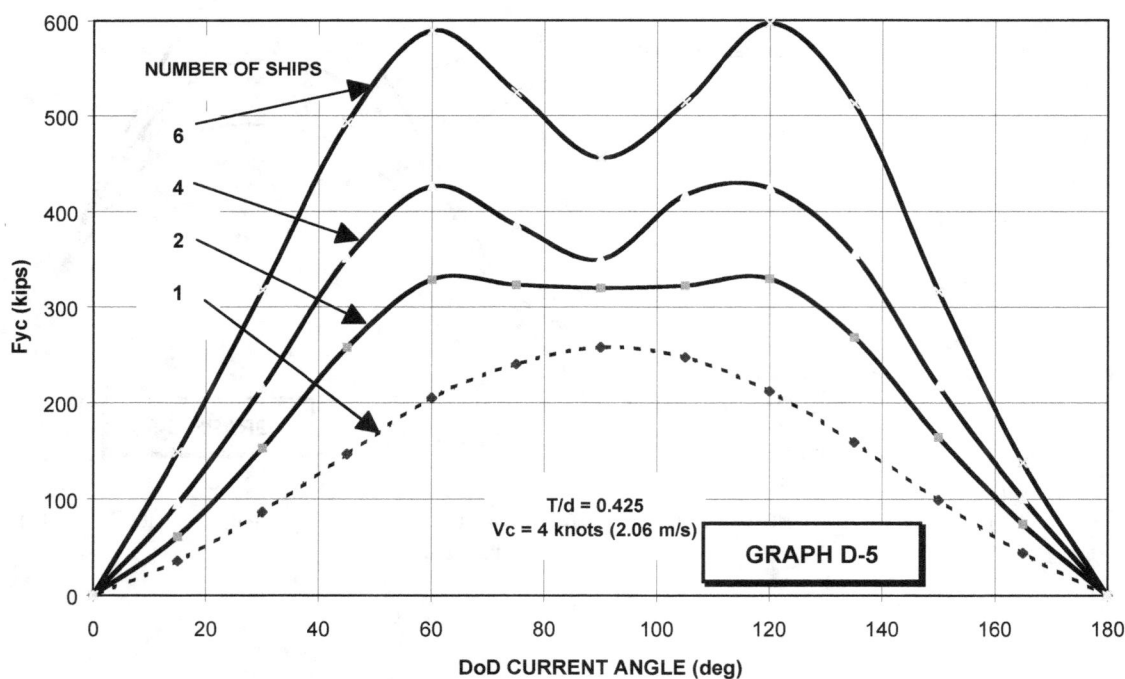

Figure 5-33 DD-692 Lateral Current Force Divided by Force on One Ship

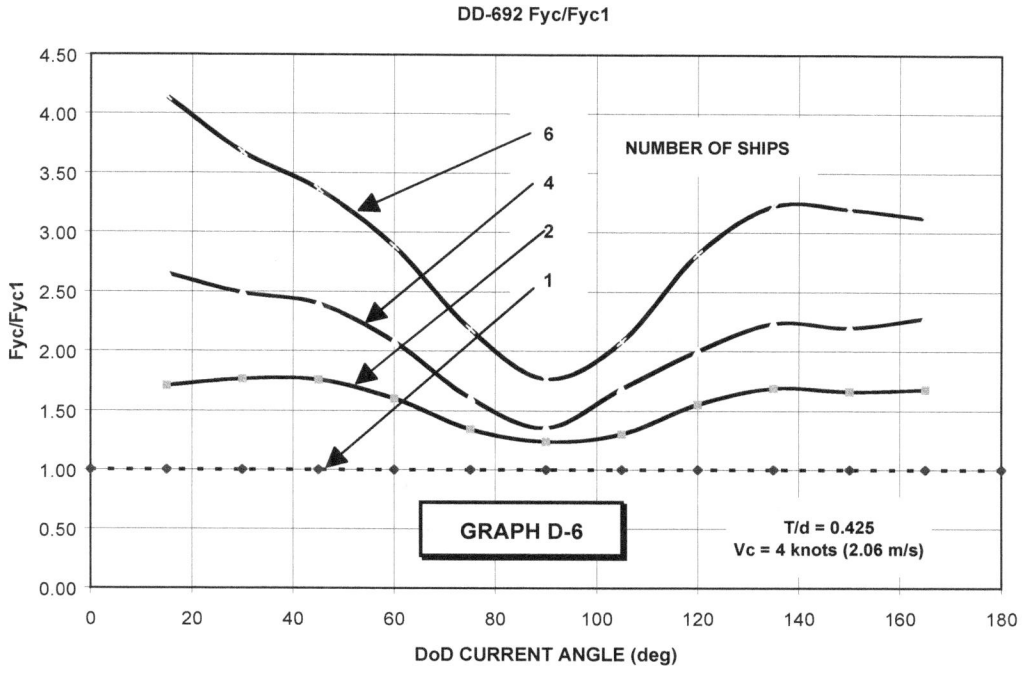

Figure 5-34 DD-692 Current Moment

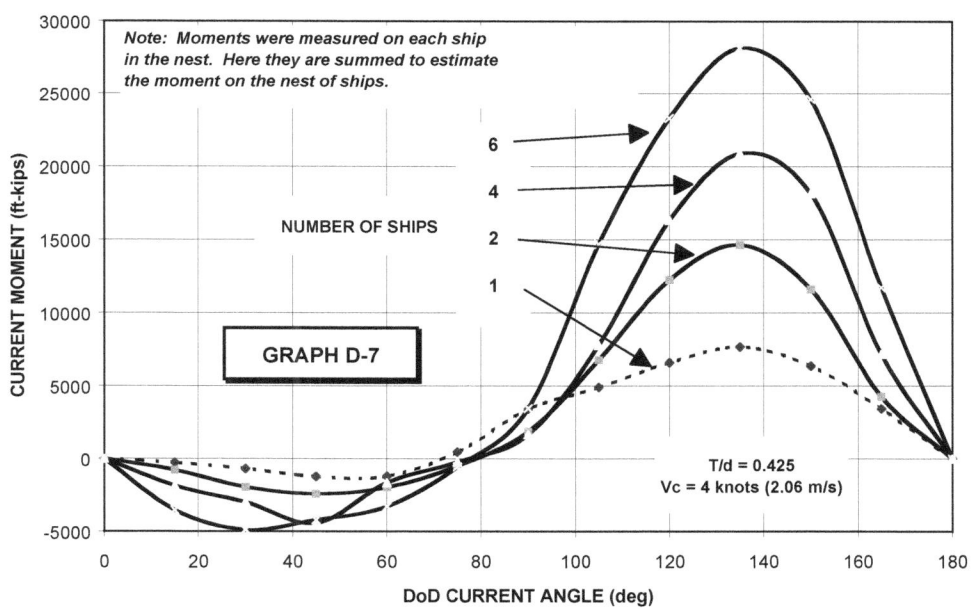

Figure 5-35 DD-692 Current Moment Arm

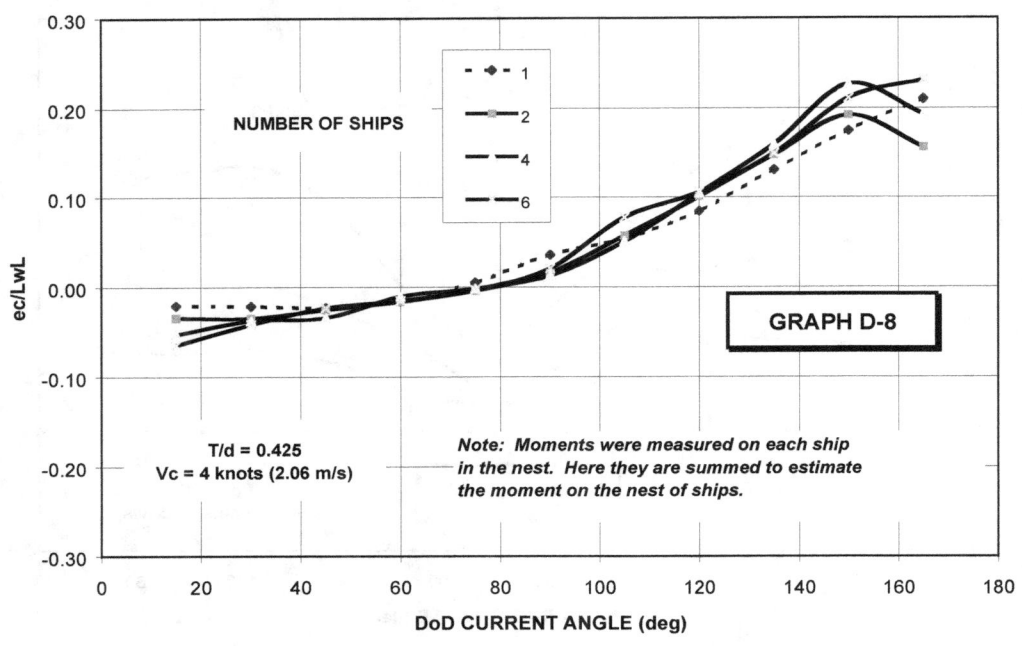

Figure 5-36 EC-2 Ship Nests Tested

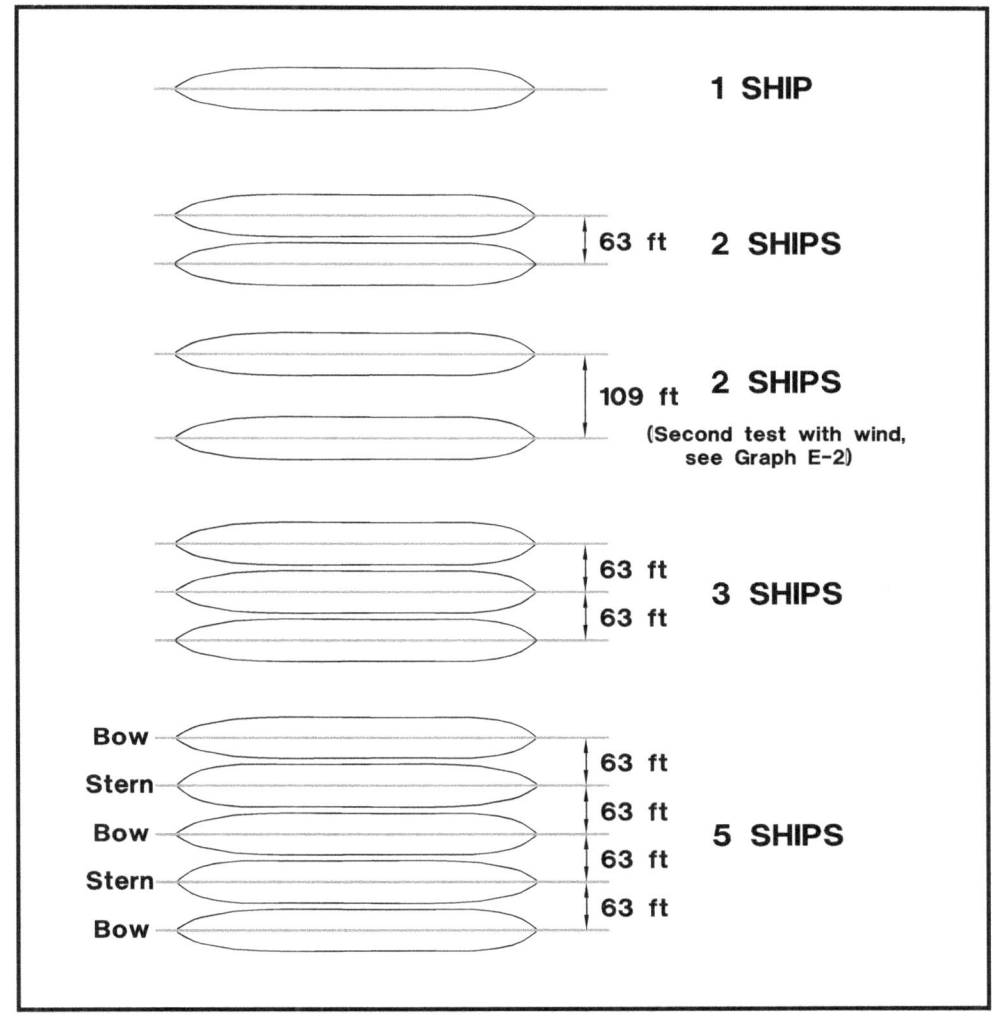

Figure 5-37 EC-2 Lateral Wind Force

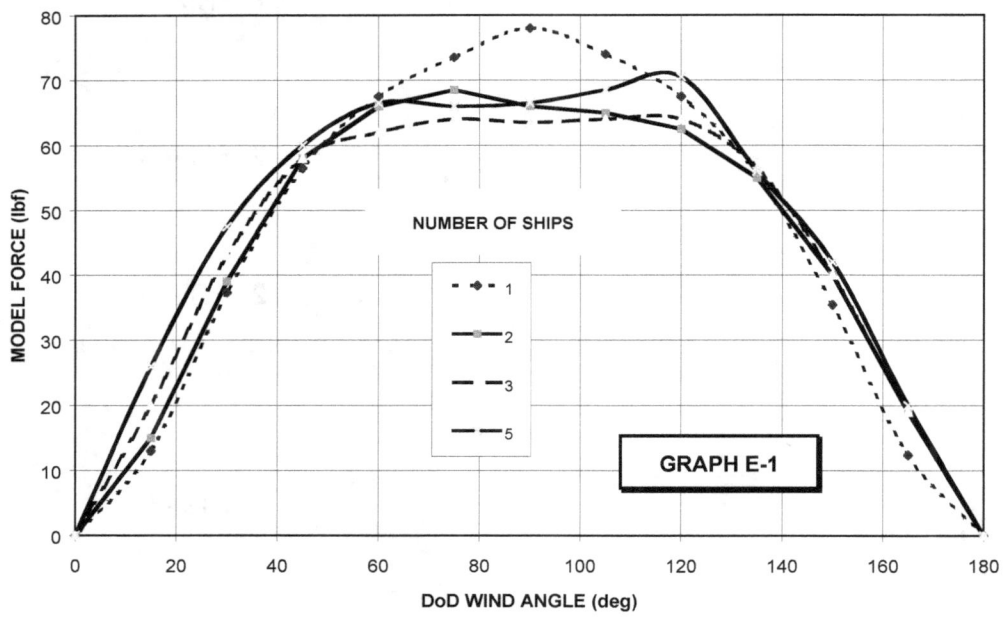

Figure 5-38 EC-2 Lateral Wind Force

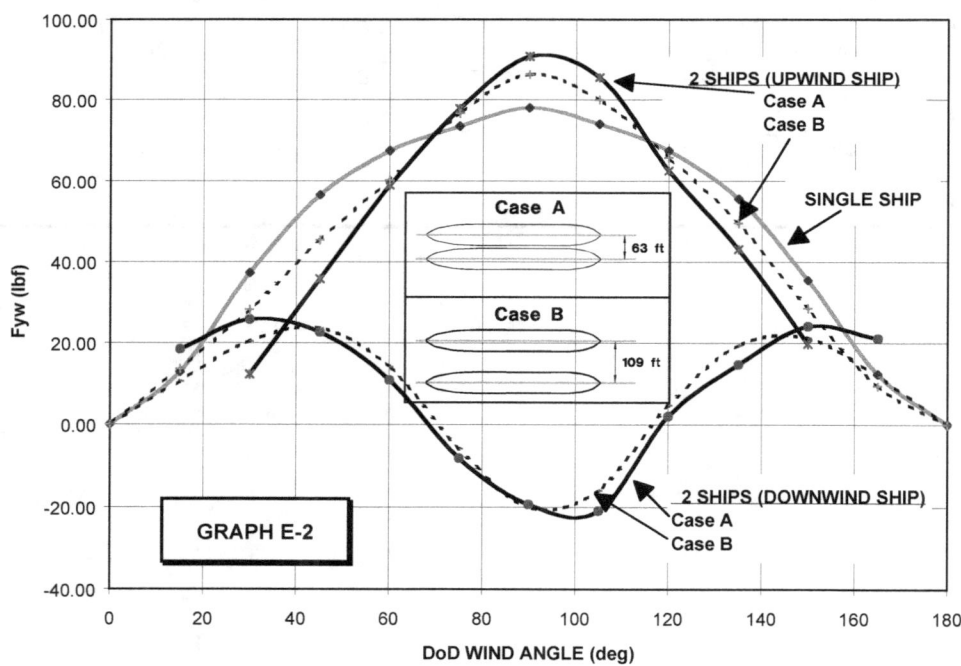

Figure 5-39 EC-2 Lateral Wind Force Divided by Force on One Ship

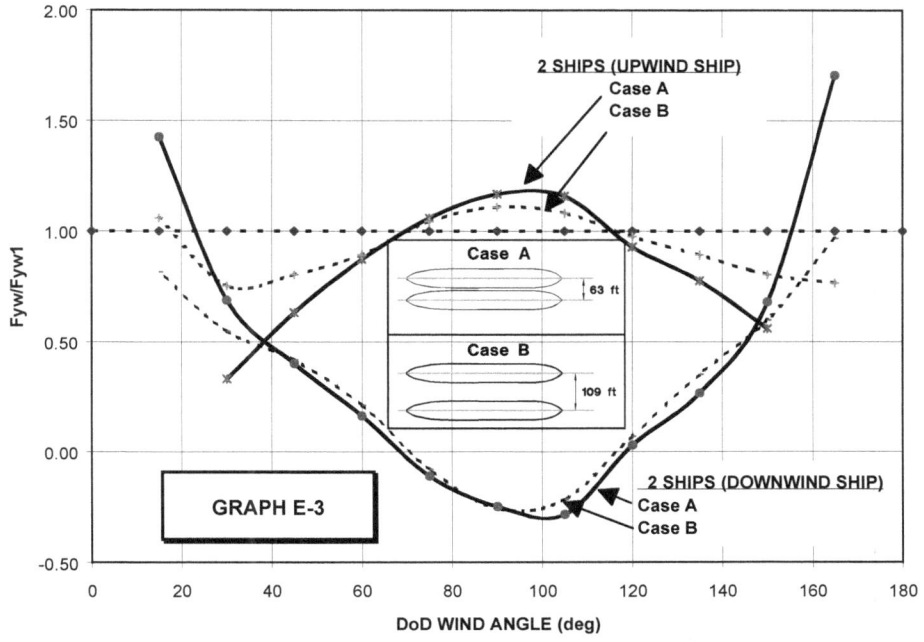

Figure 5-40 EC-2 Wind Moment

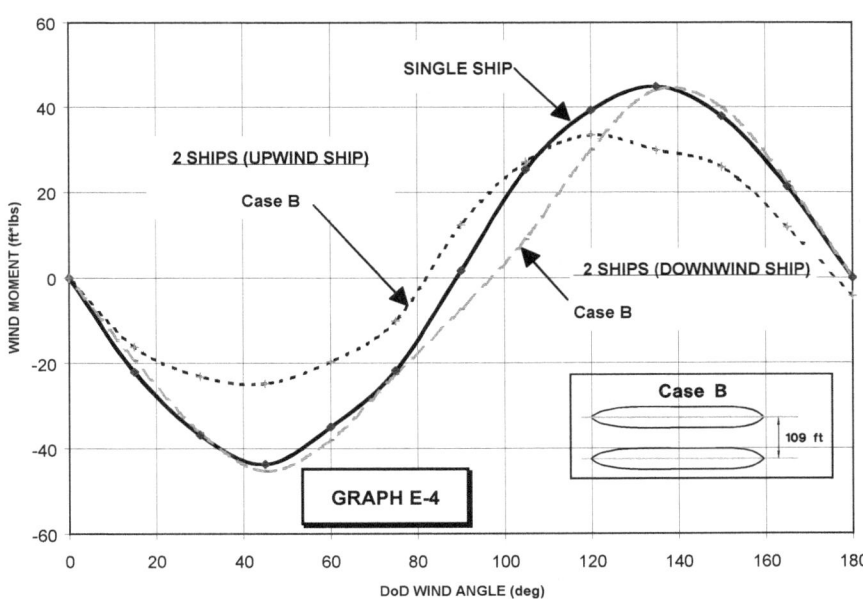

Figure 5-41 EC-2 Lateral Wind Force Divided by Force on One Ship

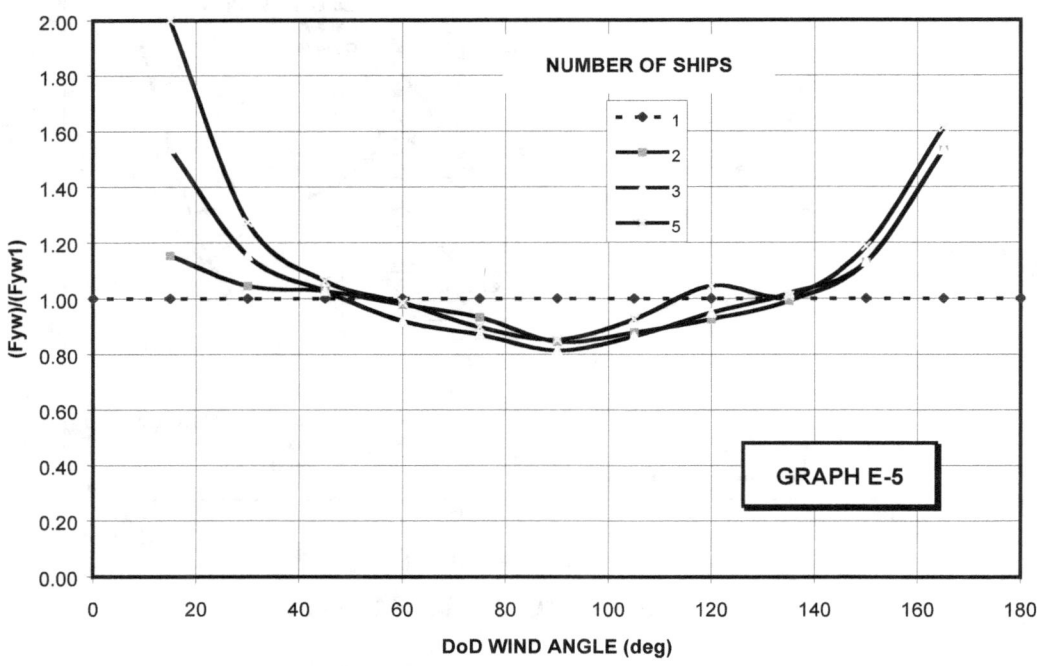

Figure 5-42 EC-2 Wind Moment

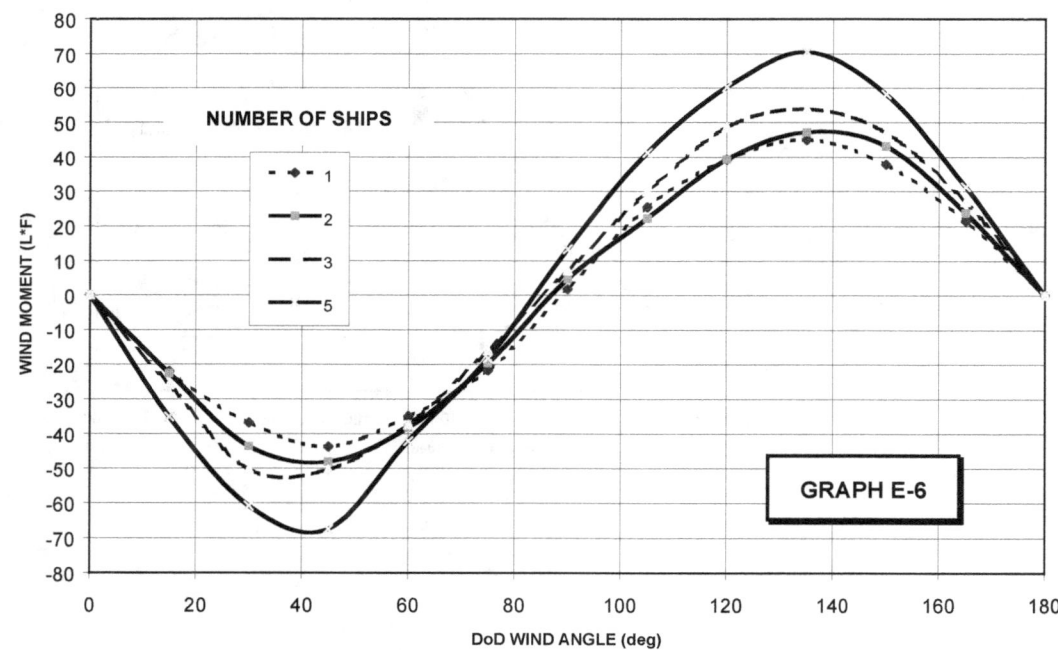

Figure 5-43 EC-2 Wind Moment Arm

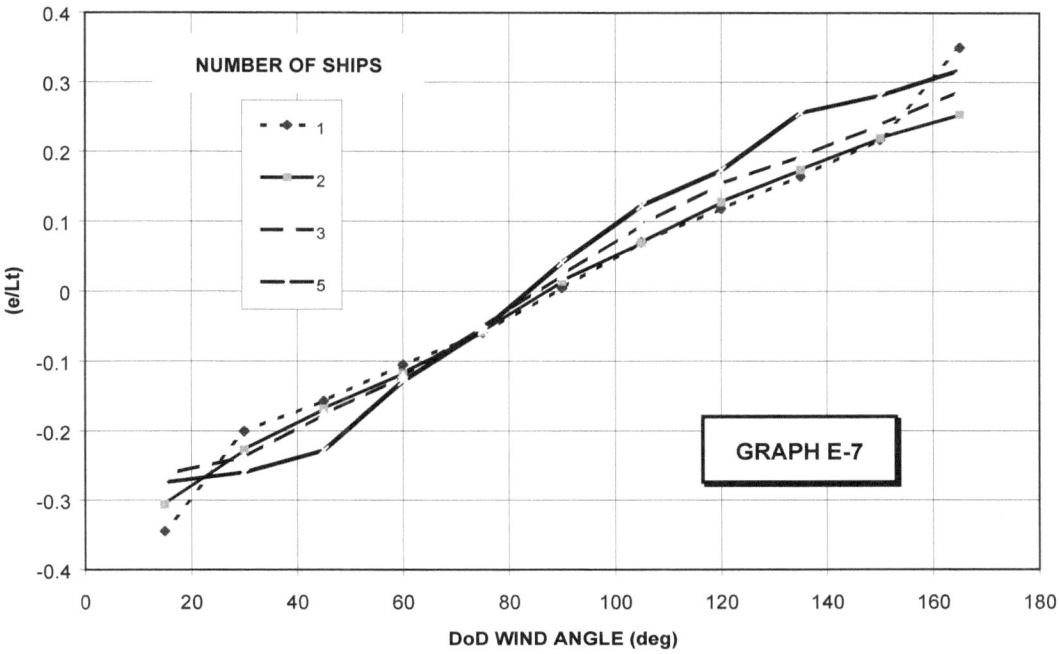

Figure 5-44 EC-2 Lateral Current Forces

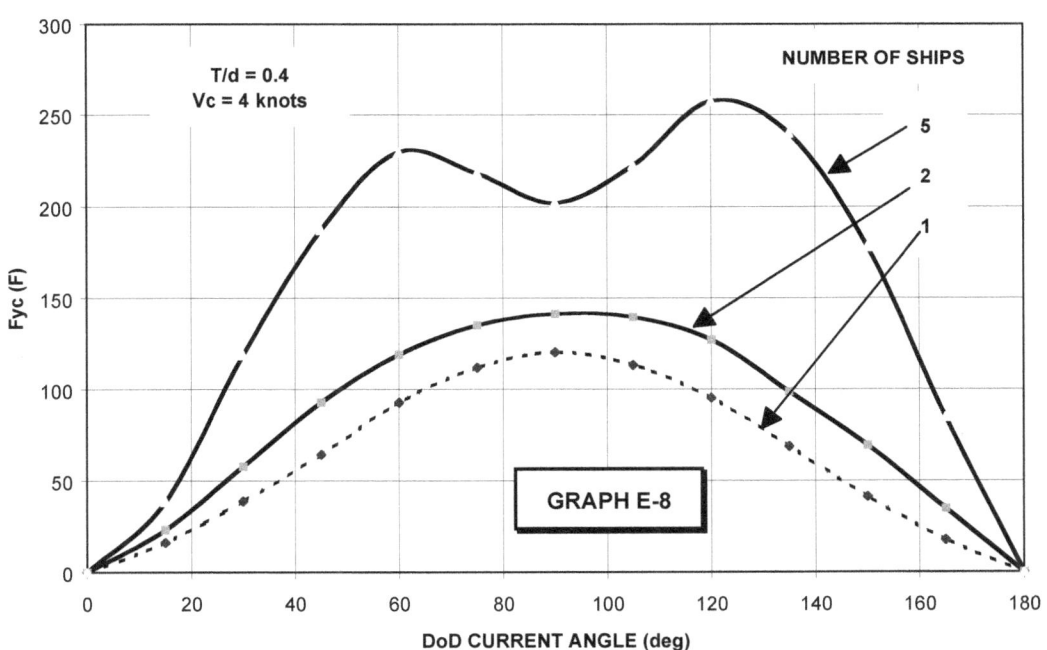

Figure 5-45 EC-2 Lateral Current Force Divided by Force on One Ship

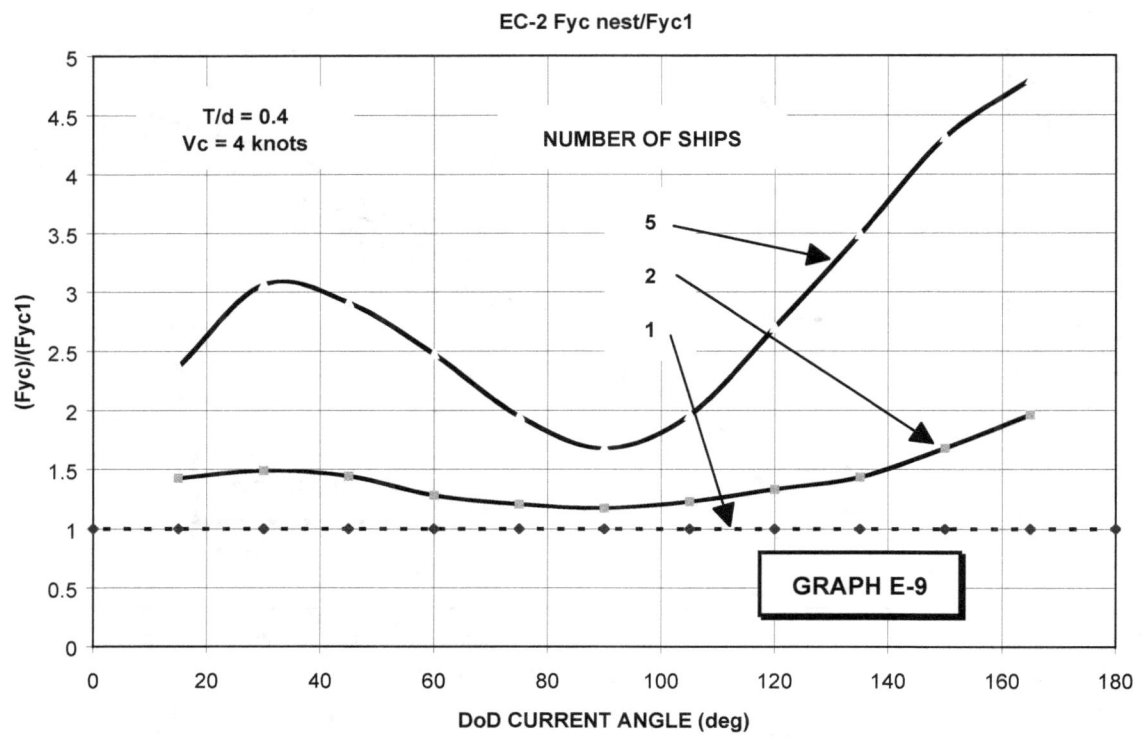

Figure 5-46 EC-2 Lateral Current Force Divided by Force on One Ship

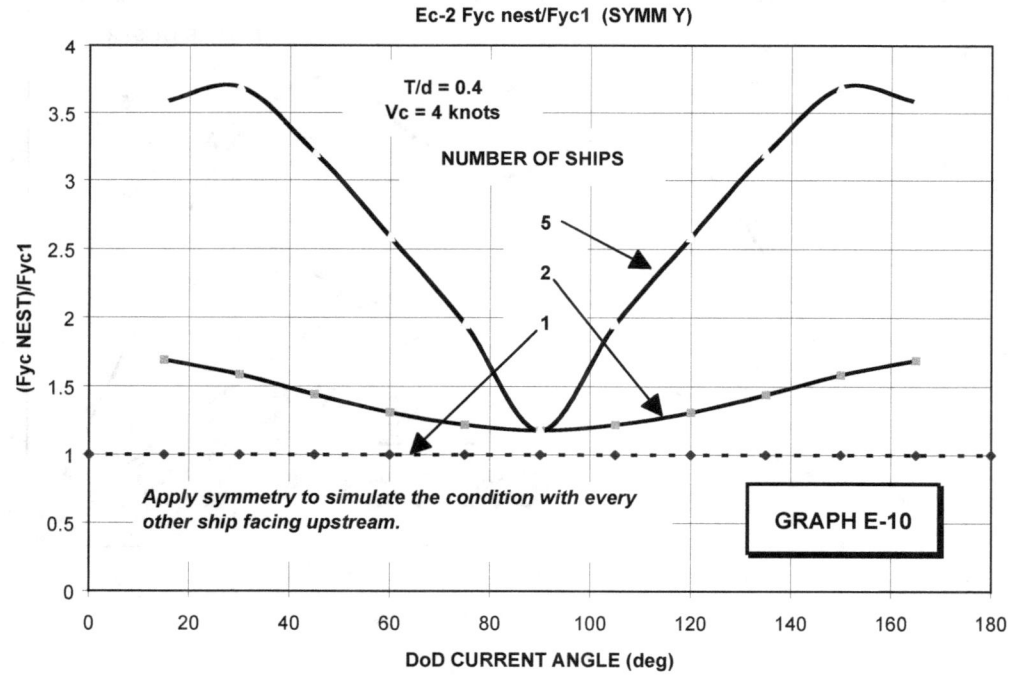

Figure 5-47 EC-2 Current Moment

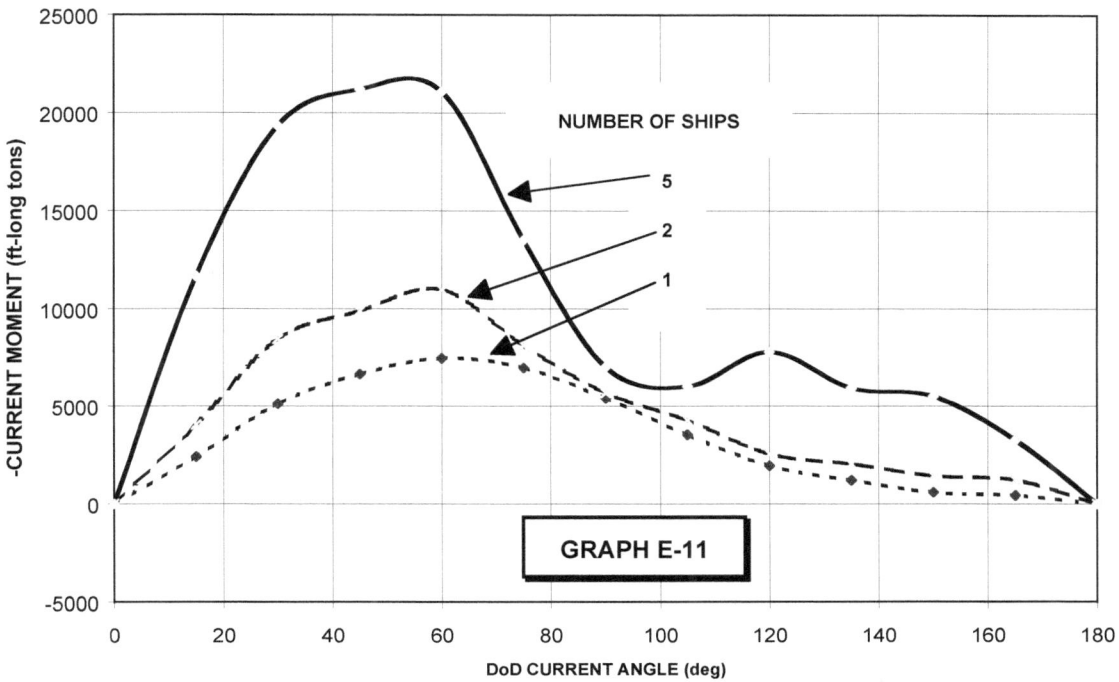

Figure 5-48 EC-2 Current Moment Arm

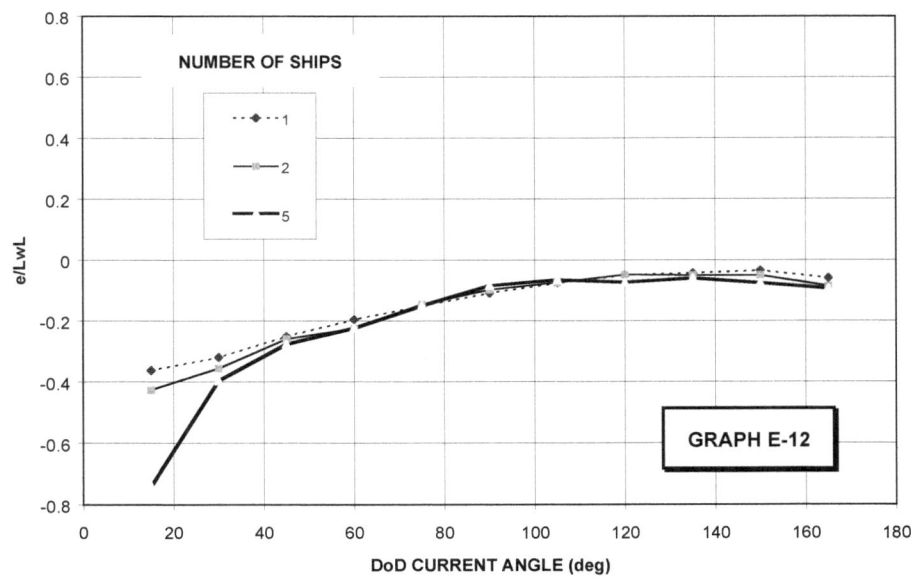

Figure 5-49 EC-2 Current Moment Arm

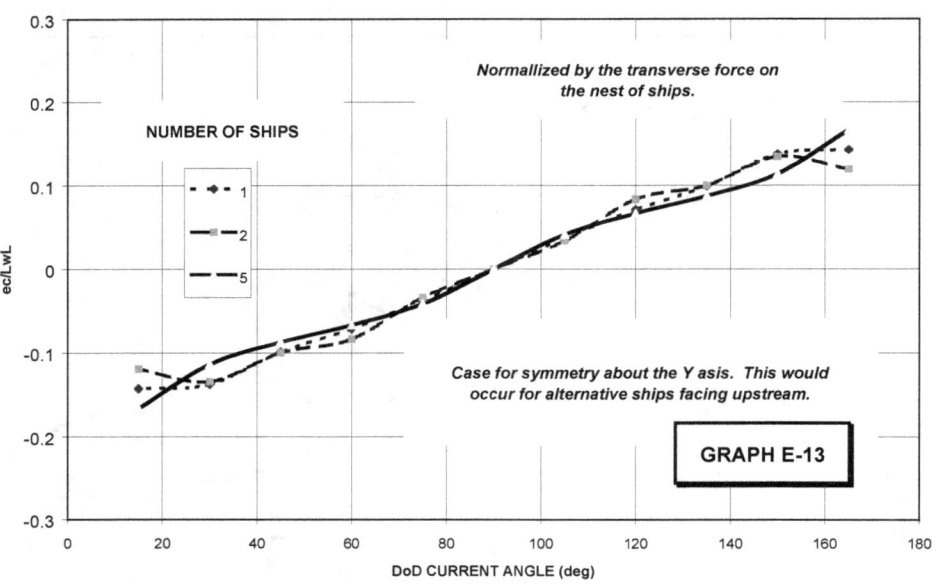

Figure 5-50 SS-212 Nests Tested

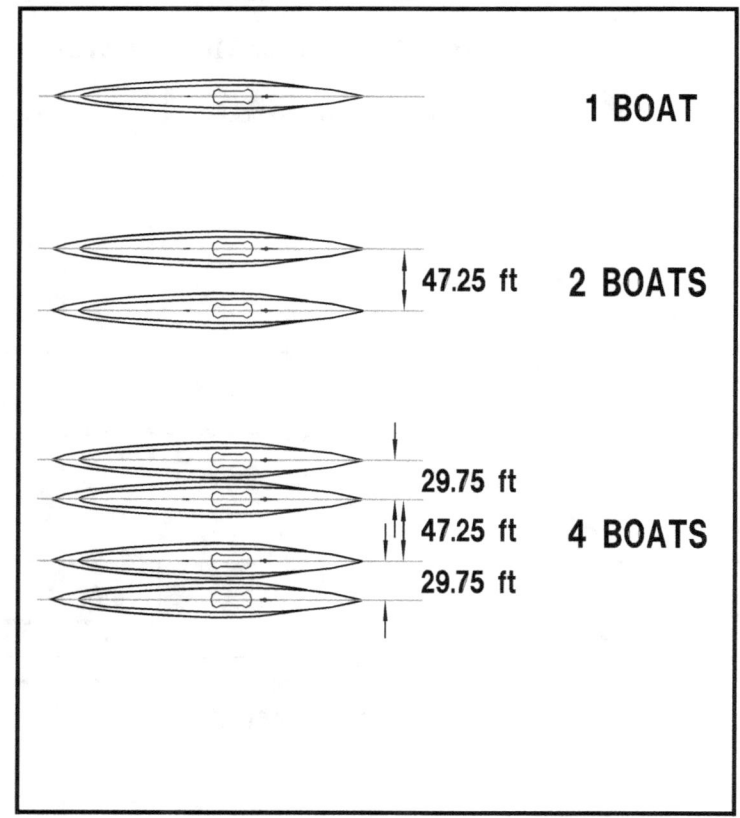

Figure 5-51 SS-212 Lateral Current Forces

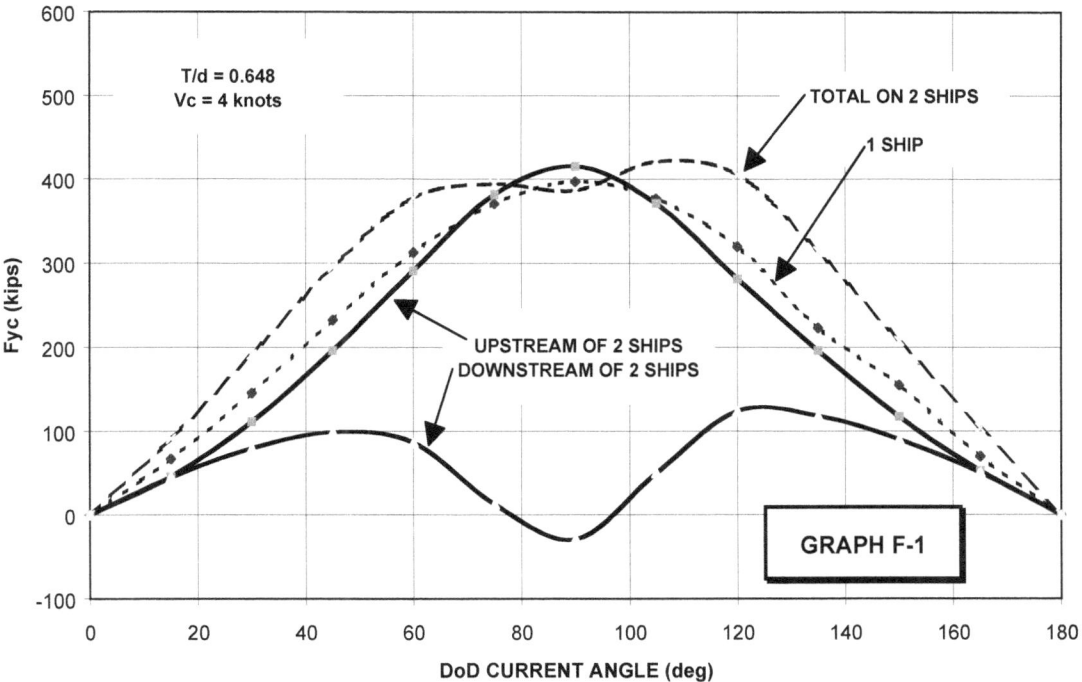

Figure 5-52 SS-212 Lateral Current Forces

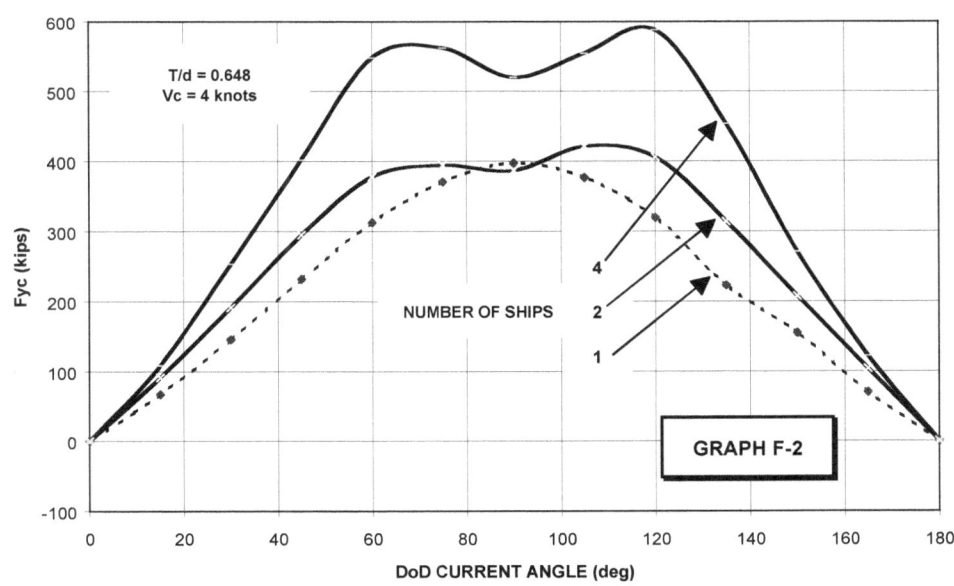

Figure 5-53 SS-212 Lateral Current Force Divided by Force on One Boat

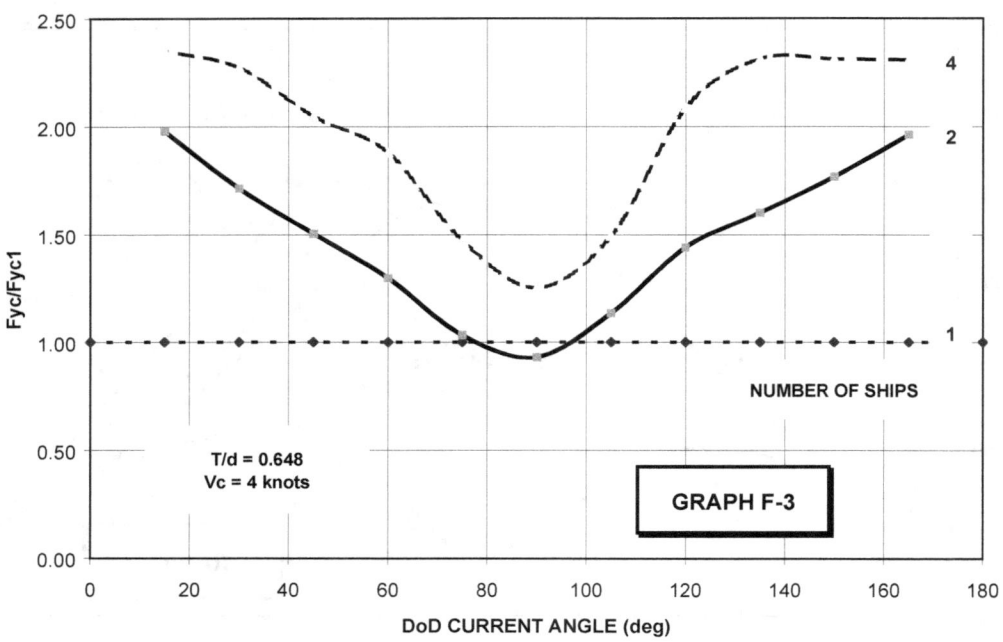

Figure 5-54 SS-212 Current Moment

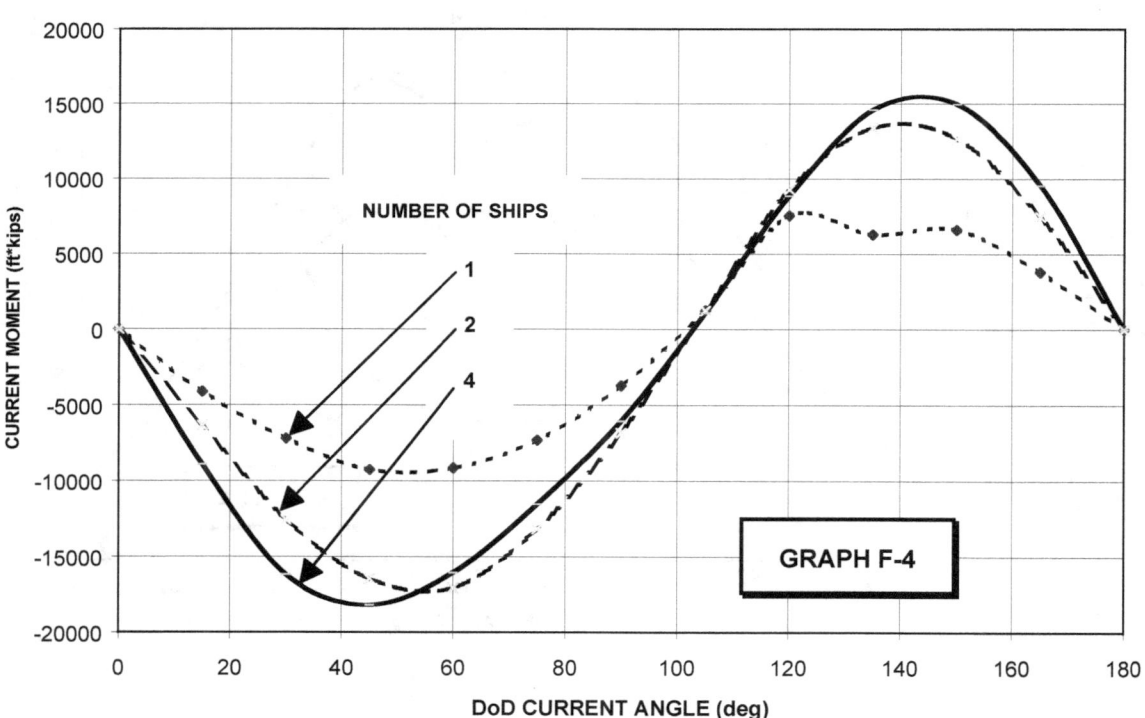

Figure 5-55 SS-212 Current Moment Arm

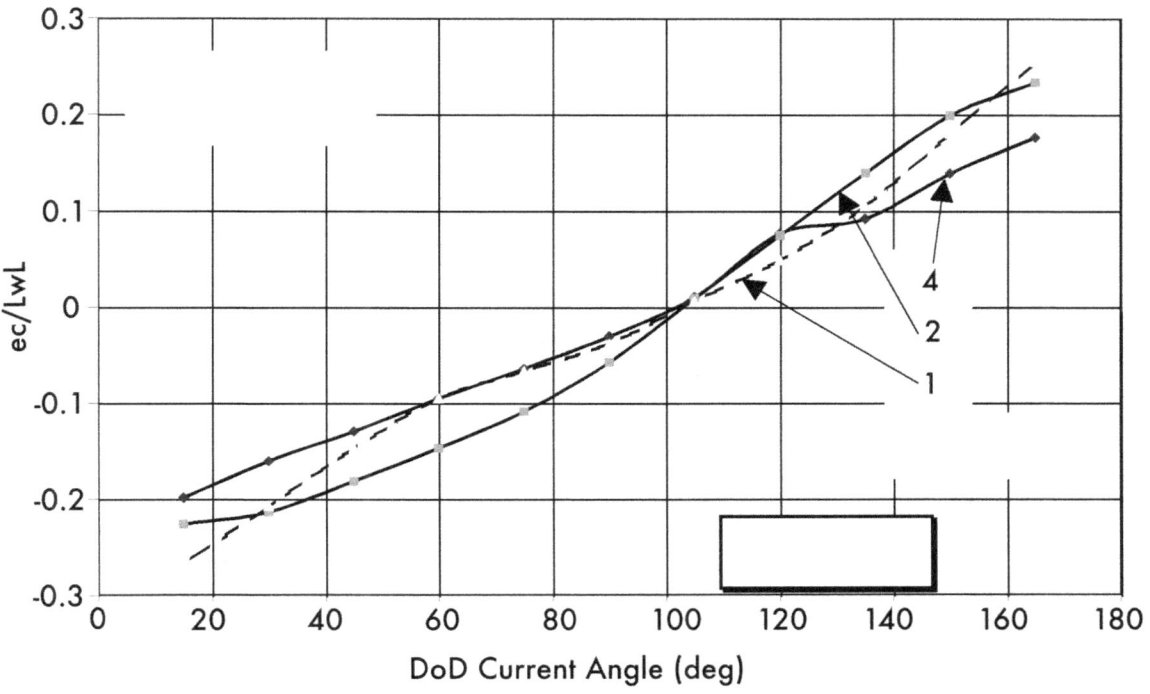

Figure 5-56 SS-212 Lateral Wind Force

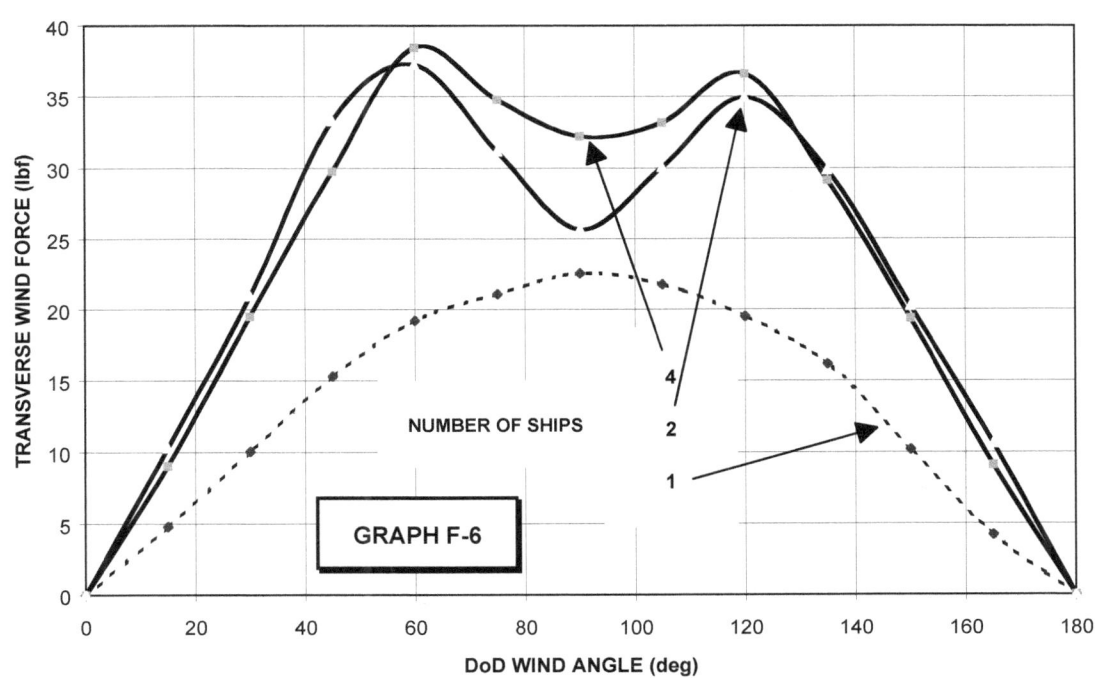

Figure 5-57 SS-212 Lateral Wind Force Divided

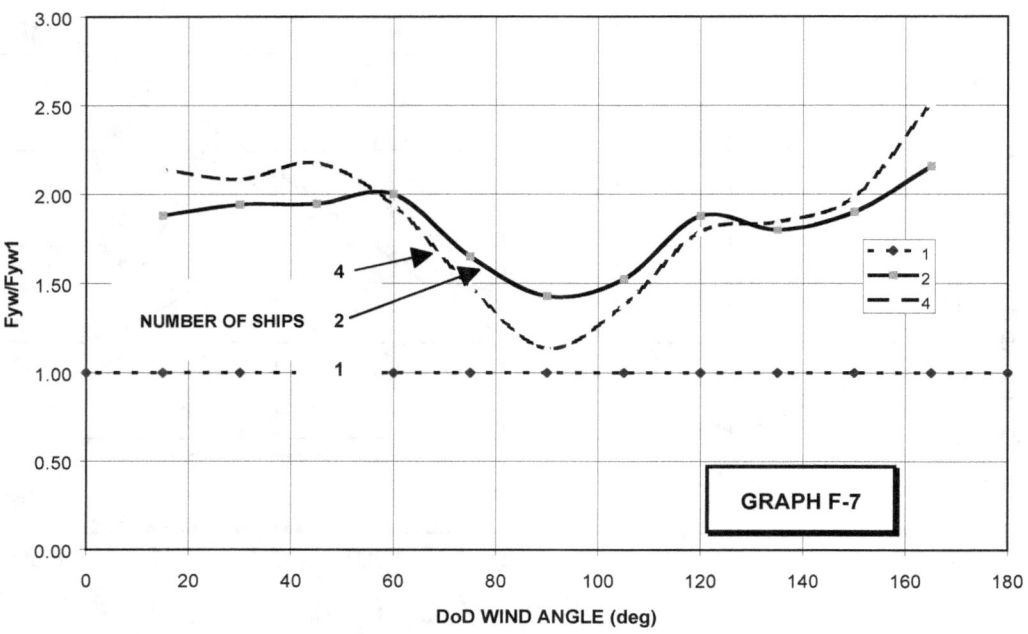

Figure 5-58 SS-212 Wind Moment

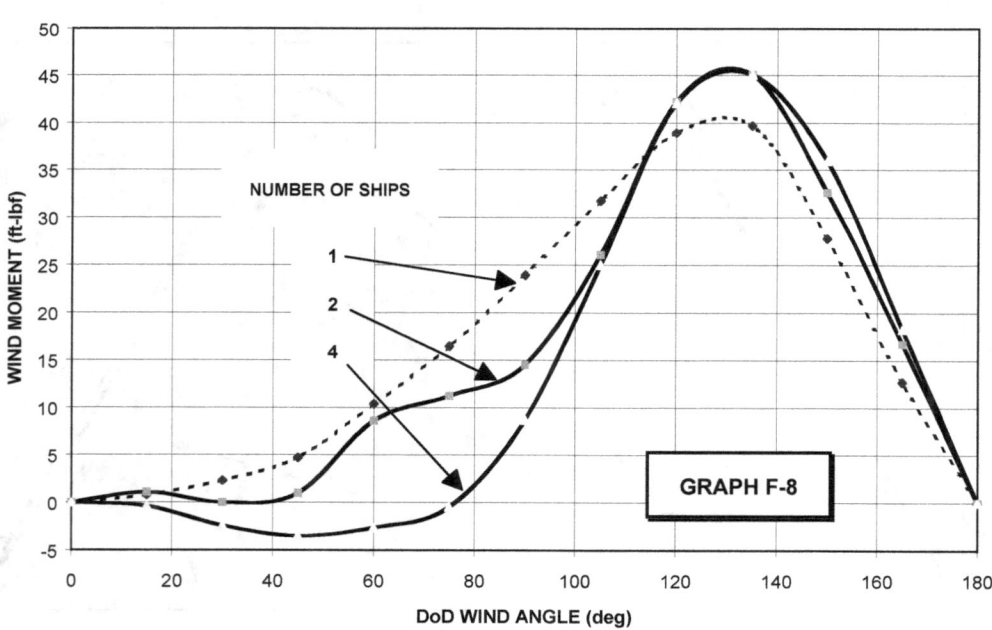

Figure 5-59 SS-212 Wind Moment Arm

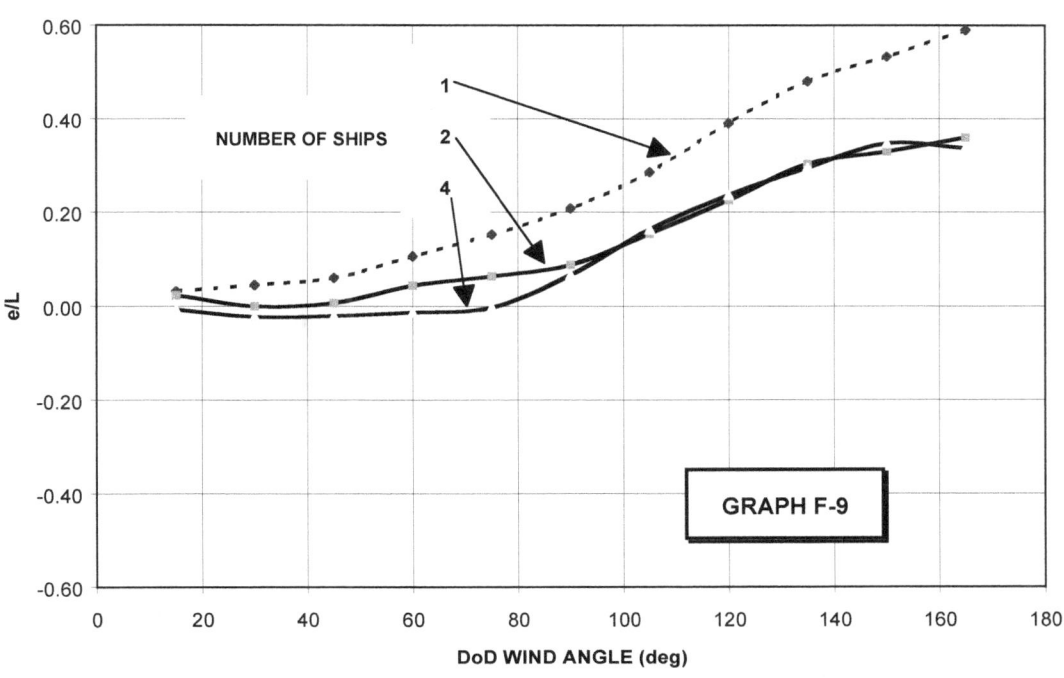

Chapter 6. Mooring System Design Procedures

6.1. Anchor Systems

6.1.1. General Anchor Design Procedure

Anchor systems ultimately hold the mooring loads in fleet mooring systems. Anchors are used on both ships and in mooring facilities, so selection and design of anchors are included in this section.

The type and size of anchor specified depends upon certain parameters, such as those shown in Table 6-1.

Table 6-1 Anchor Specification Considerations

PARAMETER	DESCRIPTION
Holding capacity	The size/type of anchor will depend on the amount of anchor holding required.
Soils	Engineering properties and sediment layer thickness influence anchor design.
Use	If anchors will be relocated, then drag anchors are most commonly used.
Weight	The amount of weight that can be handled or carried may control anchor specification.
Equipment	The size and characteristics of installation equipment are important in anchor specification.
Directionality	Drag anchors may provide little uplift capacity and primarily hold in one direction; driven plate anchors provide high omnidirectional capacity.
Performance	Whether anchor will be allowed to drag or not, as well as the amount of room available for anchors systems, will influence anchor specification.

The most commonly used anchors in DOD moorings are drag-embedment anchors and driven-plate anchors, so they will be discussed here. Other types of specialized anchors (shallow foundations, pile anchors, propellant-embedment anchors, rock bolts, etc.) are discussed in the NCEL *Handbook for Marine Geotechnical Engineering*, available from Pile Buck.

Figure 6-1 and Figure 6-2 illustrate typical drag-embedment anchors. Figure 6-3 illustrates a driven-plate anchor. Some characteristics of these two categories of anchors are given in Table 6-2.

Figure 6-1 Example of a Drag-Embedment Anchor (Stabilized Stockless Anchor)

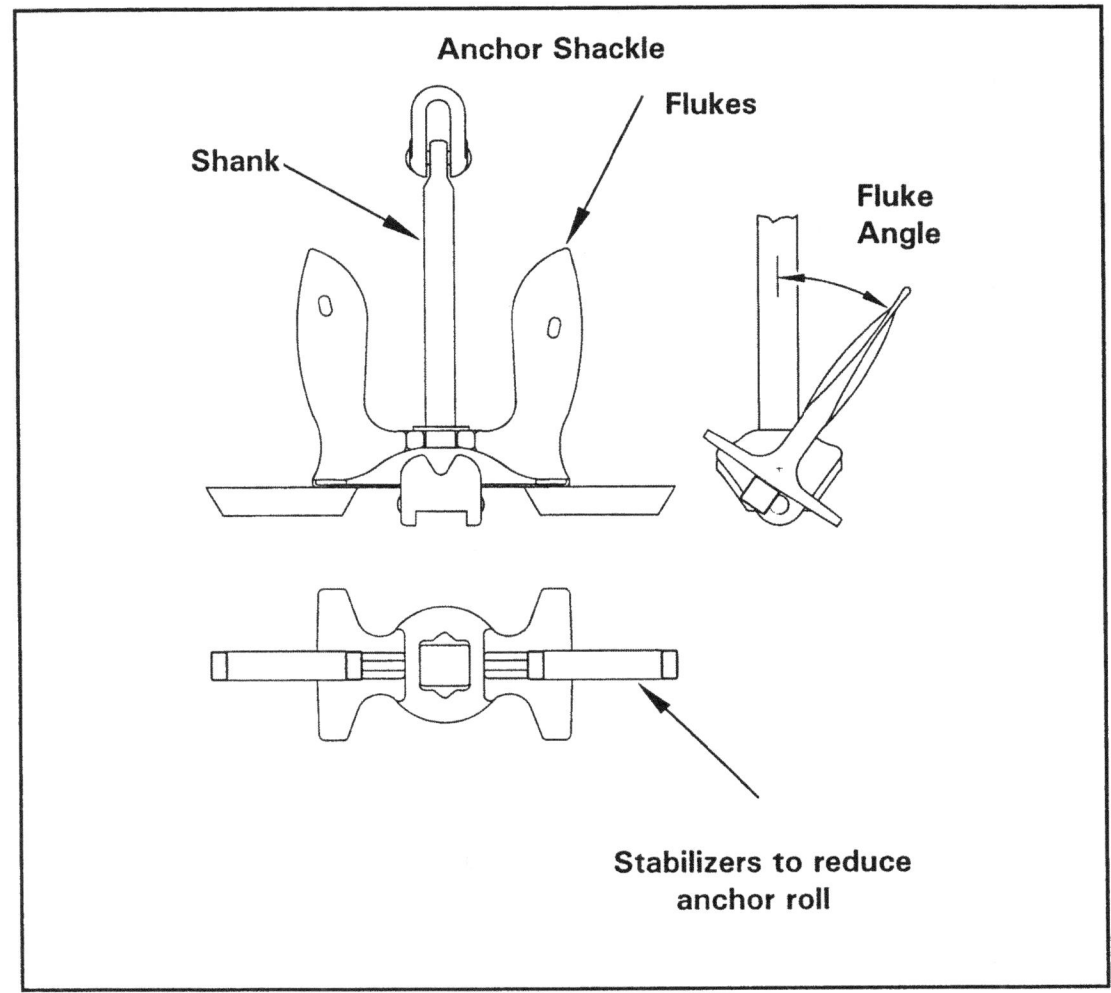

Figure 6-2 Example of a Drag-Embedment Anchor (NAVMOOR Anchor)

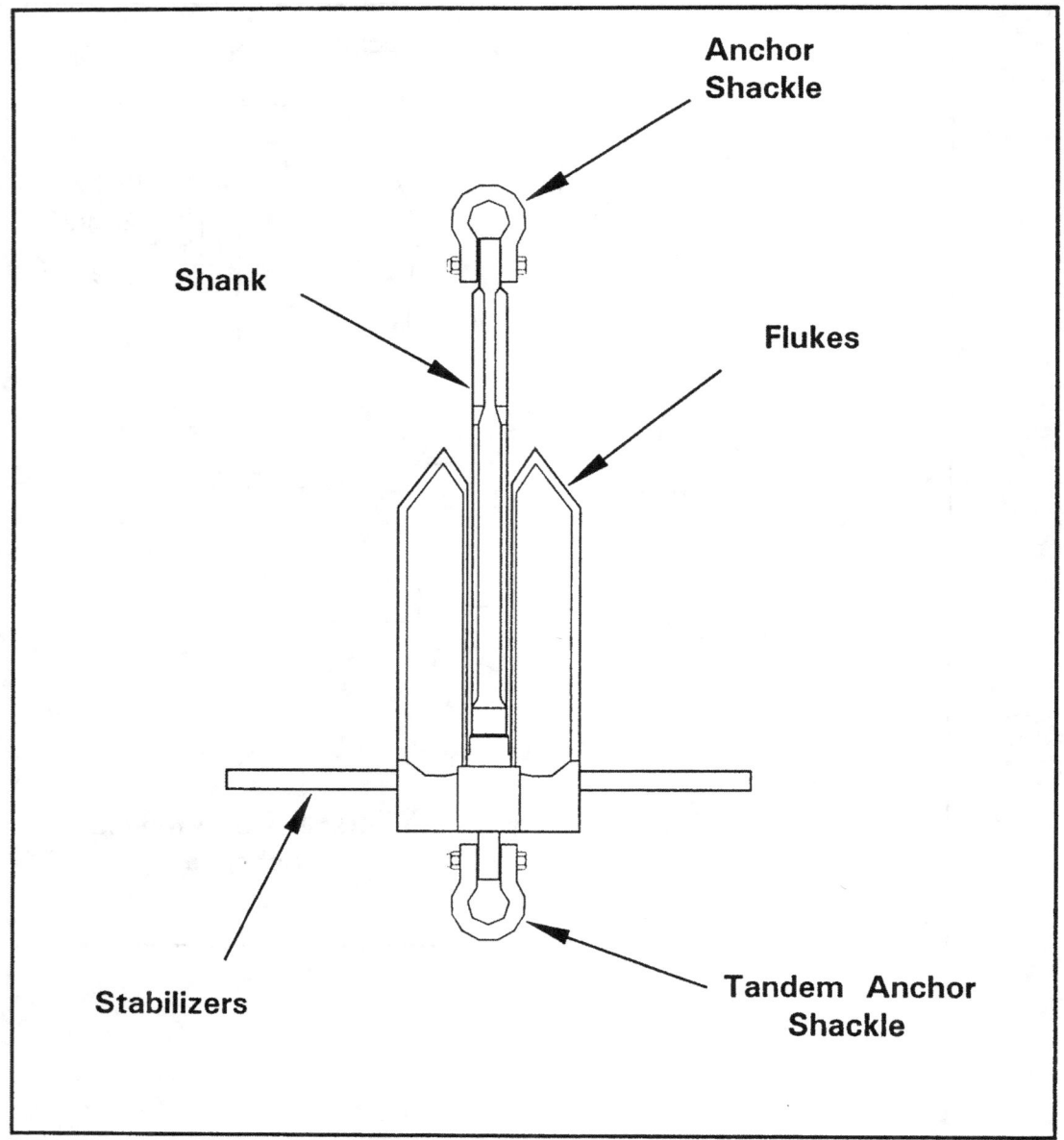

Figure 6-3 Driven-Plate Anchor

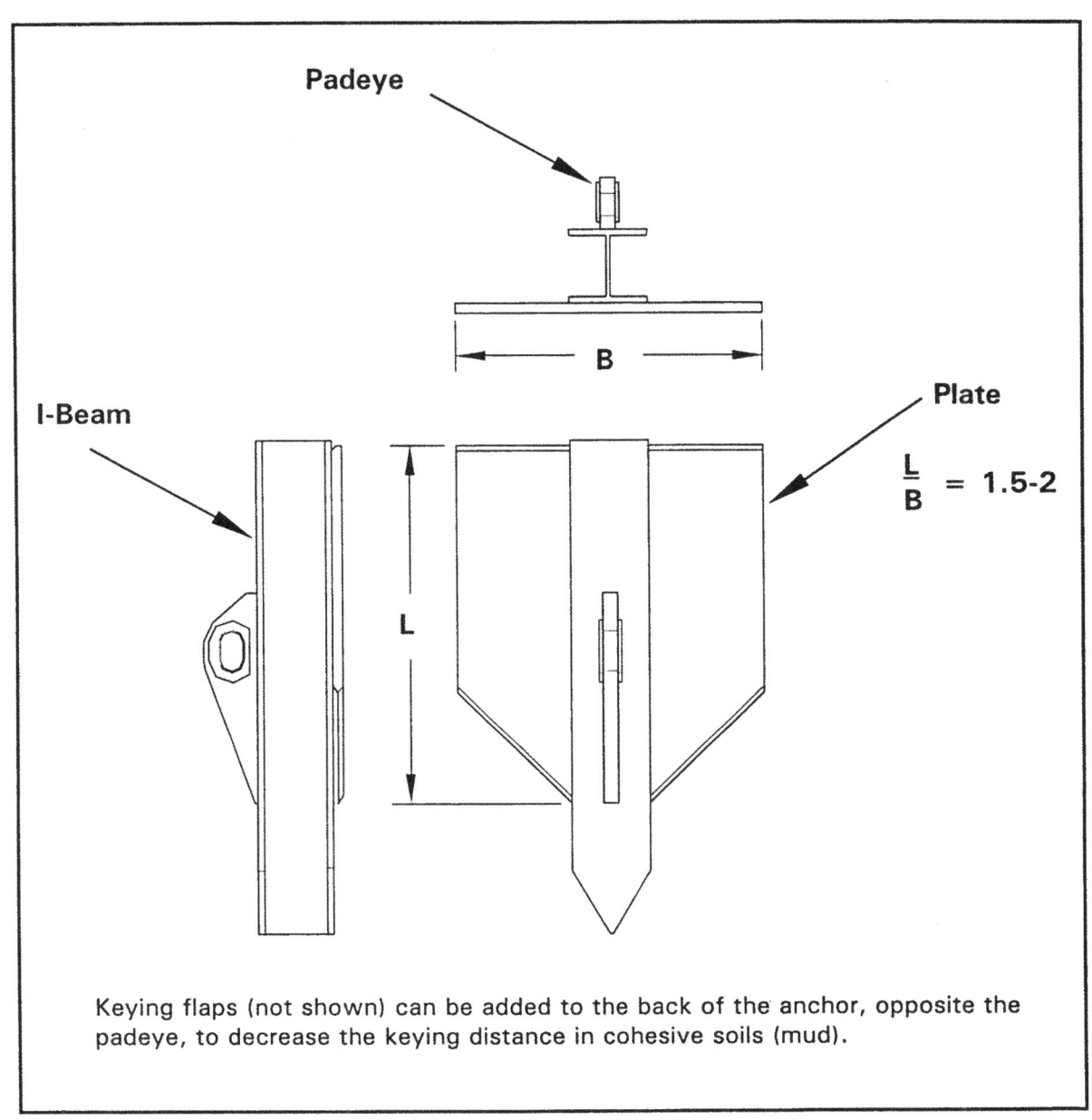

Keying flaps (not shown) can be added to the back of the anchor, opposite the padeye, to decrease the keying distance in cohesive soils (mud).

Table 6-2 Anchor Characteristics

(a) Drag-Embedment Anchors

CHARACTERISTICS	NOTES
Many basic designs and sizes are available from manufacturers.	NAVMOOR-10 & -15 and stockless of 20 to 30 kips are stocked by NFESC.
Works primarily in one horizontal direction.	Enough scope of chain and/or wire rope needs to be provided to minimize uplift forces, which can pull the

	anchor out. If a load is applied to a drag anchor at a horizontal axis off the centerline of the anchor, then the anchor may drag and reset.
Flukes should be set for the soil type.	Anchor performance depends strongly on the soil type. Fixing the maximum angle of the fluke will help ensure optimum performance. For mooring installations the flukes should be fixed open and stabilizers added for stockless anchors to help prevent overturning.
Adequate sediment required.	Sand layer thickness needs to be approximately one fluke length and mud needs to be 3 to 5 fluke lengths thick.
May not work in all seafloor types.	May be unreliable in very hard clay, gravel, coral, or rock seafloors; and in highly layered seafloors.
May not work well for sloping seafloors.	If the seafloor has a slope of more than several degrees, then the anchor may not hold reliably in the down-slope direction.
Anchor can drag.	If the anchor is overloaded at a slow enough rate, then the anchor can drag, which reduces the peak load. Anchor dragging can be a problem if the room for mooring is restricted. If adequate room is available then anchor drag can help prevent failure of other mooring components.
Anchors can be reused.	Drag-embedment anchors can be recovered and reused.
Proof loading recommended.	Pulling the anchor at the design load in the design direction will help set the anchor and assure that the soil/anchor interaction provides adequate holding.

(b) Driven-Plate Anchors

CHARACTERISTICS	NOTES
Size and design of anchor are selected to provide adequate holding, to allow driving, and to provide adequate structural capacity.	These anchors have been used in a variety of soils from soft mud to hard coral. A driving analysis is recommended for hard soil, because the anchor must be able to be driven in order to work.
Multi-directional.	Can be used on short scope, since the anchor resists uplift forces. One plate anchor may be used to replace several drag anchor legs, since the anchors are multi-directional.
Anchors designed for the soil type.	Anchors designed for the soil engineering characteristics at the site.

Adequate sediment required.	A minimum of several fluke lengths of sediment is required to provide for keying and allow the anchor to hold (NFESC TR-2039-OCN, *Design Guide for Pile-Driven Plate Anchors*).
Anchor is fixed.	The anchor will not drag, so this type of anchor is well suited to locations with limited mooring area available. The anchors cannot be recovered or inspected.
Proof loading recommended.	Pulling the anchor at the design load in the design direction will help key the anchor and assure that the soil/anchor interaction provides adequate holding.
Installation equipment.	Mobilization can be expensive, so installing a number of anchors at a time reduces the unit installation cost.

6.1.2. Drag-Embedment Anchor Specification

Drag-embedment anchors are carried on ships and used in many fleet-mooring facilities. Key considerations in selecting an anchor are: soil type, anchoring holding capacity, anchor weight, anchor stowage, cost, availability, and installation assets. Note that in SI units the anchor mass is used to characterize anchor size, while in U.S. customary units the anchor weight as a force is used.

Drag-embedment anchor holding capacities have been measured in full-scale tests, modeled in the laboratory, and derived from soil analyses. Empirical anchor holding curves were developed from this information (Naval Civil Engineering Laboratory (NCEL), TDS 83-08R, *Drag Embedment Anchors for Navy Moorings*). Predicted static ultimate anchor holding is given by

Equation 6-1: $$H_M = H_R \left(\frac{W_A}{W_R} \right)^b$$

where

- H_M = ultimate anchor system static holding capacity (kips or kN)
- H_R = reference static holding capacity
- W_A = weight of the anchor in air (for SI units use anchor weight in kilograms; for U.S. units use anchor weight in pounds force)
- W_R = reference anchor weight in air (for SI units use 4536 kg; for U.S. units use 10000 lbf)
- b = exponent

To solve for a particular anchor weight, Equation 6-1 can be solved to yield

Equation 6-2: $$W_A = W_R \left(\frac{H_M}{H_R} \right)^{\frac{1}{b}}$$

Values of H_R and b depend on the anchor and soil types. Values of these parameters are given in U.S. customary units in Table 6-3 and for SI units in Table 6-4.

Figure 6-4 and Figure 6-5 give holding capacities of selected anchors for mud and sand seafloors.

Table 6-3 Drag Anchor Holding Parameters, U.S. Customary Units

Anchor Type (a)	SOFT SOILS (Soft clays and silts)		HARD SOILS (Sands and stiff clays)	
	H_R (kips)	b	H_R (kips)	b
Boss	210	0.94	270	0.94
BRUCE Cast	32	0.92	250	0.8
BRUCE Flat Fluke Twin Shank	250	0.92	(c)	(c)
BRUCE Twin Shank	189	0.92	210	0.94
Danforth	87	0.92	126	0.8
Flipper Delta	139	0.92	(c)	(c)
G.S. AC-14	87	0.92	126	0.8
Hook	189	0.92	100	0.8
LWT (Lightweight)	87	0.92	126	0.8
Moorfast	117	0.92 (i)	60	0.8
			100 (d)	0.8
NAVMOOR	210	0.94	270	0.94
Offdrill II	117	0.92 (i)	60	0.8
			100 (d)	0.8
STATO	210	0.94	250 (e)	0.94
			190 (f)	0.94
STEVDIG	139	0.92	290	0.8
STEVFIX	189	0.92	290	0.8
STEVIN	139	0.92	165	0.8
STEVMUD	250	0.92	(g)	(g)
STEVPRIS (straight shank)	189	0.92	210	0.94
Stockless (fixed fluke)	46	0.92	70	0.8
			44 (h)	0.8
Stockless (movable fluke)	24	0.92	70	0.8
			44 (h)	0.8

(a) Fluke angles set for 50 deg in soft soils and according to manufacturer's specifications in hard soils, except when otherwise noted.
(b) "b" is an exponent constant.
(c) No data available.
(d) For 28-deg fluke angle.
(e) For 30-deg fluke angle.
(f) For dense sand conditions (near shore).
(g) Anchor not used in this seafloor condition.
(h) For 48-deg fluke angle.
(i) For 20-deg fluke angle (from API 2SK effective March 1, 1997).

Table 6-4 Drag Anchor Holding Parameters, SI Units

Anchor Type (a)	SOFT SOILS (Soft clays and silts)		HARD SOILS (Sands and stiff clays)	
	H_R (kN)	b	H_R (kN)	b
Boss	934	0.94	1201	0.94
BRUCE Cast	142	0.92	1112	0.8
BRUCE Flat Fluke Twin Shank	1112	0.92	(c)	(c)
BRUCE Twin Shank	841	0.92	934	0.94
Danforth	387	0.92	560	0.8
Flipper Delta	618	0.92	(c)	(c)
G.S. AC-14	387	0.92	560	0.8
Hook	841	0.92	445	0.8
LWT (Lightweight)	387	0.92	560	0.8
Moorfast	520	0.92 (i)	267	0.8
			445 (d)	0.8
NAVMOOR	934	0.94	1201	0.94
Offdrill II	520	0.92 (i)	267	0.8
			445 (d)	0.8
STATO	934	0.94	1112 (e)	0.94
			845 (f)	0.94
STEVDIG	618	0.92	1290	0.8
STEVFIX	841	0.92	1290	0.8
STEVIN	618	0.92	734	0.8
STEVMUD	1112	0.92	(g)	(g)
STEVPRIS (straight shank)	841	0.92	934	0.94
Stockless (fixed fluke)	205	0.92	311	0.8
			196 (h)	0.8
Stockless (movable fluke)	107	0.92	311	0.8
			196 (h)	0.8

(a) Fluke angles set for 50 deg in soft soils and according to manufacturer's specifications in hard soils, except when otherwise noted.
(b) "b" is an exponent constant.
(c) No data available.
(d) For 28-deg fluke angle.
(e) For 30-deg fluke angle.
(f) For dense sand conditions (near shore).
(g) Anchor not used in this seafloor condition.
(h) For 48-deg fluke angle.
(i) For 20-deg fluke angle (from API 2SK effective March 1, 1997).

Figure 6-4 Anchor System Holding Capacity in Cohesive Soil (Mud)

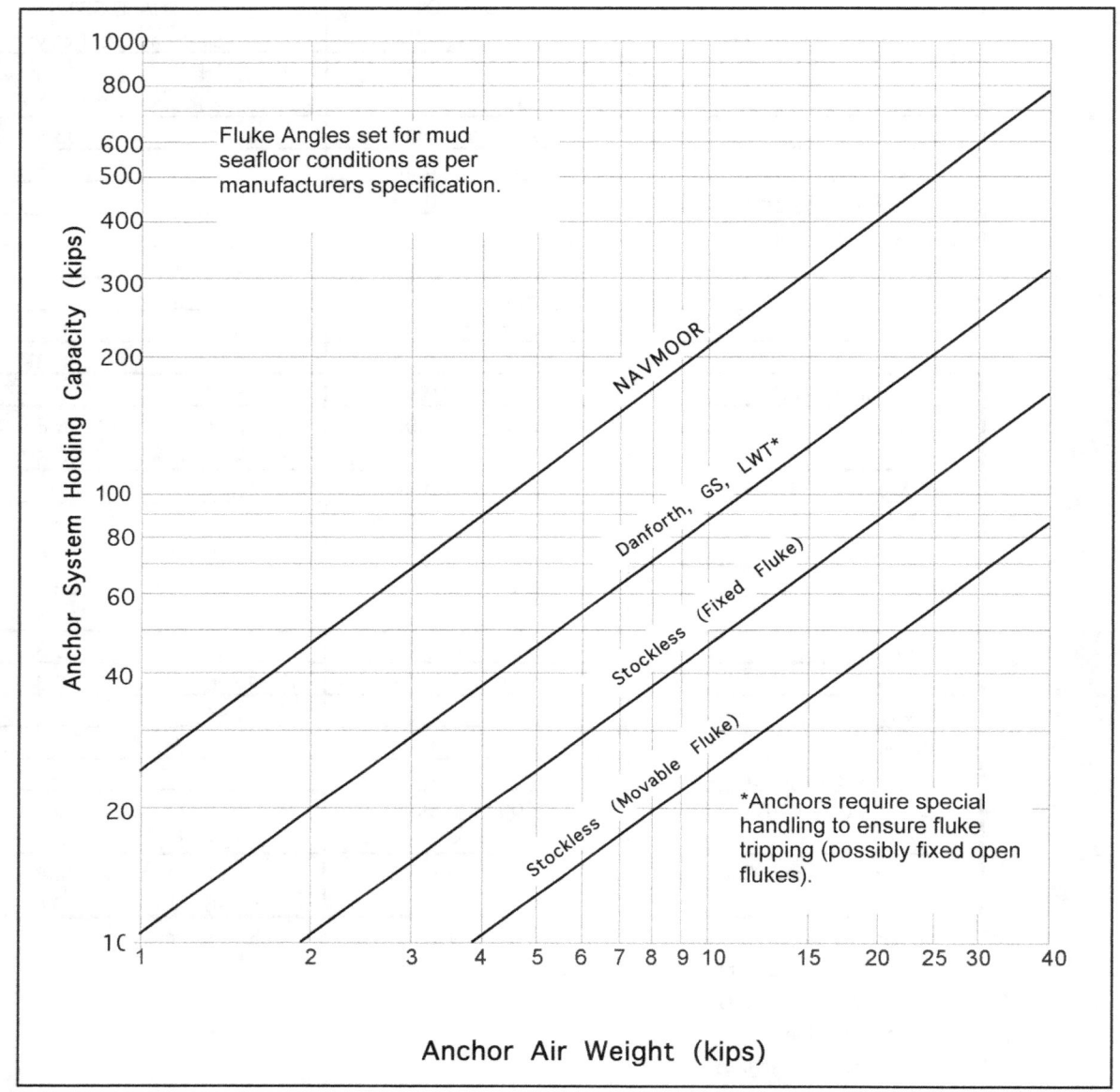

Figure 6-5 Anchor System Holding Capacity in Cohesionless Soil (Sand)

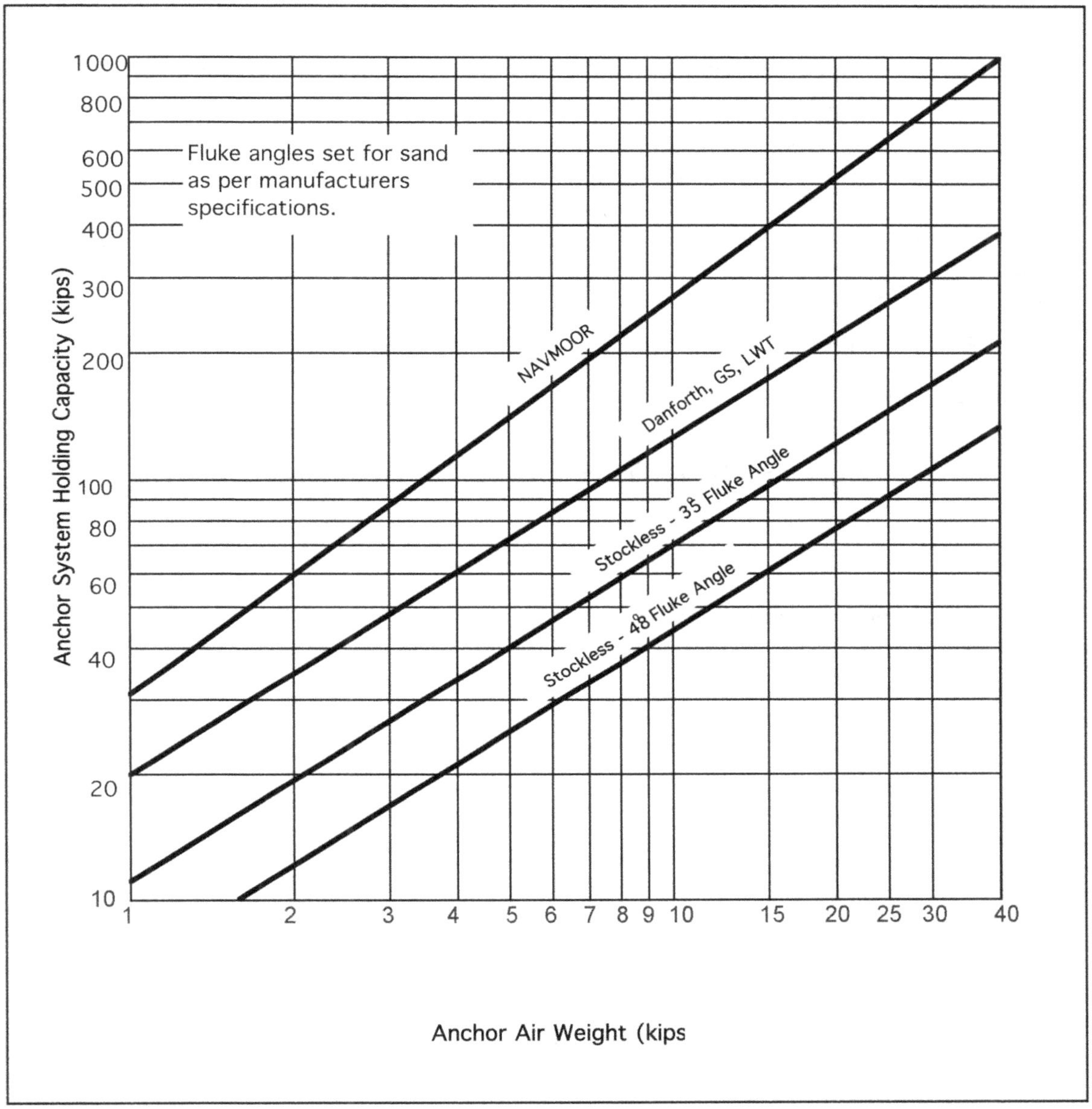

An example of anchor design is given in Example 6-1.

6.1.3. Driven-Plate Anchor Design

The U.S. Naval Facilities Engineering Service Center has found that various types of plate anchors are an efficient and cost effective method of providing permanent moorings. Detailed design procedures for these anchors are given in NFESC TR-2039-OCN, *Design Guide for Pile-Driven Plate Anchors*. Additional information is given in NCEL *Handbook for Marine Geotechnical Engineering*, available from Pile Buck. An overview of plate anchor design is given here. The recommended minimum plate anchor spacing is five times the anchor width for mud or clay and 10 times the anchor width for sand.

6.2. Catenary Behavior

It is not desirable or practical to moor a ship rigidly. For example, a ship can have a large amount of buoyancy, so it usually must be allowed to move with changing water levels. Another problem with holding a ship too rigidly is that some of the natural periods of the ship/mooring system can become short, which may cause dynamic problems.

A ship can be considered a mass and the mooring system as springs. During mooring design, the behavior of the mooring 'springs' can be controlled to fine tune the ship/mooring system behavior to achieve a specified performance. This can be controlled by the weight of chain or other tension member, scope of chain, placement of sinkers, amount the anchor penetrates the soil, and other parameters. The static behavior of catenaries can be either a) be computed by hand or b) modeled using a computer program.

6.2.1. Catenary Equations and Hand Solution

A chain mooring line supported at the surface by a buoy and extending through the water column to the seafloor behaves as a catenary. Figure 70 presents a definition sketch for use in catenary analysis.

Figure 6-6 Definition Sketch for Use in Catenary Analysis

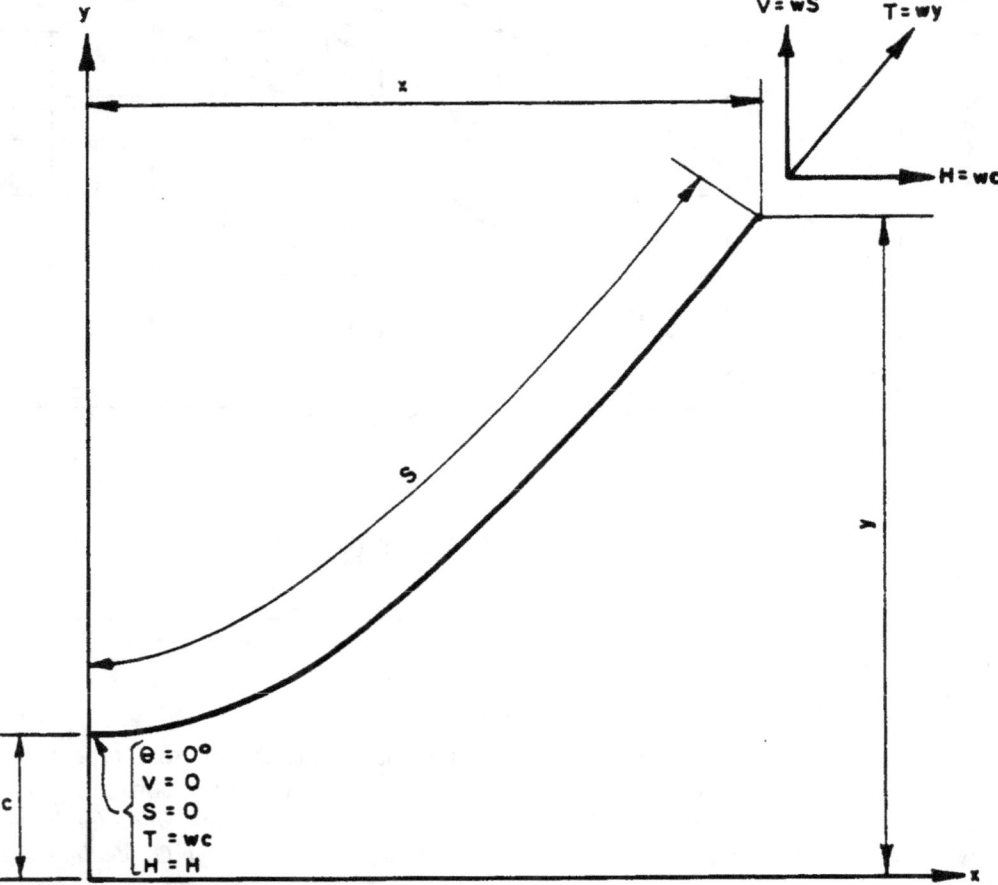

At any point (x, y) the following hold:

$$\textbf{Equation 6-3:}\quad V = wS = T\sin\theta$$

$$\textbf{Equation 6-4:}\quad H = wc = T\cos\theta$$

$$\textbf{Equation 6-5:}\quad T = wy$$

Where

- V = vertical force at point (x, y)
- w = submerged unit weight of chain
- S = length of curve (chain length) from point (O, c) to point (x, y)
- T = line tension at point (x, y)
- θ = angle of mooring line with horizontal
- H = horizontal force at point (x,y)
- c = distance from origin to y-intercept = H/w

Since

$$\textbf{Equation 6-6:}\quad \frac{V}{H} = \frac{\sin\theta}{\cos\theta} = \tan\theta$$

We can also write

$$\textbf{Equation 6-7:}\quad V = H\tan\theta$$

The following can also be derived:

$$\textbf{Equation 6-8:}\quad S = \frac{V}{w}$$

$$\textbf{Equation 6-9:}\quad c = \frac{H}{w}$$

The shape of the catenary is governed by the following:

$$\textbf{Equation 6-10:}\quad y = \sqrt{S^2 + c^2} = c\cosh\left(\frac{x}{c}\right)$$

$$\textbf{Equation 6-11:}\quad S = c\sinh\left(\frac{x}{c}\right)$$

Equation 6-11 may be more conveniently expressed by solving for x:

$$\textbf{Equation 6-12:}\quad x = c\ln\left(\frac{S}{c} + \sqrt{\left(\frac{S}{c}\right)^2 + 1}\right)$$

Note that, in the above equations, the horizontal load in the chain is the same at every point and that all measurements of x, y, and S are referenced to the catenary origin.

When catenary properties are desired at point (x_m, y_m), as shown in Figure 71, the following equations are used:

Equation 6-13: $x_{ab} = 2\ln\left(\dfrac{\sqrt{S_{ab}^2 - w_d^2} + \sqrt{S_{ab}^2 - w_d^2 + 4c^2}}{2c}\right)$

Equation 6-14: $x_m = \operatorname{arctanh}\left(\dfrac{w_d}{S_{ab}}\right) c$

Equation 6-15: $x_a = x_m - \dfrac{x_{ab}}{2}$

Equation 6-16: $x_b = x_m + \dfrac{x_{ab}}{2}$

Equation 6-17: $y_b = y_a + w_d$

Terms are defined in Figure 6-7.

Figure 6-7 Definition Sketch for Catenary Analysis at point (x_m, y_m)

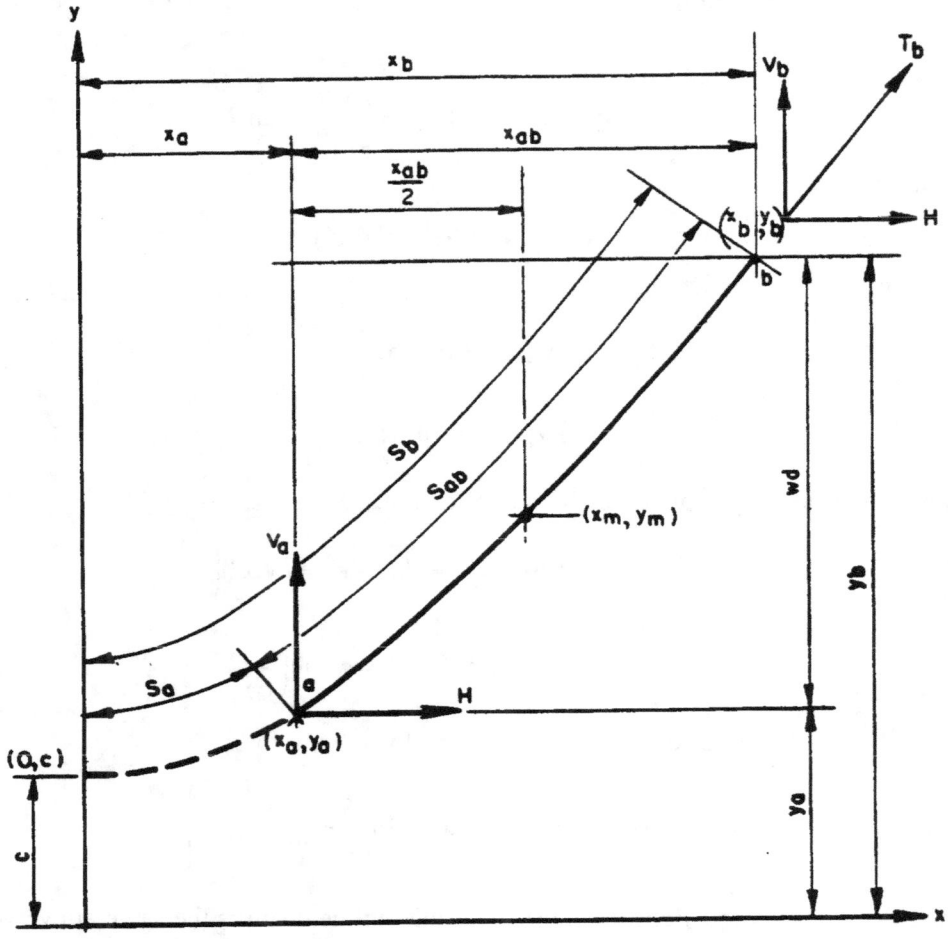

Some applications of the catenary equations are shown in Table 6-5.

Table 6-5 Applications of Catenary Equations

Case	Known Variables	Method	Figure with Procedure
I	<ul><li>mooring-line angle at the anchor, θ_a (which is zero: $\theta_a = 0°$)</li><li>water depth, w_d</li><li>the horizontal load, H</li><li>the submerged unit weight of the chain, w.</li></ul>	The known variables are the A zero anchor angle is often specified because drag-anchor capacity is drastically reduced as the angle of the chain at the seafloor is increased. The length of mooring line, S_{ab}, the horizontal distance from the anchor to the buoy, x_{ab}, and the tension in the mooring line at the buoy, T_b, are desired. Check to determine if the entire chain has been lifted off the bottom by comparing the computed chain length from anchor to buoy, S_{ab}, to the actual chain length, S_{actual} If the actual chain length is less than the computed, then Case I cannot used and Case V must be used.	Figure 6-8(a)
II	<ul><li>mooring-line angle at the anchor, θ_a (or, equivalently, a specified vertical load at the anchor, V_a)</li><li>water depth, w_d</li><li>horizontal load at the surface, H</li><li>submerged unit weight of the chain, w.</li></ul>	This situation arises when a drag anchor is capable of sustaining a small prescribed angle at the anchor, or an uplift-resisting anchor of given vertical capacity, $V_a = H \tan \theta_a$, is specified. The origin of the catenary is not at the anchor, but is some distance below the bottom. The length of the chain from anchor to buoy, S_{ab}, the tension in the mooring line at the buoy, T_b, and the horizontal distance from the anchor to the surface, X_{ab} are desired.	Figure 6-8(b)
III	<ul><li>horizontal distance from the anchor to the buoy, X_{ab}</li><li>water depth, w_d</li><li>horizontal load at the surface, H</li><li>submerged unit weight of the chain, w.</li></ul>	This situation arises when it is necessary to limit the horizontal distance from buoy to anchor due to space limitations. The length of chain from anchor to buoy, S_{ab}, the tension in the mooring line at the buoy, T_b, and the vertical load at the anchor, V, are required.	Figure 6-8(c)
IV	<ul><li>water depth, w_d</li><li>horizontal load at the surface, H</li><li>submerged unit weight of the</li></ul>	The mooring consists of a chain of constant unit weight with a sinker attached to it. The total length of chain, S_{ac} the distance of the top of the sinker off the bottom, y_s, and the tension the mooring line at the buoy, T_{C2}, are desired.	Figure 6-9(a)

	- chain, w. - angle at the anchor, θ_a - sinker weight, W_s - unit weight of the sinker, s - unit weight of water, w - length of chain from anchor to sinker, S_{ab}		
V	- water depth, w_d - horizontal load at the surface, H - submerged unit weight of the chain, w. - length of chain from anchor to buoy, S_{ab}	The horizontal load, H, is sufficiently large to lift the entire chain off the bottom, resulting in an unknown vertical load at the anchor, V_a. This situation arises when one is computing points on a load-deflection curve for higher values of load. Solution involves determining the vertical load at the anchor, V, using the trial-and-error procedure. The problem is solved efficiently using a Newton-Raphson iteration method (Gerald, 1980); this method gives accurate solutions in two or three iterations, provided the initial estimate is close to the final answer.	Figure 6-9(b)

Figure 6-8 Solution Procedures for Cases I, II and III

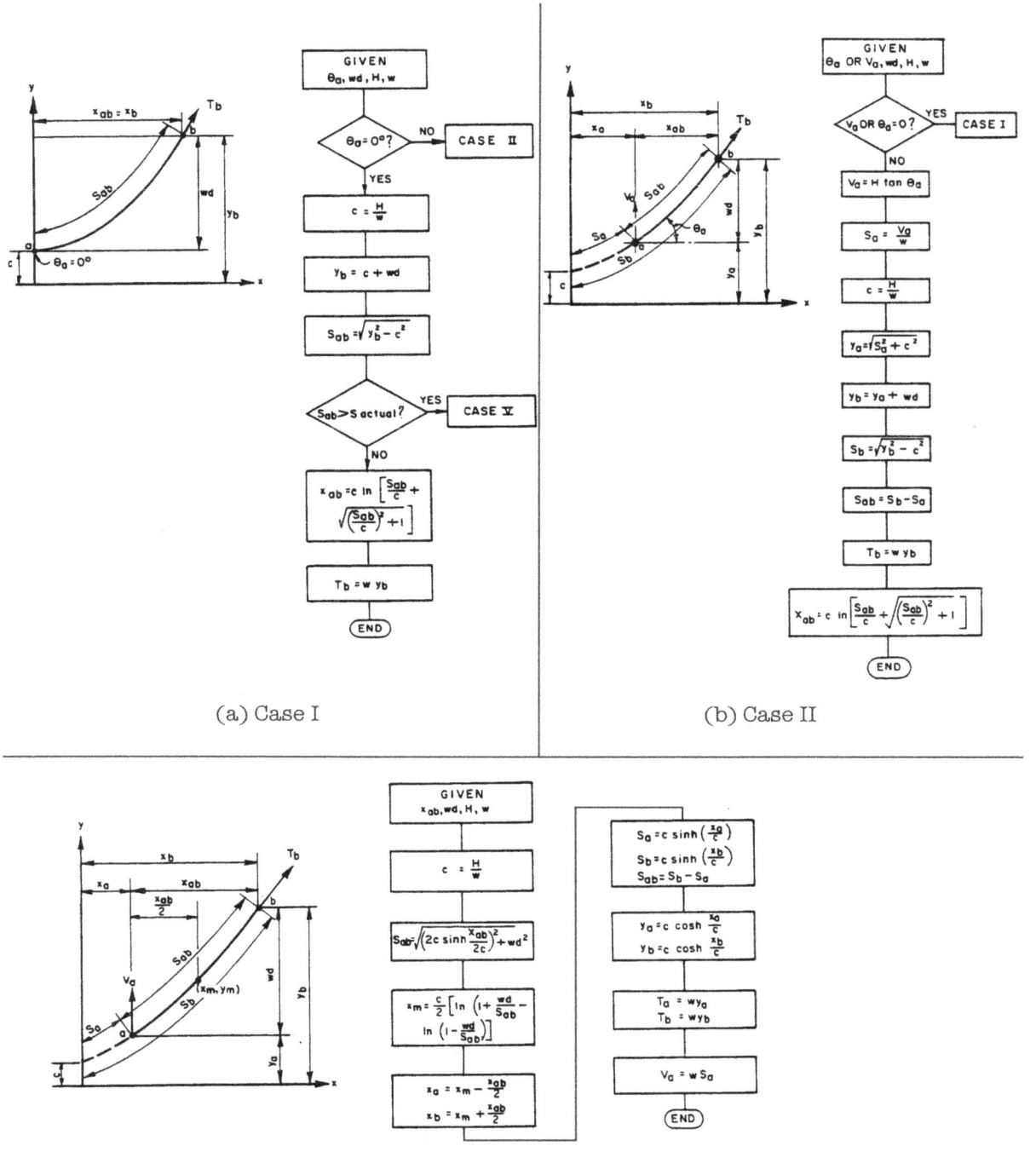

(a) Case I

(b) Case II

(c) Case III

Figure 6-9 Solution Procedure for Cases IV and V

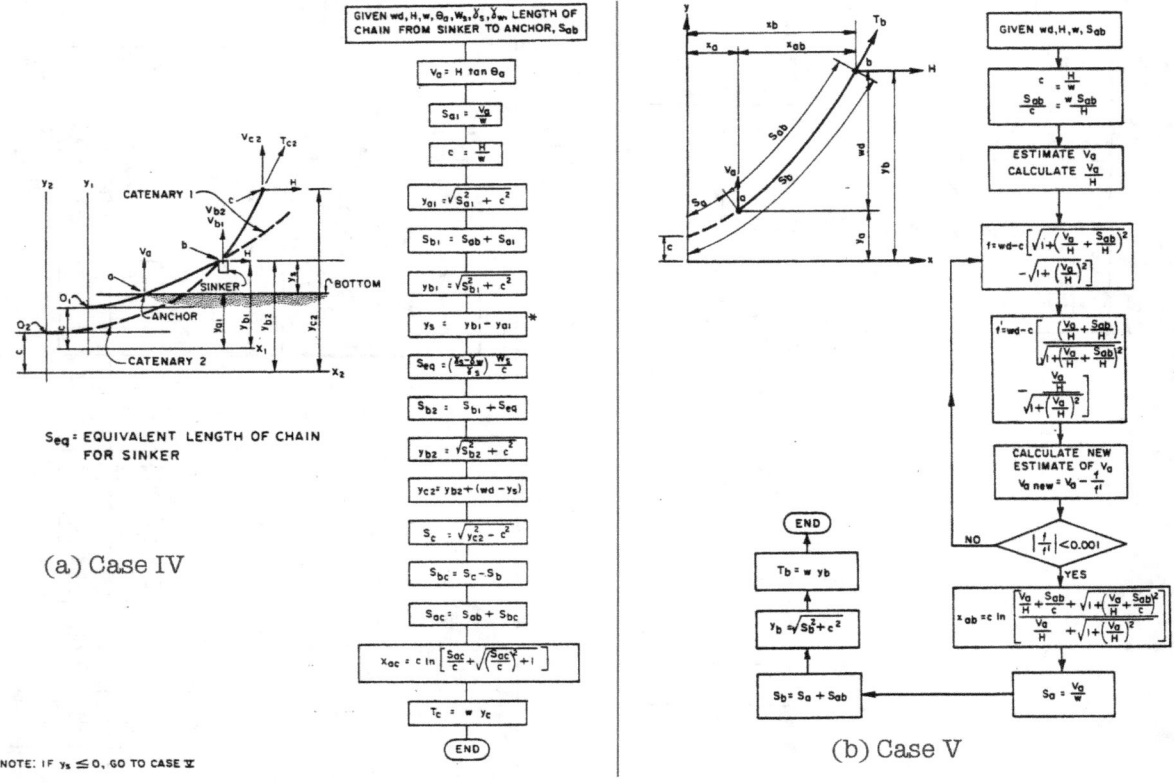

(a) Case IV

(b) Case V

Example 6-1 Catenary Analysis and Anchor Design—Hand Solution

Consider the following mooring scenario for a single point ship's anchoring:

- Water depth w_d at high tide = 41'
- Chain angle at bottom $\theta_a = 2°$
- Horizontal load $H = 21.396$ kips
- Proposed chain: 1", breaking strength of 84.5 kips
- Anchored into sand.

Compute the chain length and tension; also specify the anchor.

The first thing we must do is to determine the submerged weight of the chain. In the absence of other data, the weight in air of stud link chain can be estimated by the equation

Equation 6-18: $w_{air} = 9.5d^2$

where d is the diameter of the chain and wair is the weight of the chain in pounds per foot of chain. The submerged weight is

Equation 6-19: $w_s = 8.26d^2$

For our chain, $w_s = (8.26)(1^2) = 8.26$ lb/ft = 0.00826 kips/ft of chain.

The known parameters suggest that we use the Case II procedure to solve the problem; however, we will reference the basic equations. To determine the length of the chain, we proceed as follows:

$$V_a = H \tan\theta_a = 21.396 \times \tan 2° = 0.7472 \text{ kips} \textbf{ (Equation 6-7)}$$

$$S_a = \frac{V_a}{w_s} = \frac{0.7472}{0.00826} = 90.46' \textbf{ (Equation 6-8)}$$

$$c = \frac{H}{w_s} = \frac{21.396}{0.00826} = 2{,}590.3' \textbf{ (Equation 6-9)}$$

$$y_a = \sqrt{S_a^2 + c^2} = \sqrt{90.46^2 + 2590.3^2} = 2{,}591.9' \textbf{ (Equation 6-10)}$$

$$y_b = y_a + w_d = 2591.9 + 41 = 2{,}632.9' \textbf{ (Equation 6-17)}$$

$$S_b = \sqrt{y_b^2 - c^2} = \sqrt{2{,}632.9^2 - 2{,}590.3^2} = 471.7' \textbf{ (Equation 6-10)}$$

$$S_{ab} = S_b - S_a = 471.7 - 90.46 = 381.25' \textbf{ (see Figure 6-7)}$$

Since chain is generally supplied in 90' shots, we can estimate the number of shots required, thus 381.25/90 = 4.24, or use 4.5 shots.

Turning to the tension of the chain,

$$T_b = wy_b = (0.00826)(2{,}632.9) = 21.7478 \text{ kips} \textbf{ (Equation 6-5)}$$

According to Table 4-7, a factor of safety of 3 should be applied to chain (other than around bends.) Thus, the design load of the chain is 84.5/3 = 28.2 kips > 21.7478 kips, so the tension of the chain is within the design load.

Turning to the anchor design, we plan to use a stockless anchor. According to Table 4-7, we can use a factor of safety of 1.5. The simplest way to implement this is to multiply the design load by the factor of safety. For sand, from Table 6-3, H_R = 70 kips and b = 0.8 for stockless anchors with 35° flukes. For an anchor force of 21.396 kips, using Equation 6-2,

$$W_A = 10\left(\frac{(21.396)(1.5)}{70}\right)^{\frac{1}{0.8}} = 3.77 \text{ kips}$$

Thus any of the anchors shown in Table 3-1 are satisfactory.

6.2.2. Catenary Analysis—Computer Solution

For computer analysis of catenary problems, one can use the computer program CSAP2 (NFESC CR-6108-OCN, *Anchor Mooring Line Computer Program Final Report, User's Manual for Program CSAP2*). This program includes the effects of chain and wire rope interaction with soils, as well as the behavior of the catenary in the water column and above the water surface.

Example 6-2 Catenary Analysis—Computer Solution

Consider the catenary shown in Figure 6-10. This mooring leg consists of four sections. The segment next to the anchor, Segment 1, consists of wire rope, followed by three segments of chain. Sinkers with the shown in-water weight are located at the ends of Segments 2 and 3. In this example, a plate anchor is driven 55 feet (16.8 meters) into mud below the seafloor. The chain attachment point to the ship is 64 feet (19.5 meters) above the seafloor. The mooring leg is loaded to its design horizontal load of H = 195 kips (870 kN) to key and proof load the anchor soon after the anchor is installed. The keying and proofing corresponds to a tension in the top of the chain of approximately 210 kips. Figure 6-10 shows the shape of the chain catenary predicted by CSAP2 for the design load.

The computed load/deflection curve for the design water level for this mooring leg, after proofing, is shown in Figure 6-11. The shape of this and the other mooring legs in this mooring, which are not shown, will strongly influence the static and dynamic behavior of the ship/mooring system during forcing.

Figure 6-10 Sample Catenary

SEGMENT	TYPE	DIA (inches)	WEIGHT (lbf/ft)	LENGTH (ft)	SINKER (kips)
1	W	3.00	13.15	30	0
2	C	2.75	62.25	156	13.35
3	C	2.75	62.25	15	17.8
4	C	2.75	62.25	113	0

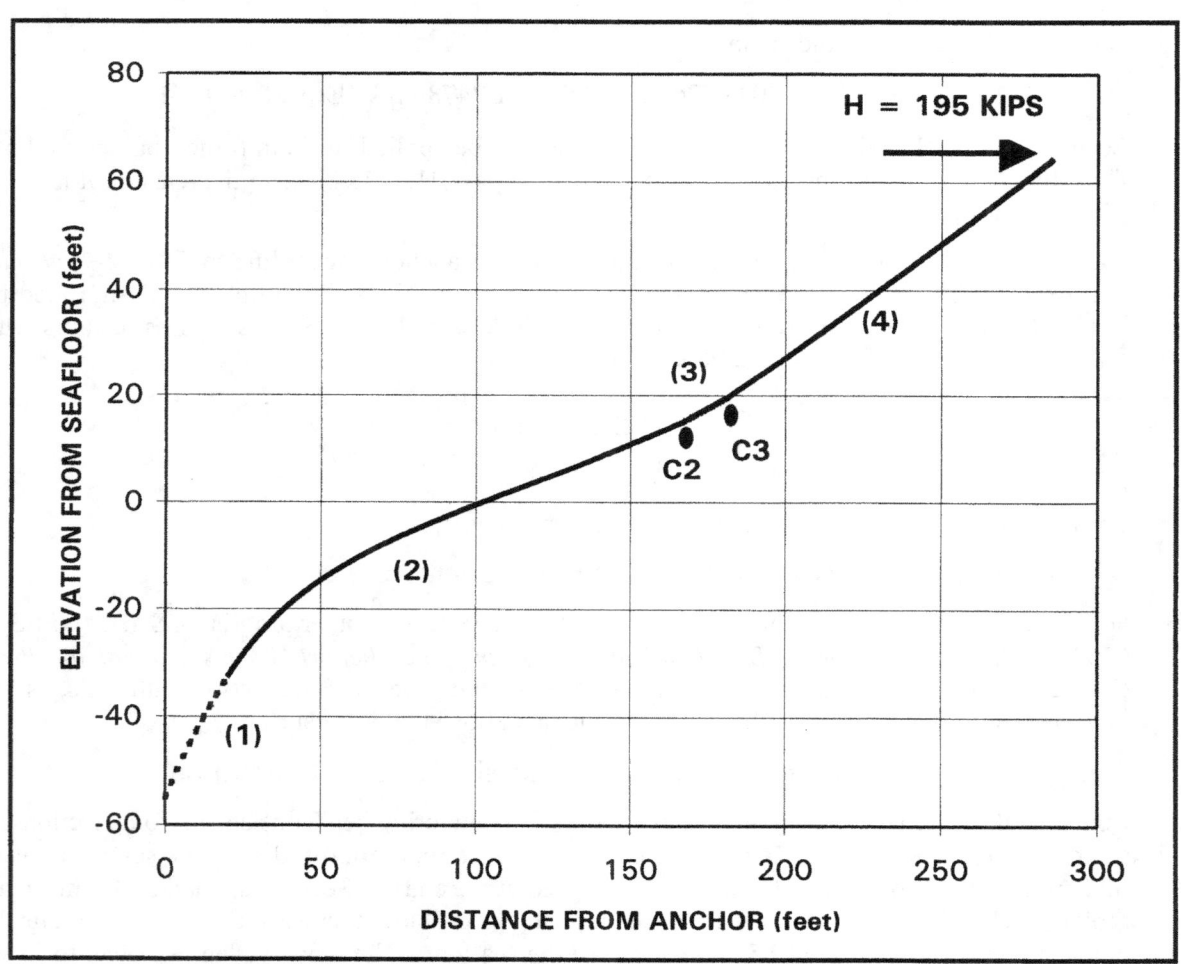

Figure 6-11 Load/Deflection Curve for the Example Mooring Leg

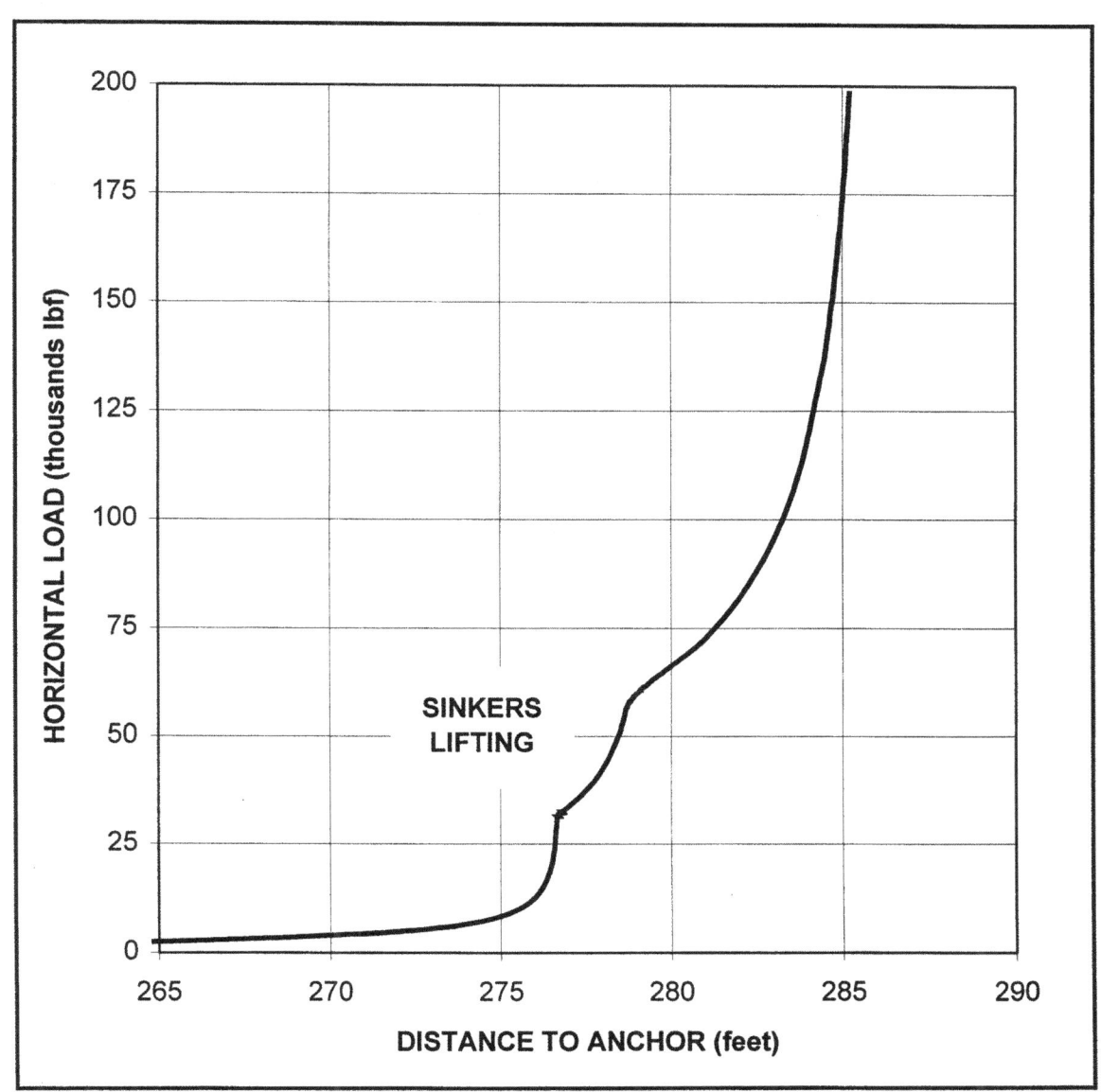

6.3. Example Problems

Example 6-3 Single Point Mooring - Basic Approach.

Design of single point fleet moorings (SPMs) is illustrated here. In this example two moorings were designed and installed. The original designs were based on quasi-static methods. Ships moored to these buoys broke their mooring hawsers when a wind gust front struck the ships. In this example, the design and hawser failures are reviewed. The effects of wind dynamics on a single point mooring are illustrated.

Let us first assume that the wind is coming from a specified direction and has stationary statistical properties. The current speed and direction are constant. In this case there are three common types of ship behavior, shown in Figure 6-12, that a vessel at a single point mooring can have:

a) Quasi-static. In this case the ship remains in approximately a fixed position with the forces and moments acting on the ship in balance. For quasi-static behavior, the tension in the attachment from the ship to mooring will remain approximately constant. Quasi-static analyses can be used for design in this case.

b) Fishtailing. In this case the ship undergoes significant surge, sway, and yaw with the ship center of gravity following a butterfly-shaped pattern. The mooring can experience high dynamic loads, even though the wind and current are constant.

c) Horsing. In this case the ship undergoes significant surge and sway with the ship center of gravity following a U-shaped pattern. The mooring can experience high dynamic loads.

Figure 6-12 Some Types of Behavior of Ships at Single Point Moorings

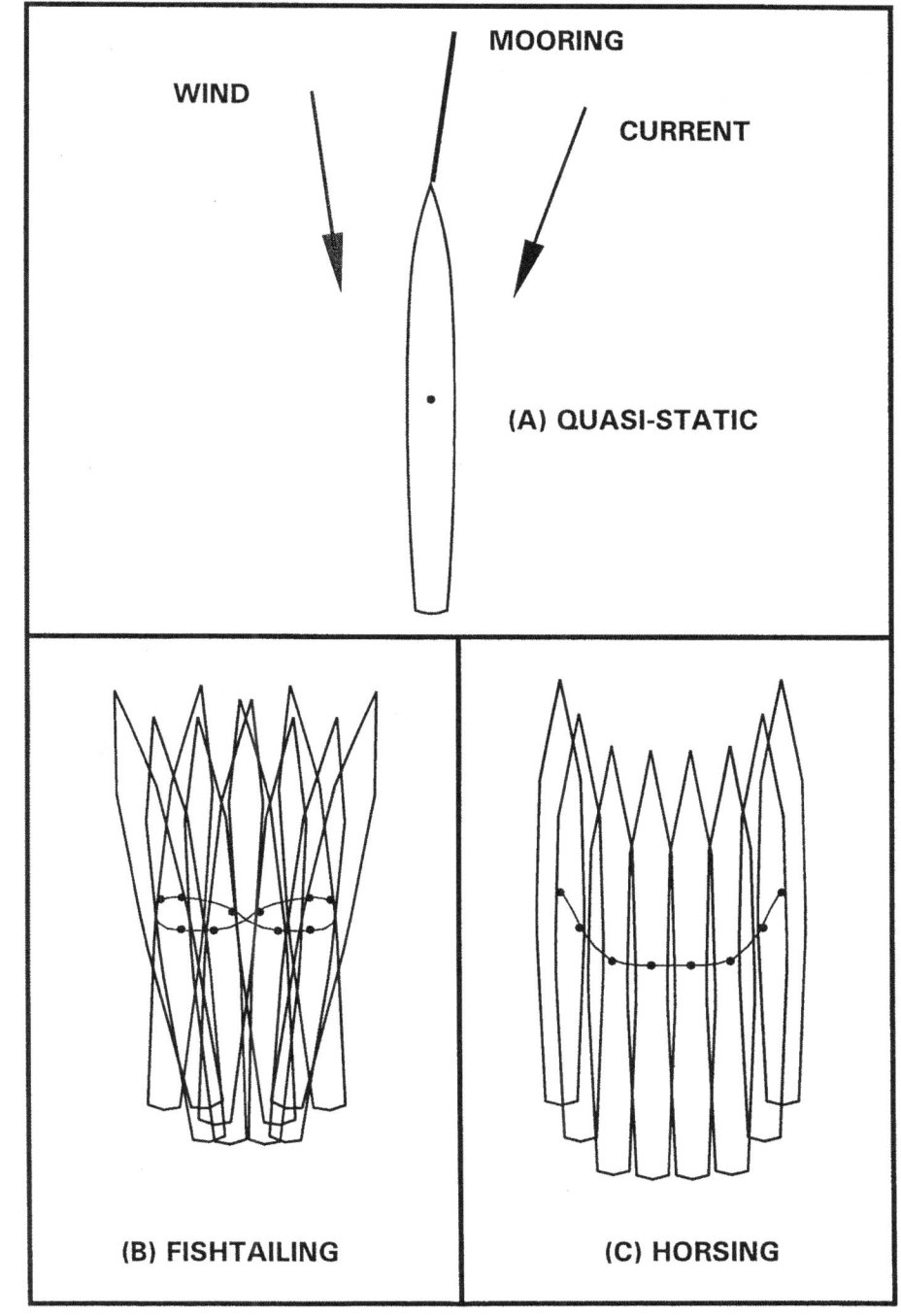

These cases show that the type of behavior of a given ship at a given single point mooring in a given environment can be very complex (Wichers, 1988), even though the wind and current are steady. It is recommended that a dynamic stability analysis first be conducted (Wichers, 1988) at the early stages of single point mooring design. Then the type of analysis required can be determined. The results from this analysis will suggest what type of method should be used to design a single point mooring. These methods are complex and beyond the scope of this handbook. Behavior of single point moorings is illustrated by example.

1. *Ship.* A single 2nd LT JOHN P. BOBO (T-AK 3008) class ship was moored at each of two fleet mooring buoys. Table 6-6 gives basic characteristics of the ships.

Table 6-6 2nd LT JOHN P BOBO Parameters (Fully Loaded)

PARAMETER	DESIGN BASIS (SI units)	DESIGN BASIS (English units)
Length Overall At Waterline Between Perpendiculars	 193.2 m 187.3 m 187.3 m	 633.76 ft 614.58 ft 614.58 ft
Beam @ Waterline	9.80	32.15 ft
Draft	9.75 m	32 ft
Displacement	4.69E7 kg	46111 long tons
Line Size (2 nylon hawsers)	-	12 inches

2. *Forces/Moments.* In this case the design wind speed is 45 knots (23 m/s). Currents, waves, and tidal effects are neglected for these 'fair weather' moorings. The bow-on ship wind drag coefficient is taken as the value given for normal ships of 0.7, plus 0.1 is added for a clutter deck to give a drag coefficient of 0.8. Methods in Section 4 are used to compute the forces and moments on the ship. The computed bow-on wind force is 68.6 kips (300 kN) for 45-knot (23-m/s) winds, as shown in Figure 6-13.

Figure 6-13 Example Single Point Mooring

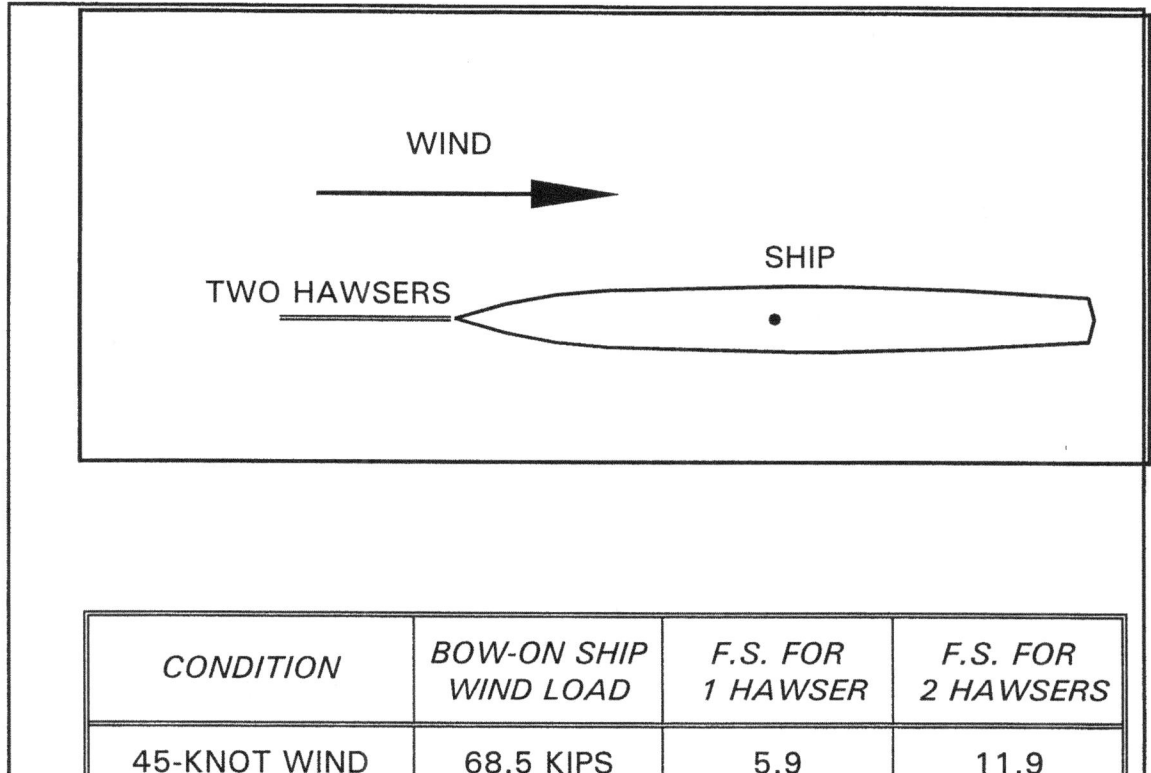

3. *Quasi-Static Design.* Quasi-static design procedures place the ship parallel to the wind for this example, because in this position the forces and moments on the ship are balanced out. Two mooring hawsers were specified for this design. Extra factor of safety was specified for the two 12" nylon mooring hawsers, which had a new wet breaking strength of 406 kips (1.8 MN), to account for poor load sharing between the two hawsers.

4. *Mooring Hawser Break.* The ships were moored and faced into 15-knot winds. The weather was unsettled, due to two nearby typhoons, so the ships had their engines in idle. A wind gust front struck very quickly with a wind speed increase from 15 to 50 knots. As the wind speed increased, the wind direction changed 90°, so the higher wind speed hit the ships broadside. The predicted peak dynamic tension on the mooring hawsers was 1140 kips (5.07 MN), (Seelig and Headland, 1998). Figure 6-14 is a simulation predicting the dynamic behavior of the moored ship and hawser tension. In this case, the mooring hawsers broke and the predicted factor of safety dropped to less than 1. In this event, the peak dynamic tension on the mooring hawser is predicted to be 13.5 times the bow-on wind force for 50-knot (25.7-m/s) winds.

Mooring Systems: Design and Maintenance
Pile Buck International Inc.

Figure 6-14 Example Mooring Failure Due to a Wind Gust Front

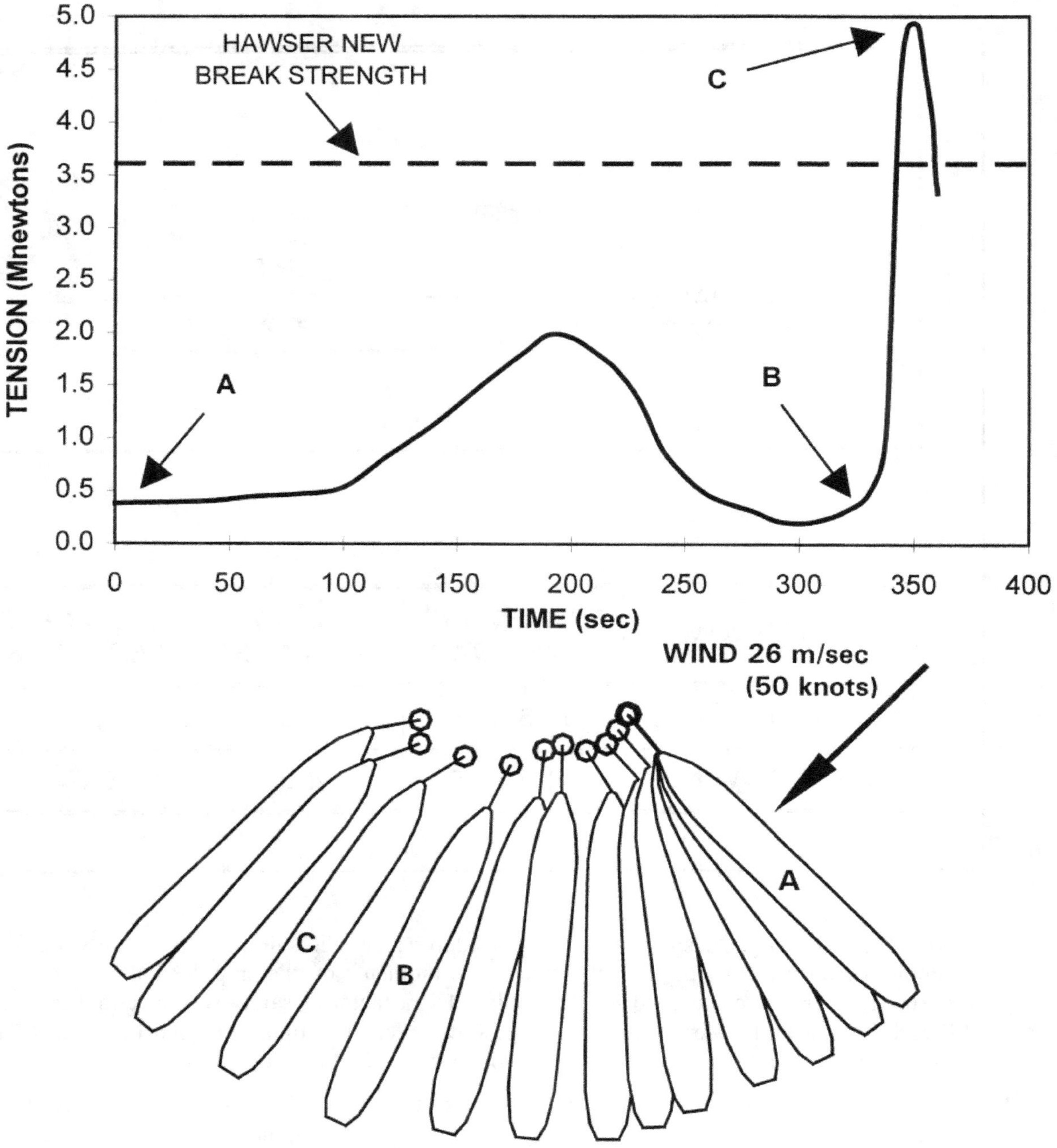

Predicted ship positions at 30 second intervals.

This example shows that single point moorings can be susceptible to dynamic effects, such as those caused by wind gust fronts or other effects.

Example 6-4 *Fixed Mooring - Basic Approach.*

Development of a design concept for a fixed mooring, a mooring that includes both tension and compression members, is illustrated here. Several new aircraft carrier berthing wharf facilities are being programmed. Users expressed concerns regarding the possibility of excessive ship movement. Wind is the major environmental parameter of concern. Assume the proposed sites have small tidal ranges and tidal currents.

1. *Goal.* Develop a concept to moor USS NIMITZ (CVN-68) class ships at newly constructed wharves. Assume the Mooring Service Type is II and the design wind speed is 75 mph (33.5 m/s).

2. *Ship.* Fully loaded USS NIMITZ (CVN-68) class ships are used in this example. Table 6-7 gives some ship parameters. Additional information is found in the Naval Facilities Engineering Command (NAVFAC) *Ships Characteristics Database*.

Table 6-7 CVN-68 Criteria (Fully Loaded)

PARAMETER	DESIGN BASIS (SI units)	DESIGN BASIS (English units)
Length Overall At Waterline Between Perpendiculars	 249.9 m 233.2 m 237.1 m	 1115 ft 1056 ft 1040 ft
Beam @ Waterline	32.3	134 ft
Draft	11.55 m	37.91 ft
Displacement	96,440 Mg	94,917 long tons
Bitt Size	-	12 inches
Line Size (nylon)	-	8 and 9 inches

3. *Forces/Moments.* Methods in Section 4 are used to compute the forces and moments on the ship. These values are summarized in Figure 6-15.

Figure 6-15 Wind Forces and Moments on a Single Loaded CVN-68 for a 75-mph (33.5-m/s) Wind

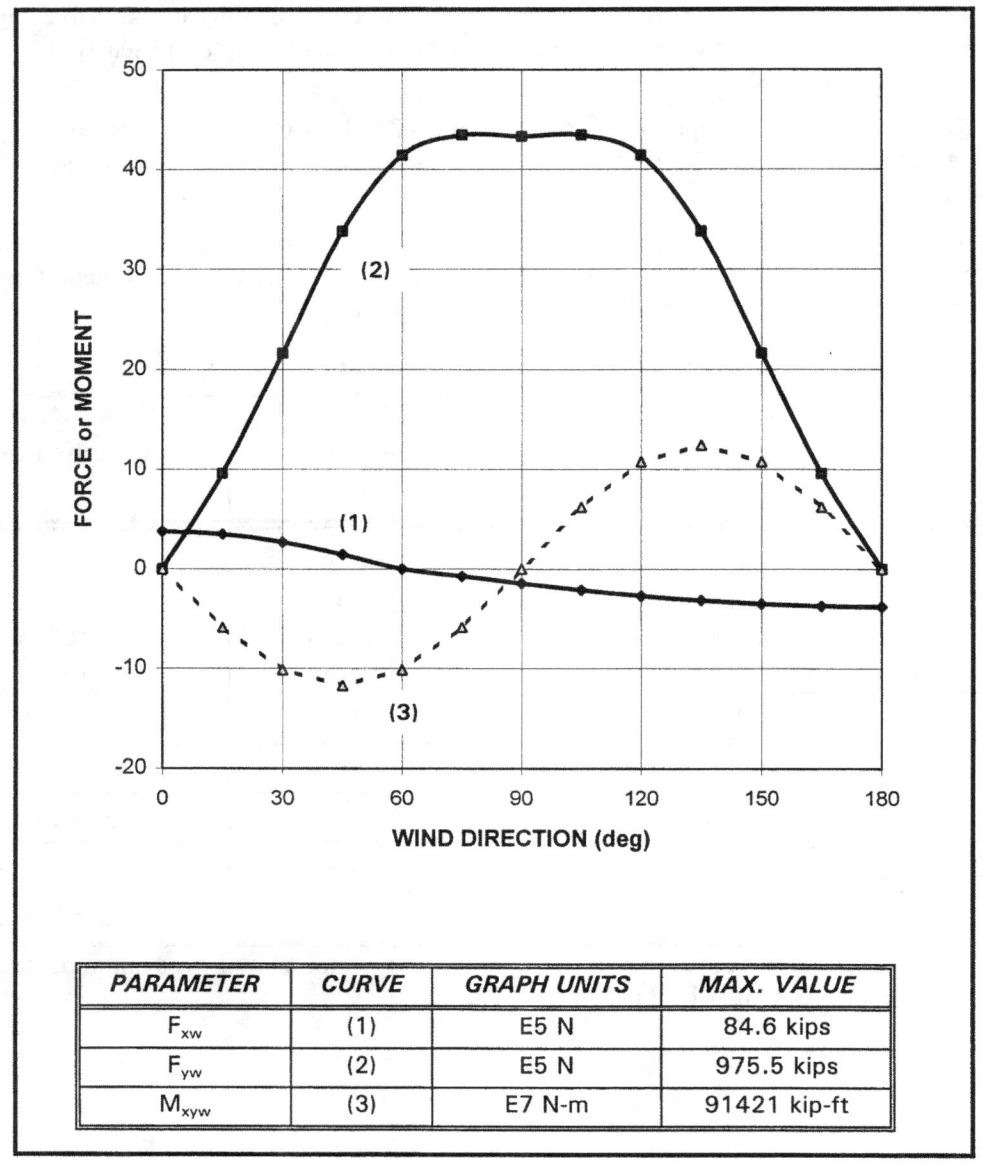

PARAMETER	CURVE	GRAPH UNITS	MAX. VALUE
F_{xw}	(1)	E5 N	84.6 kips
F_{yw}	(2)	E5 N	975.5 kips
M_{xyw}	(3)	E7 N-m	91421 kip-ft

4. *Definitions.* In this example we define a global coordinate system with "X" parallel to the wharf, as shown in Figure 55. Then "Y" is a distance perpendicular to the wharf in a seaward direction and "Z" is a vertical distance. Let "Pt 2" be the ship chock coordinate and "Pt 1" be the pier fitting. A spring line is defined as a line whose angle in the horizontal plane is less than 45° and a breasting line whose angle in the horizontal plane is greater than or equal to 45°, as shown in Figure 6-16.

Figure 6-16 Definitions

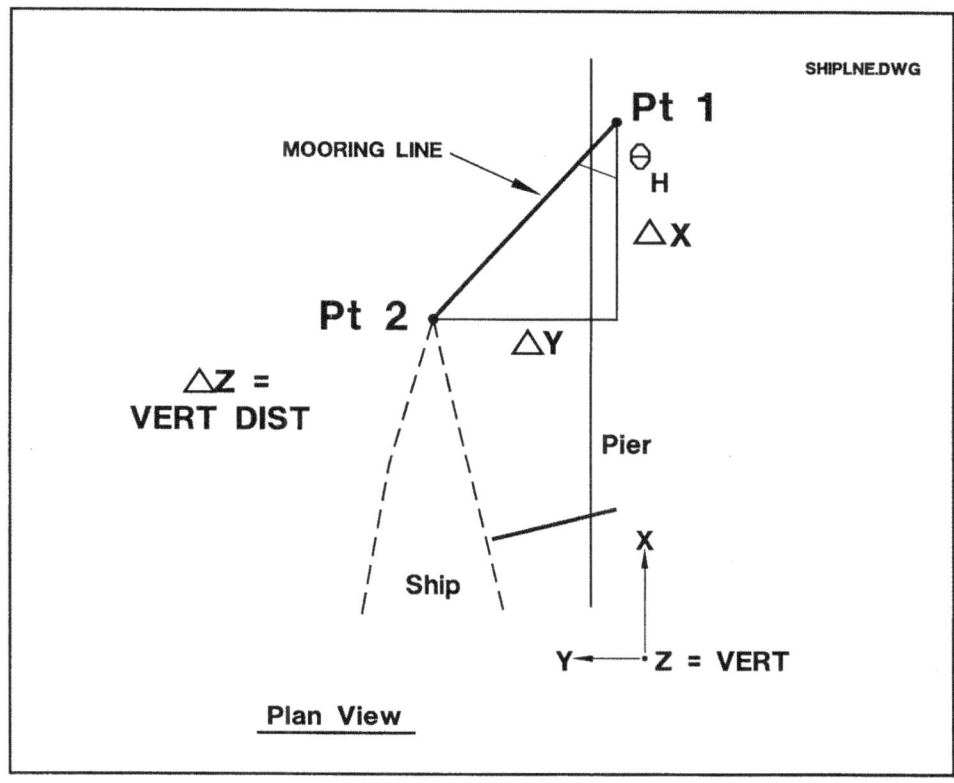

$$\theta_H = \tan^{-1} \left| \frac{\Delta Y}{\Delta X} \right|$$

$$\theta_V = \tan^{-1} \left| \frac{\Delta Z}{\Delta Y} \right|$$

IF $|\theta_H| < 45$ deg → SPRING LINE

IF $|\theta_H| \geq 45$ deg → BREASTING LINE

5. *Preliminary Analysis.* The first step for fixed mooring design is to analyze the mooring requirements for the optimum ideal mooring shown in Figure 6-17. Analyzing the optimum ideal arrangement is recommended because: (1) calculations can be performed by hand and; (2) this simple arrangement can be used as a standard to evaluate other fixed mooring configurations (NFESC TR-6005-OCN, *EMOOR - A Quick and Easy Method for Evaluating Ship Mooring at Piers and Wharves.*

Figure 6-17 Optimum Ideal Mooring (Lines are parallel to the water surface and breasting lines are spaced one-half ship's length from midships)

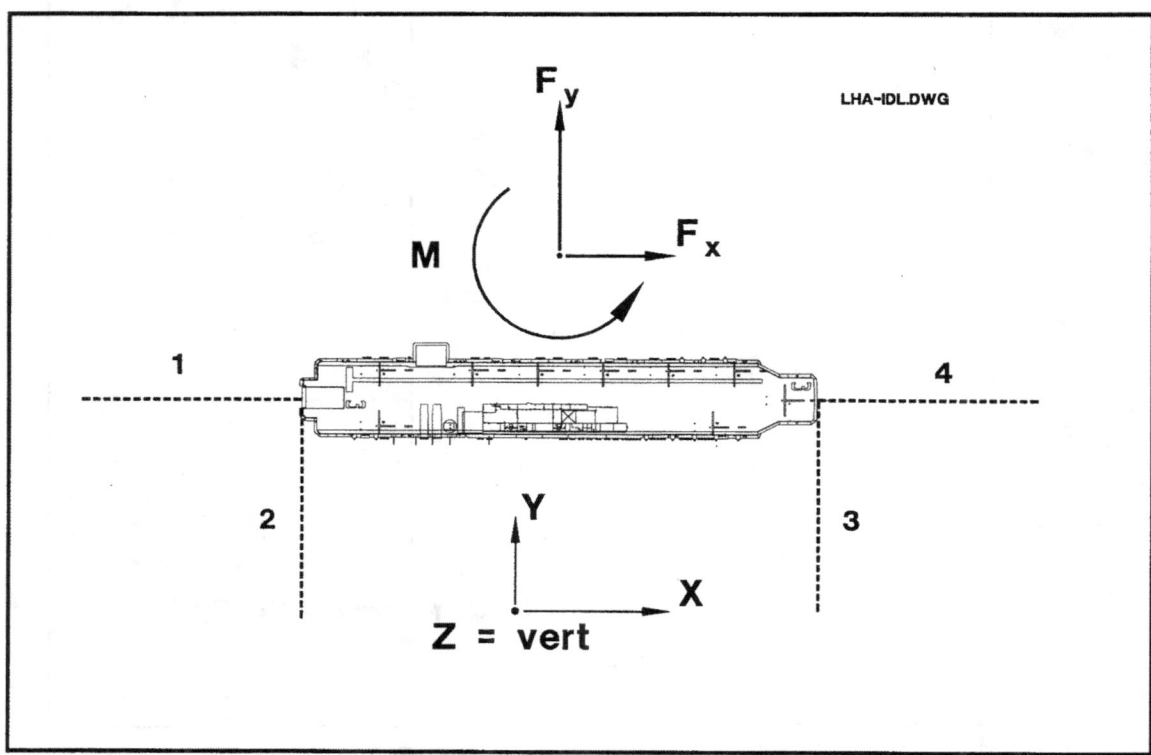

The optimum ideal mooring shown in Figure 6-16 consists of two spring lines, Lines 1 and 4, which are assumed to resist longitudinal forces. There are two breast lines, Lines 2 and 3, which are assumed to resist lateral forces and moments for winds with directions from 0 to 180°. Fenders are not shown. All lines are assumed to be parallel to the water surface in the ideal mooring.

A free body diagram is made of the optimum ideal mooring for a loaded CVN-68 in 75-mph (33.5-m/s) winds. It is found that the sum of the working mooring capacity required for Lines 1 and 4 is 174 kips (770 kN) and the sum of the working mooring capacity required for Lines 2 and 3 is 1069 kips (4.76 MN), as shown in Figure 6-18. Note that no working line capacity is required in the 'Z' direction, because the ship's buoyancy supports the ship. The sum of all the mooring line working capacities for the optimum ideal mooring is 1243 kips (5.53 MN).

Figure 6-18 Required Mooring Capacity Using the Optimum Ideal Mooring

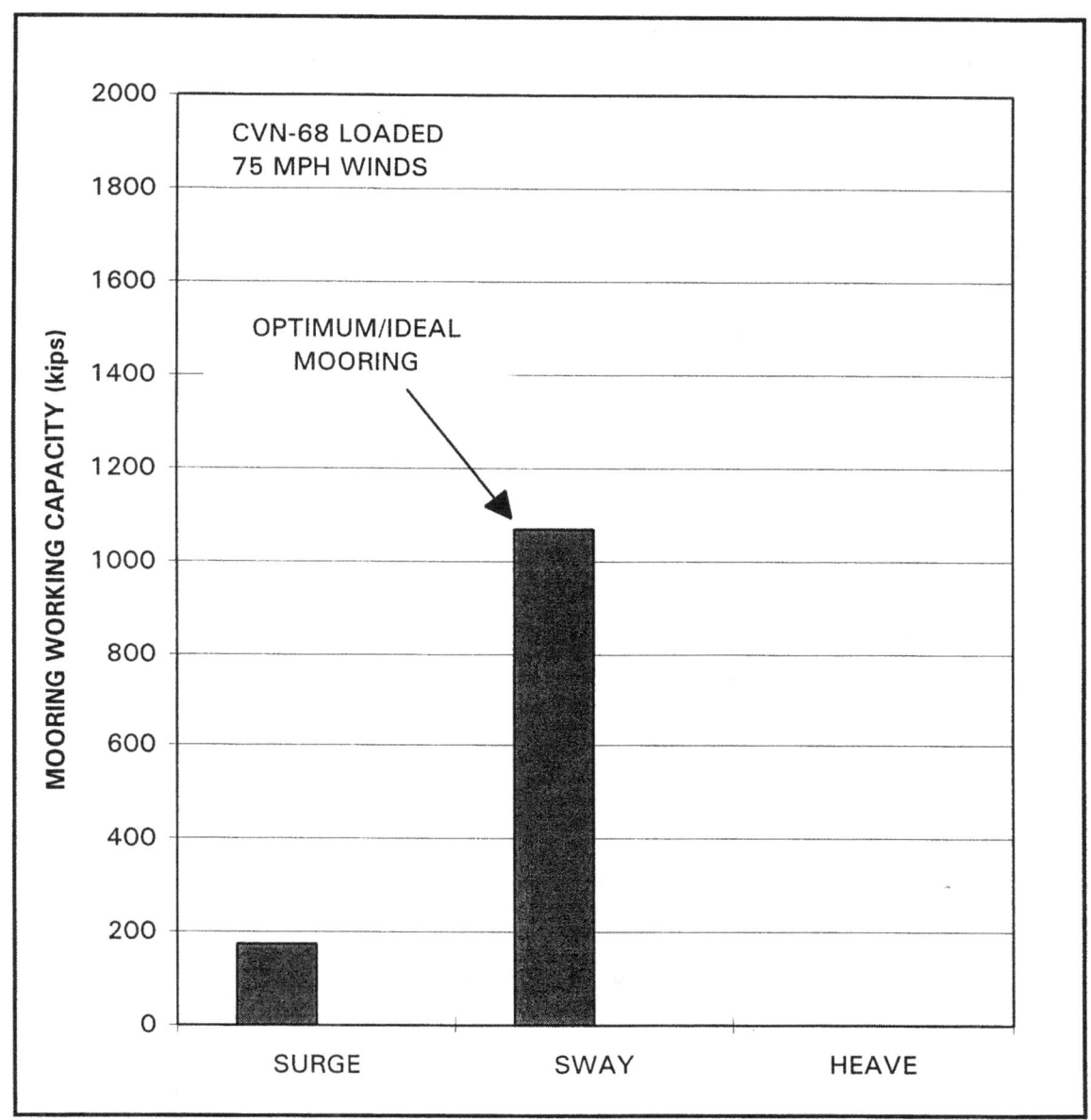

6. *Wharf Mooring Concept.* Camels and fenders are located between the wharf and ship to offset the ship in this design. Also, the wharf breasting line bollards are set back from the face of the wharf, so that the vertical angles of the breasting lines are approximately 10°. Figure 6-19, from a study of a number of ship moorings at piers and wharves (NFESC TR-6005-OCN) is used to estimate that a mooring system using synthetic lines will have an efficiency of approximately 0.67 for the case of breasting lines with a 10-degree vertical angle. The estimated total required working mooring line capacity is the working line capacity of the optimum ideal mooring divided by the efficiency. In this case, the estimated working line capacity required is 1243 kips/0.67 or approximately 1855 kips.

Figure 6-19 Efficiency of Ship Moorings Using Synthetic Lines at Piers and Wharves (after NFESC TR-6005-OCN)

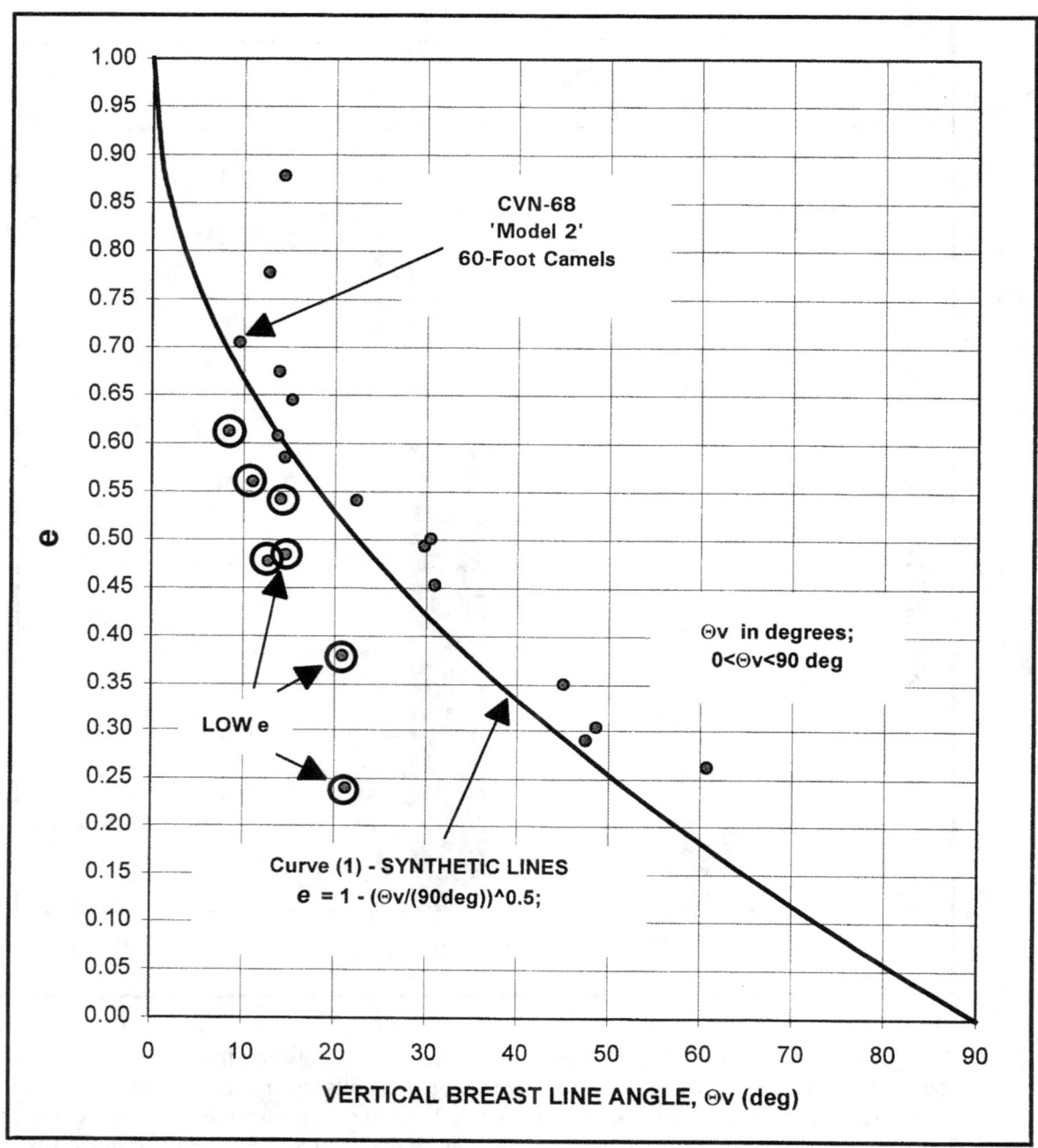

For extra safety, the selected concept 'Model 2' is given 11 mooring lines of three parts each of aramid mooring line, as shown in Figure 6-20. A single part of line is taken as having a break strength of 215 kips (920 kN). These lines have a combined working strength of 11*3*215/3 = 2365 kips with a factor of safety of 3. These lines are selected to provide extra safety. A component analysis, Figure 6-21, suggests that this mooring concept has adequate mooring line capacity in the surge and sway directions.

Figure 6-20 CVN-68 Wharf Mooring Concept ('Model 2')

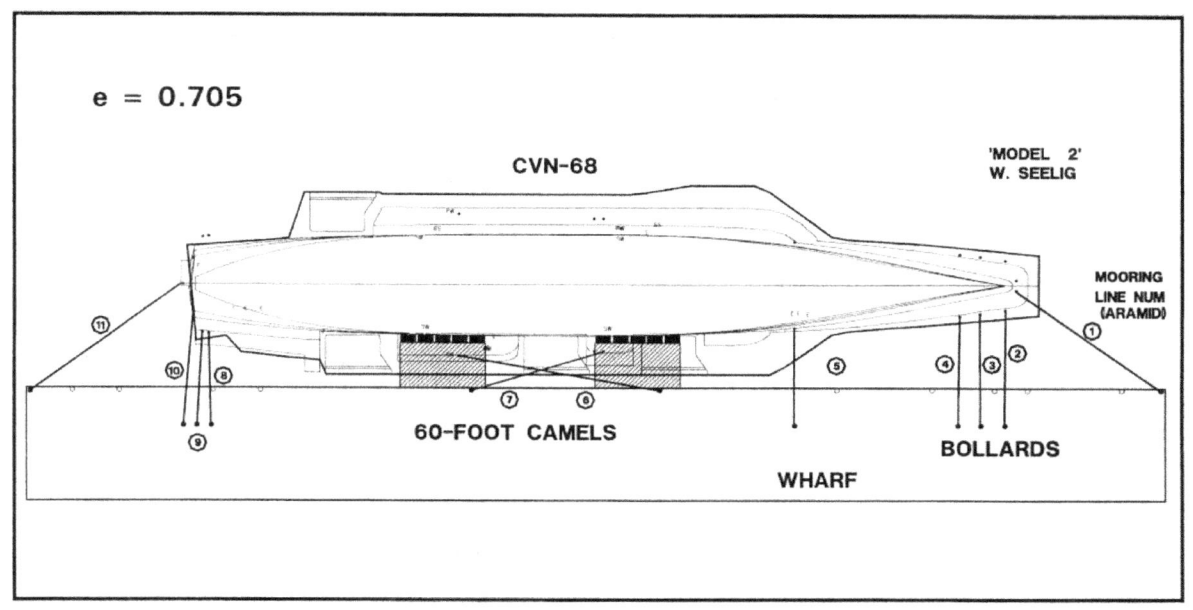

Figure 6-21 Component Analysis of Mooring Working Capacity

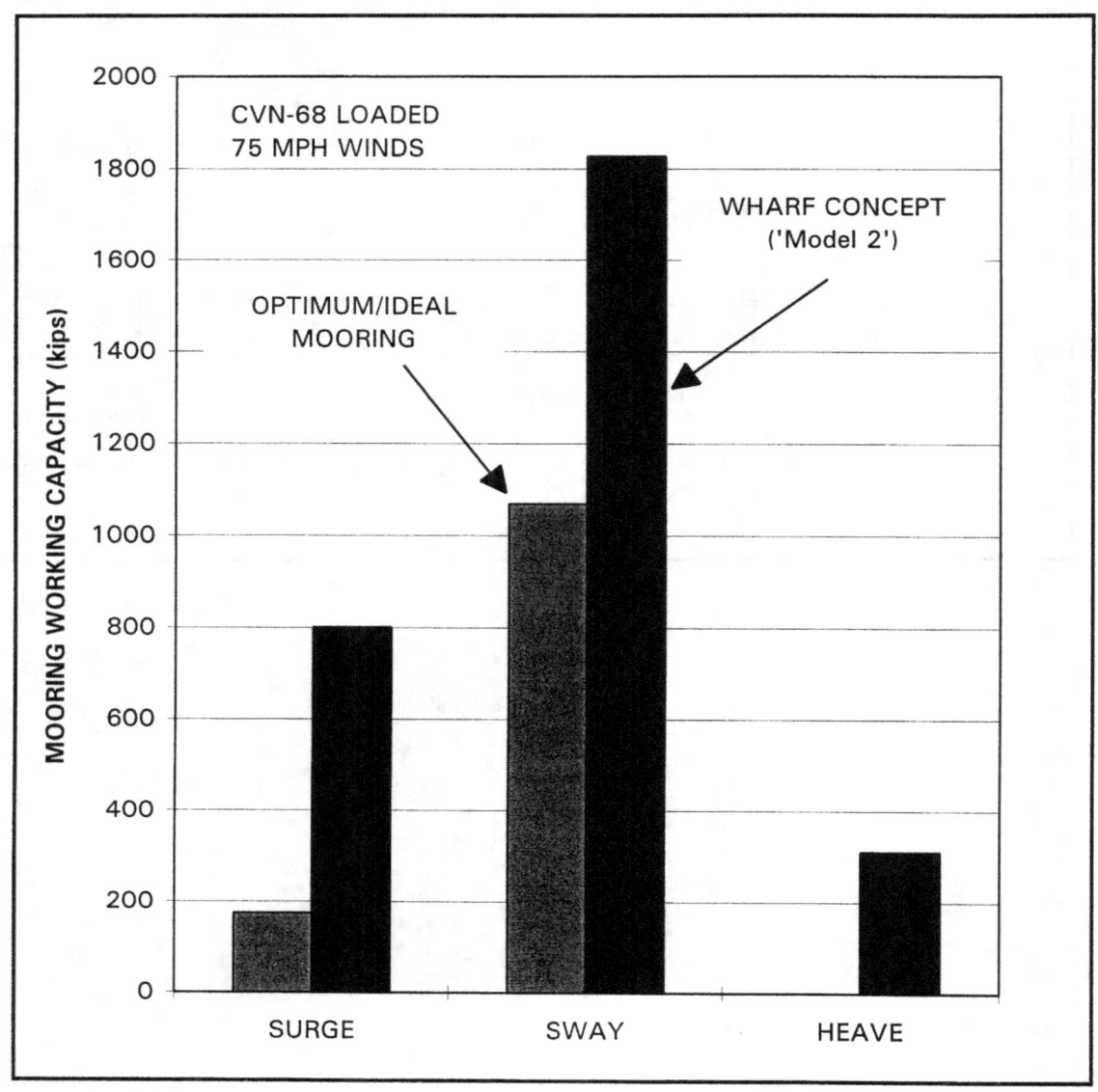

A computer software program (W.S. Atkins Engineering Sciences, *AQWA Reference Manual*) using a fixed mooring performs quasi-static analyses. Analyses are performed for various wind directions around the wind rose. Results show that the mooring line factors of safety are larger than the required minimum of 3 (i.e., line tensions divided by the new line break strength is less than 0.33), as shown in Figure 6-22. This extra safety is justified, because the ship is nuclear powered. In this concept the spring lines are especially safe with a factor of safety of about 10. These analyses show ship motions of approximately 1 foot (0.3 meter) under the action of the 75-mph (33.5-m/s) design winds.

Figure 6-22 Mooring Line Tensions for a CVN-68 Moored at a Wharf With 75 mph (33.5 m/s) Winds ('Model 2')

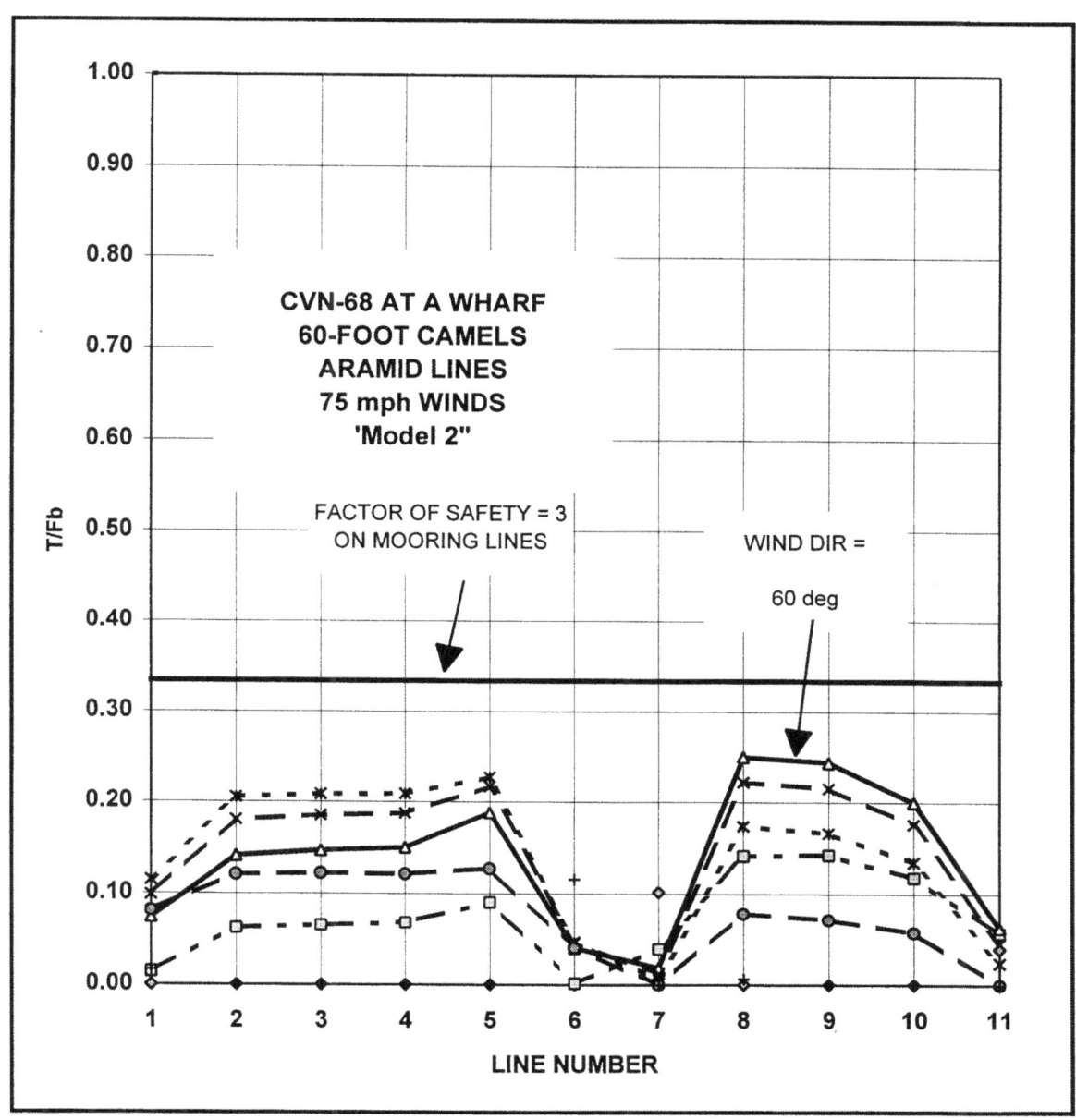

Further quasi-static analyses show this concept is safe in up to 87-mph (38.9-m/s) winds with a factor of safety of 3 or more on all the mooring lines. The computed mooring efficiency for 'Model 2' at this limiting safe wind speed is 0.705, which is slightly higher than the estimated value of 0.67, as shown in Figure 6-19.

These preliminary calculations show that this fixed mooring concept could safely secure the ship. Figure 6-23 illustrates the mooring concept in perspective view. Further information on this example is provided in NFESC TR-6004-OCN, *Wind Effects on Moored Aircraft Carriers*.

Figure 6-23 Aircraft Carrier Mooring Concept (perspective view)

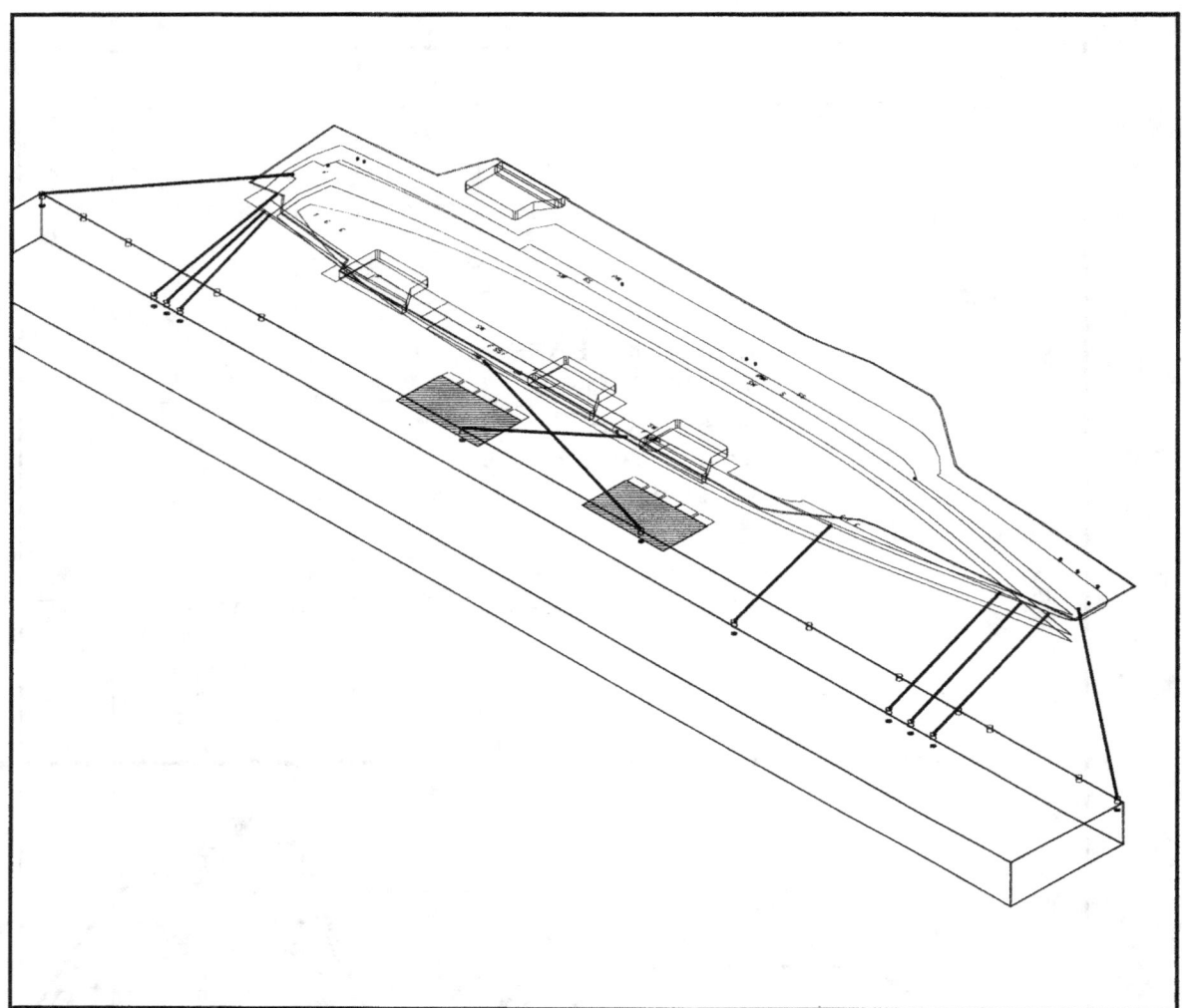

Example 6-5 *Spread Mooring - Basic Approach.*

Design of a spread mooring for a nest of ships is illustrated in this section. SPRUANCE class (DD 963) destroyers are scheduled for inactivation and a mooring is required to secure four of these vessels (NFESC SSR-6119-OCN, *D-8 Mooring Upgrade Design Report*). These ships are inactive and cannot go out to sea, so the mooring must safely secure the vessels in a hurricane using Mooring Service Type IV design criteria. At this location, wind is the predominant environmental factor of concern. At this site the tidal range and tidal current are small. Soil conditions at the site consist of an upper soft silty layer between 50 to 80 feet in depth (15 to 24 meters) over a stiff clay underneath. Water depth at the site ranges between 31 to 35 feet (9.4 to 10.7 meters) MLLW.

1. *Goal.* Develop a concept to moor four DD 963 class destroyers in a spread mooring. Use Mooring Service Type IV criteria and a design wind speed of 78.3 mph (68 knots or 35 m/s).

2. *Ship.* The ships are assumed to be at one-third stores/cargo/ballast condition, since DD-963 vessels are unstable in the light condition. Table 6-8 gives some ship parameters. Additional information is found in the NAVFAC *Ships Characteristics Database*.

Table 6-8 DD 963 Criteria (1/3 Stores)

PARAMETER	DESIGN BASIS (SI units)	DESIGN BASIS (English units)
Length Overall At Waterline	171.9 m 161.2 m	564 ft 529 ft
Beam @ Waterline	16.8 m	55 ft
Average Draft Draft at Sonar Dome	6.5 m 8.8 m	21.2 ft 29 ft
Displacement	9070 Mg	8928 long tons
Chock Height From Baseline	10.7m stern 15.9m bow	35 ft 52 ft

6 *Forces/Moments*. Methods described elsewhere in this book are used to compute the forces and moments on the ships. These values are summarized in Figure 6-24. Wind angles are based on the local coordinate system for a ship shown in Figure 5-3.

Figure 6-24 Wind Forces and Moments on a Nest of Four DD 963 Class Vessels for a Wind Speed of 78 mph (35 m/s)

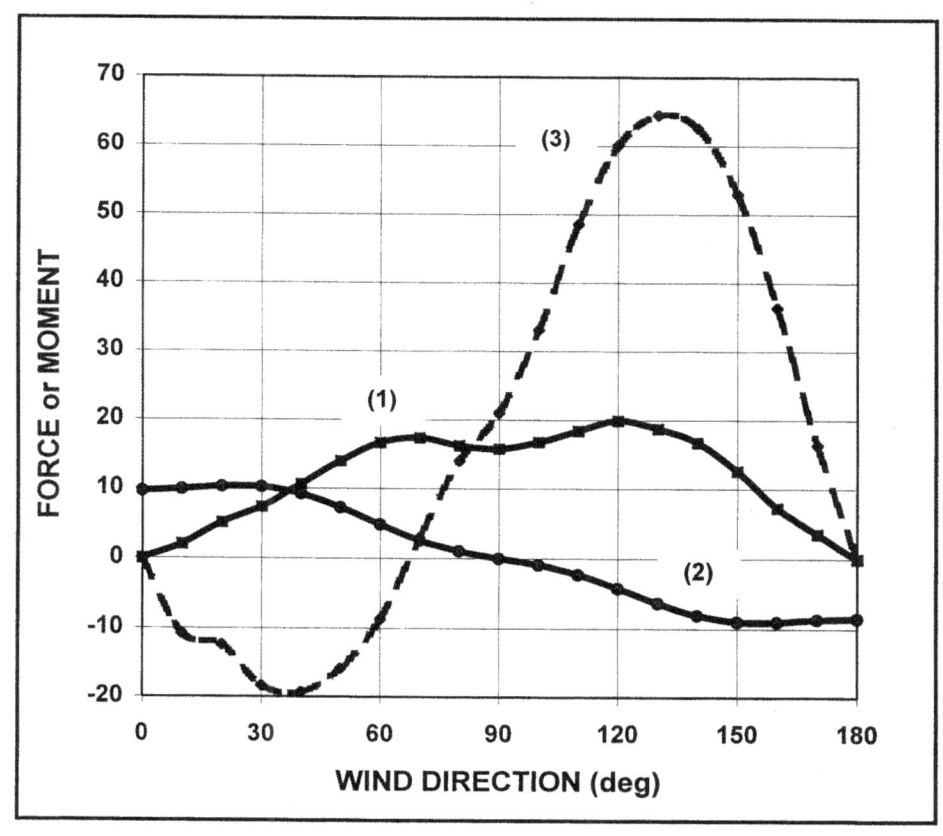

	CURVE	GRAPH LIMITS	MAX. VALUE
(1)	Transverse Wind Force	E5 N	447.4 kips
(2)	Longitudinal Wind Force	E5 N	231.3 kips
(3)	Yaw Moment	E6 N	47643 kip-ft

Note that wind tunnel model tests show that there is significant sheltering in the transverse direction of downwind ships in this nest of identical ships, as shown in Appendix A. However, there is little wind sheltering in the longitudinal direction. Table 6-9 summarizes the environmental force calculations used for this example.

Table 6-9 Environmental Forces

Condition	Load (Metric)	Load (US)	Comments
Single DD 963	1663.8 kN	374 kips	Transverse Wind
	257 kN	57.82 kips	Longitudinal Wind
	35972 m-kN	26531 ft-kips	Wind-Yaw Moment
	104.2 kN	23.4 kips	Transverse Current
	2.5 kN	0.56 kips	Longitudinal Current
	1216 m-kN	863.1 ft-kips	Current Yaw Moment
4 ea DD 963	1989.9 kN	447.4 kips	Transverse Wind
	1028.7 kN	231.3 kips	Longitudinal Wind
	64595 m-kN	47643 ft-kips	Wind-Yaw Moment
	190.6 kN	42.8 kips	Transverse Current
	9.8 kN	2.2 kips	Longitudinal Current
	3342 m-kN	2372.7 ft-kips	Current Yaw Moment

7   *Anchor Locations.* Driven-plate anchors are selected as a cost-effective method to safely moor the nest of ships. The soils at the site are soft harbor mud of depths between 50 to 80 feet (15 to 24 meters), so a chain catenary will form below the seafloor (in the mud) as well as it the water column, as illustrated in Figure 6-10. A horizontal distance of 100 feet (30 meters) between the anchor location and the chain daylight location (point where the anchor leg chain exits the seafloor) is estimated based on Chain Soil Analysis Program (CSAP) modeling of the chain catenary in the soil and in the water column.

To ensure the mooring legs are efficient in resisting the imposed environmental horizontal forces, a target horizontal distance of 170 feet (52 meters) is chosen between the predicted daylight location (where the chain exits the soil) and the attachment point on the ship for each of the mooring legs. Therefore, anchor locations are established at a horizontal distance of 270 feet (82 meters) away from the vessel.

8   *Definitions.* In this example, a local ship and a global coordinate system are defined. The local ship coordinate system is used to determine environmental loads at the various wind and current attack angles, as shown in Figure 27, with the origin of the "Z" direction at the vessel keel. A global coordinate system for the entire spread mooring design is selected with the point (0,0,0)

defined to be at a specific location. For this example, the origin is selected to be in the middle of the vessel nest and 164 feet (50 meters) aft of the stern of the vessels. The origin for the "Z" direction in the global coordinate system is at the waterline. The various analysis programs use this global coordinate system to define the "chain daylight" locations and the location of the vessel center of gravity within the spread mooring footprint.

9 *Number of Mooring Legs.* It is estimated that eight 2.75" chain mooring legs are required, based on the safe working load the chain (289 kips or 1.29 MN) and the applied environmental forces and moments on the nest of ships. Four legs are situated on both sides of the nest and each mooring leg is angled to be effective in resisting the longitudinal wind forces, as well as lateral wind forces and moments, from winds approaching at angles other than broadside. Legs are also placed toward the ends of the nest to be effective in resisting the yaw moment. To help control ship motions, two 20-kip (9000-kg) sinkers are placed on each mooring leg approximately midway between the vessel attachment point and the predicted chain daylight location. A schematic of the planned spread mooring arrangement is shown in Figure 6-25.

Figure 6-25 Spread Mooring Arrangement for a Nest of Four Destroyers

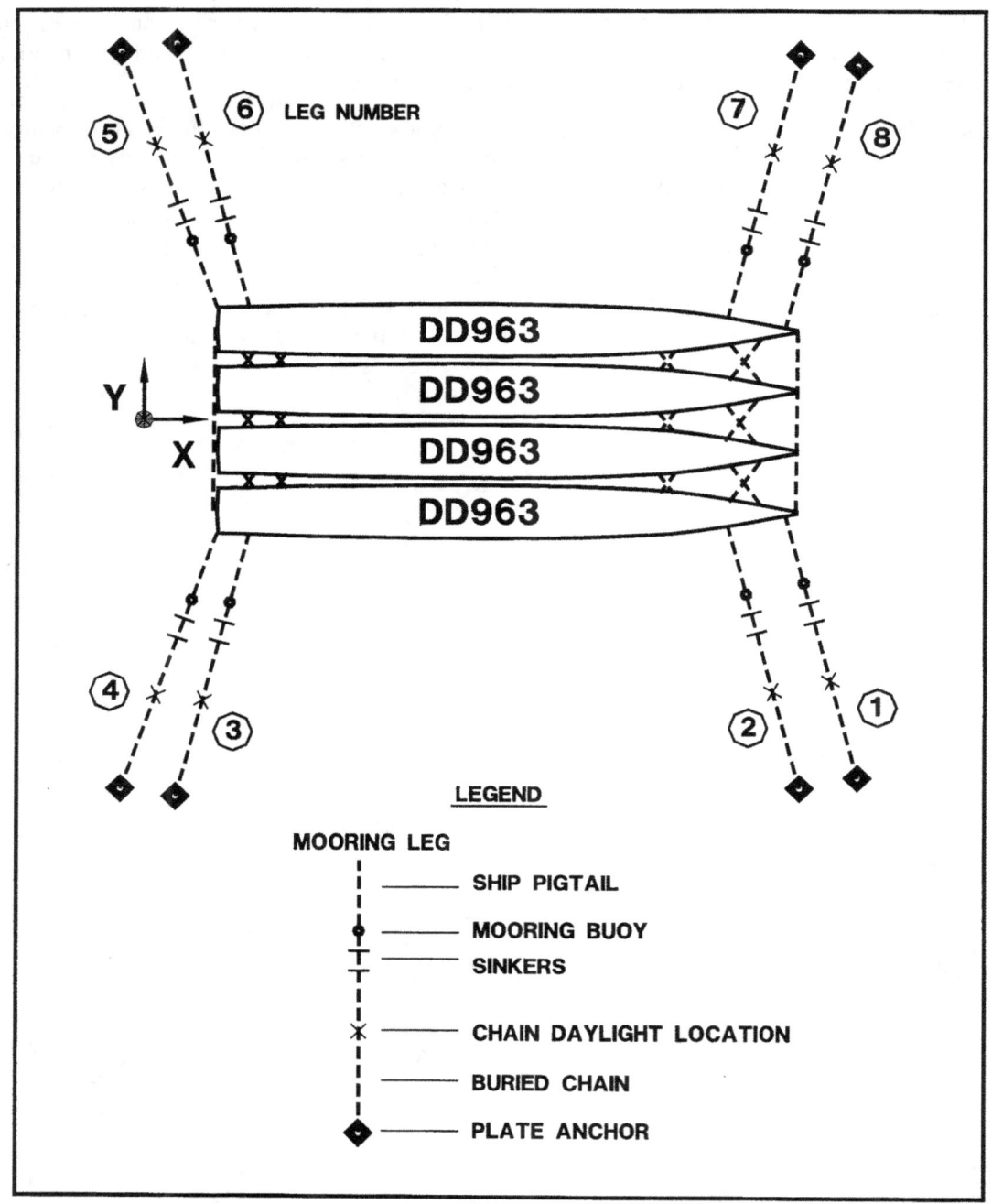

10 *Static Analysis*. A quasi-static analysis is performed on the mooring system using a mooring analysis program (W.S. Atkins Engineering Sciences, *AQWA Reference Manual*). Each mooring leg is initially pretensioned to a tension of 3.6 knots (10 kips). Quasi-static analysis is performed for various combinations of wind and current directions. Quasi-static results for various wind directions in conjunction with a 60-degree flood tidal current of 0.6 knots (0.31 m/s) are shown in Table 6-10.

Table 6-10 Quasi-Static Leg Tensions for the Spread Mooring at Various Wind Directions With a Flood Tidal Current

	Wind Direction						
LEG	0	30°	60°	90°	120°	150°	180°
	kN						
1	52.49	214.99	447.01	609.02	**945.05**	866.04	541.00
2	-	62.50	347.99	486.02	769.03	927.03	571.03
3	693.00	941.04	844.02	588.02	560.00	343.99	93.50
4	668.00	808.04	611.02	387.00	255.02	45.60	-
5	622.01	490.03	84.52	-	-	-	-
6	563.02	454.00	64.72	-	-	-	-
7	-	-	-	-	-	220.99	449.01
8	-	-	-	-	-	309.02	564.00
	Kips						
1	11.8	48.33	100.49	136.91	**212.45**	194.69	121.62
2	-	14.05	78.23	109.26	172.88	208.4	128.37
3	155.79	211.55	189.74	132.19	125.89	77.33	21.02
4	150.17	181.65	137.36	87	57.33	10.25	-
5	139.83	110.16	19	-	-	-	-
6	126.57	102.06	14.55	-	-	-	-
7	-	-	-	-	-	49.68	100.94
8	-	-	-	-	-	69.47	126.79

\- Indicates that the leg does not get loaded

A maximum load of 945 knots (212 kips) occurs on Leg 1 at a wind direction of 120°. This provides a quasi-static factor of safety of approximately 4 to the breaking strength of 2.75" FM3 chain.

11 *Dynamic Analysis.* A dynamic analysis is performed on the mooring system to evaluate peak mooring loads and vessel motions using a mooring analysis program (W.S. Atkins Engineering Sciences, *AQWA Reference Manual*). The initial location of the vessel nest is based on the equilibrium location of the vessel nest determined in the quasi-static analysis. An Ochi-Shin wind spectrum is used to simulate the design storm (*Wind Turbulent Spectra for Design Consideration of Offshore Structures*, Ochi-Shin, 1988). This simulation is performed for a 60-minute duration at the peak of the design storm.

Figure 6-25 shows that the four vessels in the nest are close together and Figure 6-26 shows that the ships have a large ratio of ship draft to water depth. In this case it is estimated that the ships will capture the water between them as the ships move. Therefore, the nest of moored ships was modeled as a rectangular box having a single mass with the dimensions of 161.2 meters (length of each ship at the waterline), 71.62 meters wide (four ship beams + 5 feet spacing between ships), and 6.5 meters deep (average vessel draft). Added mass for sway and surge was computed as if the nest was cylindrical in shape with a diameter equal to the average draft. Damping as a function of frequency was estimated from a diffraction analysis (W.S. Atkins Engineering Sciences, *AQWA Reference Manual*).

Figure 6-26 End View of DD 963 Mooring Nest

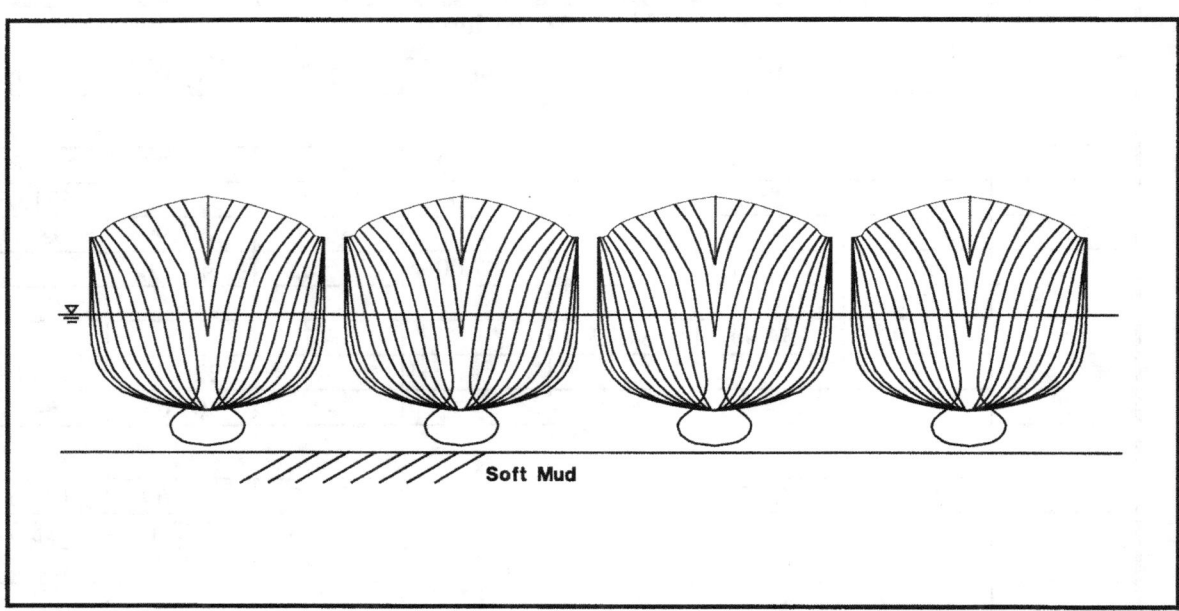

Dynamic analyses were performed for various combinations of wind and current directions using a wind speed time history that simulated the design storm. Results showing the instantaneous peak tensions for various wind directions in conjunction with a flood tidal current of 0.6 knots (0.31 m/s) are shown on Table 6-11.

Table 6-11 Peak Dynamic Chain Tensions for DD 963 Nest for Various Wind Directions and a Flood Tidal Current

	Wind Direction						
LEG	0	30^0	60^0	90^0	120^0	150^0	180^0
	kN						
1	167.05	288.9	634.03	828.68	**2246.2**	1848.5	731.73
2	55.089	174.58	430.31	545.27	1067.2	1152.1	720.54
3	1202.5	1625	995.98	818.81	1370	647.78	210.62
4	1362.2	1651.7	653.82	480.82	486.77	240.56	-
5	1284.2	1356.4	219.12	-	-	-	-
6	938.06	901.87	217.04	-	-	-	-
7	-	-	-	-	55.019	374.91	514.26
8	-	-	-	-	170.54	485.43	834.54
	Kips						
1	37.55	64.95	142.53	186.29	**504.95**	415.55	164.50
2	12.38	39.25	96.74	122.58	239.91	259.00	161.98
3	270.33	365.31	223.90	184.07	307.98	145.62	47.35
4	306.23	371.31	146.98	108.09	109.43	54.08	-
5	288.69	304.92	49.26	-	-	-	-
6	210.88	202.74	48.79	-	-	-	-

| 7 | - | - | - | - | 12.37 | 84.28 | 115.61 |
| 8 | - | - | - | - | 38.34 | 109.13 | 187.61 |

Modeling shows that the instantaneous peak chain tension of 2246 knots (505 kips) is predicted on Leg 1 as the moored vessel nest responds to wind gusts. This provides a peak instantaneous factor of safety of 1.5 on the breaking strength of the selected chain size. For this example, the peak dynamic chain tension during the 1 hour at the peak of the design storm is 2.4 times the quasi-static tension in the mooring leg with the highest tension, Leg 1.

Nest motions for surge, sway, and yaw are provided in Table 6-12. This table shows that the maximum surge of the vessel nest is approximately 7.4 meters (24.3 feet) from its equilibrium condition at no loading. Maximum sway and yaw of the vessel nest is 3.2 meters (10.5 feet) and 1.59° clockwise, respectively. During a dynamic analysis simulation, nest motions oscillated up to 5.4 meters (17.7 feet) in surge (for a wind direction coming from the stern), 1.9 meters (6.2 feet) in sway (for a wind direction 30° aft of broadside), and 2.1° in yaw (for a wind direction 30° off the stern).

Table 6-12 DD 963 Nest Motions for Surge, Sway, and Yaw at Various Wind Directions with a Flood Tidal Current

Motion	Wind Direction						
	$0°$	$30°$	$60°$	$90°$	$120°$	$150°$	$180°$
Surge (meters)							
Origin	98.17	98.17	98.17	98.17	98.17	98.17	98.17
Start	105.6	105.4	103.6	98.1	93.7	89.2	88.1
Max	106.9	106.8	103.9	98.8	95.1	93.4	93.5
Min	102.3	102.3	102.4	98.1	93.7	89.2	88.1
Diff	4.6	4.5	1.5	0.7	1.4	4.2	5.4
Sway (meters)							
Origin	0.0	0.0	0.0	0.0	0.0	0.0	0.0
Start	0.84	1.49	2.39	2.97	1.27	2.02	1.14
Max	0.84	1.49	2.65	3.13	3.22	2.50	1.45
Min	0.52	0.83	0.93	1.35	1.27	1.43	1.11
Diff	0.32	0.66	1.72	1.78	1.93	1.07	0.34
Yaw (degrees)							
Origin	0.0	0.0	0.0	0.0	0.0	0.0	0.0
Start	0.76	1.09	1.43	0.64	-0.08	-0.74	-0.89
Max	0.76	1.18	1.59	0.80	-1.22	-1.49	-1.12
Min	0.38	0.27	0.43	-0.25	0.76	0.54	-0.83
Diff	0.38	0.91	1.16	1.05	1.96	2.03	0.29

12 *Anchor Design.* Using the quasi-static design mooring leg tension, anchor capacity and loads on the embedded plate anchor are calculated using procedures outlined in NFESC TR-2039-OCN, *Design Guide for Pile-Driven Plate Anchors* and NFESC CR-6108-OCN, *Anchor Mooring Line Computer Program Final Report, User's Manual for Program CSAP2*. Due to the lower shear strengths of the soft silty upper layers at the site, a 6-foot by 11-foot mud plate anchor, as described in Table 36 (par. 5.3) is specified. A design keyed depth of 55 feet is selected for the plate anchor. This will provide an estimated static holding capacity of 1913 kN (430 kips).

CSAP is used to predict the mooring leg tension at the anchor. Input requirements of CSAP include: (1) mooring leg configuration between the anchor and the buoy or chock; (2) water depth or height of chock above the seafloor; (3) soil profiles and strength parameters; (4) location and size of sinkers; (5) horizontal tension component of the mooring leg at the buoy or chock; (6) horizontal distance or total length of the mooring leg between anchor and buoy or chock; and (7) anchor depth. Output provided by CSAP includes: (1) chain catenary profile from the anchor to the buoy or chock attachment point; (2) angle of the mooring leg from the horizontal at the anchor, the seafloor, and the buoy or chock; (3) tension of the mooring leg at the anchor, seafloor, and at the buoy or chock; (4) predicted daylight location for the mooring leg; and (5) length of mooring leg required or horizontal distance between anchor and buoy or chock.

For this example, a keyed anchor depth of 55 feet was selected. Input data included: (1) configuration of the mooring leg (30 feet of 3" wire attached to 2+ shots of 2.75" chain); (2) height of seafloor to vessel chock (46 feet stern and 64 feet bow); (3) soil profile and strength for the site (shear strength increases linearly at 10 pounds per ft^2 per foot of depth); (4) information on the sinkers (2 each 20-kip sinkers placed a horizontal distance of 170 feet away from the anchor; (5) horizontal tension component of the mooring leg from the quasi-static results (195 kips); (6) horizontal distance between anchor and chock (280 feet) from the quasi-static results; and (7) depth of anchor (55 feet).

CSAP results for this design leg at this anchor depth indicate that the predicted daylight location of the mooring leg is approximately 99 feet (30 meters) from the anchor location and the leg tension at the anchor is 166 kips. A profile of this leg is shown in Figure 6-10. Note that the interaction between the chain and the soil accounts for a 25 percent reduction in tension on the mooring leg at the anchor. This gives a predicted quasi-static anchor holding factor of safety of 2.6.

Based on the CSAP results, 6-foot by 11-foot plate anchors are specified. Based on predicted keying distances required for this anchor, as outlined in NFESC TR-2039-OCN, *Design Guide for Pile-Driven Plate Anchors*, the anchors should be installed to a tip depth of 70 feet (21 meters) below the mudline to ensure that the anchor is keyed at a minimum depth of 55 feet (16.8 meters). Figure 6-27 provides a comparison between tip depth, keyed depth and ultimate capacity for this size anchor.

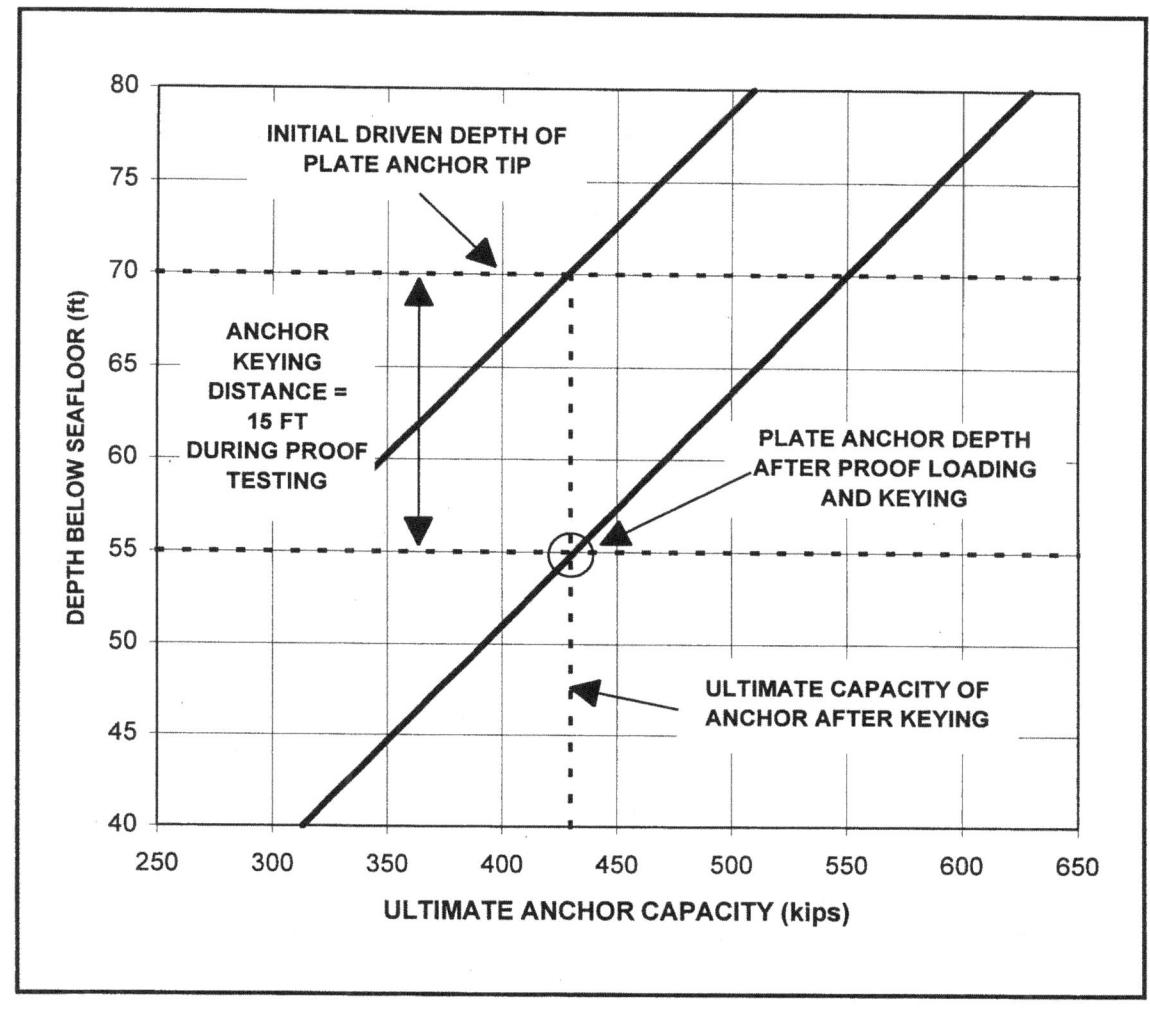

Figure 6-27 Plate Anchor Holding Capacity (6-foot x 11-foot anchor with keying flaps in soft mud)

Chapter 7. Overview Of Fleet Mooring Maintenance

7.1. Introduction

Fleet mooring maintenance includes all actions taken to ensure that moorings are safe, reliable, and in satisfactory condition. Maintenance may range from a simple annual surface inspection of a buoy to a complex operation involving recovery, refurbishment, and reinstallation of an entire mooring.

Mooring maintenance actions will include the following:

- Minor above-water repairs of buoys and topside mooring components.
- Removal of buoys for painting or repairs ashore.
- Replacement of the riser chain because of wear or corrosion.
- Replacement of depleted zinc anodes.
- Inspection and preservation of ashore inventory.
- Underwater inspections of installed moorings.
- Repair or relaying of a mooring damaged by a collision or a severe storm.

7.2. Inspections

Inspections are perhaps the most important, but often the most neglected, of all maintenance performed. One of the primary purposes of inspections is to detect any deficient conditions that require immediate remedial attention. Often overlooked is that inspection results should be used to plan future maintenance of both installed moorings and inventory ashore. If maintenance is performance on a routine basis, costs may be higher than need be because action is taken sooner than required. On the other hand, if maintenance is performed without regard for inspection results, the importance of a particular inspection may be diminished in the inspector's view. If this happens, even critical repair requirements may be overlooked.

The importance of inspection results as a maintenance-planning tool will increase as more long-life upgraded moorings are introduced into the system. These moorings are designed to remain in service for extended periods of time, thus making possible a reduction in expenditures for both maintenance and new material. If the improved components are regularly replaced on a schedule, based on the maintenance needs of older components, then overall mooring maintenance costs will surely decline. A well planned and aggressively managed maintenance program, supported by a comprehensive inspection plan, is the most effective way to keep costs down while continuing to provide reliable service to the fleet.

Chapter 8. Mooring Installation And Recovery

8.1. Preparation

8.1.1. General.

This chapter describes methods for installing and recovering moorings in sheltered waters and provides the general procedures for accomplishing these tasks. Installation and recovery of both riser and non-riser buoy systems are covered. Procedures set forth may be modified to suit local environmental conditions and/or material constraints, or to accommodate a special mooring design. The procedures developed herein assume that the tasks will be accomplished using a floating crane or crane barge for installation and recovery.

8.1.2. Planning And Preparation.

Development of a plan is a necessary requirement needed and the procedures necessary for accomplishment of the task. Design drawings, as well as materials required, should also be included. Recovery plans should take into account the as-built drawings. The plan should be prepared well enough in advance to allow riggers, crane and barge operators, design engineers, and others to review and comment on the details. Prior to commencement of offshore operations, all persons involved should understand the general sequence of events that will occur and their specific responsibilities in the effort. The plan should also take into consideration the following:

8.1.2.1. Preparation

Thorough preparation is one of the principal keys to a successful mooring installation or recovery operation. Proper ashore preparations will ensure that offshore operations will be expeditiously and efficiently carried out. As examples:

1. A floating crane will be provided for the installation or recovery of mooring components. The size and placement of required temporary padeyes must be determined and action taken to install these padeyes on the deck of the barge.
2. Physical fit checks of all mooring materials should be accomplished to ensure proper assembly of the mooring in accordance with design specifications. If being installed, chain stud anodes should be attached to the chain prior to moving the material offshore.
3. After anode installation, caution must be taken in handling the chain to preclude damage to or loss of the anodes.

8.1.2.2. Environmental Factors

Since offshore operations can be strongly influenced by environmental conditions, plans should be made flexible enough to allow for unexpected delays due to inclement weather or rough seas.

8.1.3. Tools And Equipment Required

The floating crane or other suitable platform will be stocked with all tools needed not only for the planned operation, but also for any field modifications or in-place repairs that may be required. Tools, equipment, and materials for the effort should include the following items:

1. Welding and cutting equipment.
2. High-pressure seawater pump (100 psi) or a water blaster.
3. Spare parts (e.g., locking pins, punches, retaining pellets, etc.). A good source is BUSHIPS No. 52603-840327.
4. Lubricants, preservative greases, coatings, etc.

8.1.4. Pre-Installation Layout

Mooring components will be laid out on the crane barge, prior to offshore operations, in an arrangement governed by procedures established for installation of the mooring. Chains shall be placed so that they will readily pay out during a controlled installation. Figure 8-1 illustrates a typical mooring material/component layout on a crane barge. The mooring should be assembled on the barge to the maximum extent possible prior to departing for installation operations to minimize assembly activity offshore.

Figure 8-1 Typical Layout of Mooring Equipment Aboard a Crane Barge

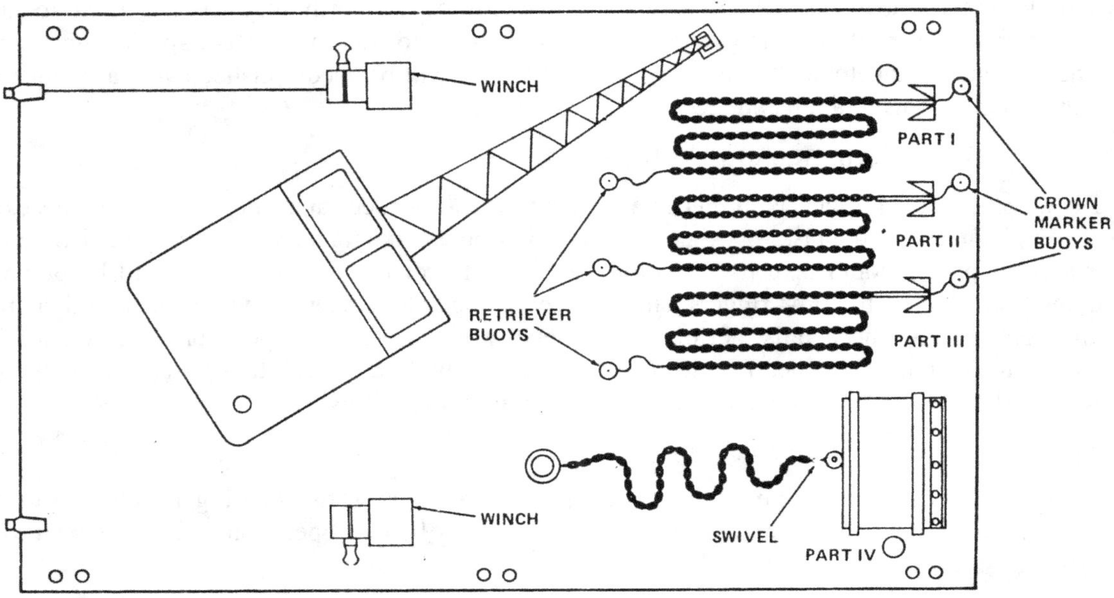

8.1.5. Pre-Installation Positioning

A mooring is designed for installation at a specific site. The characteristics of the site (e.g., water depth, bottom conditions) will determine the mooring design features such as length of riser and anchor chain subassemblies and fluke angle of the anchor. It is important, therefore, that the mooring be placed in the location for which it was designed, since a misplaced mooring would require the expense and effort of reinstalling the entire system. Equipment used for positioning the mooring is described below.

8.1.5.1. Marker Buoys

To ensure that the mooring is installed in the right location and that the anchor chain assemblies are properly orientated, marker buoys should be placed at the desired points at the site before installation begins. A center marker buoy will be used to indicate the position of the buoy; ring markers will be used to indicate anchor locations; and range markers will be used to assist in anchor chain subassembly orientation (see Figure 8-2). Correct placement of anchors will ensure that the anchor legs are straight and taut. Range buoys may be used in conjunction with the ring markers to aid crane barge movement along proper lines of bearing as it lays the anchor chain subassemblies.

Figure 8-2 Typical Center, Ring, and Range Marker Buoy Configuration

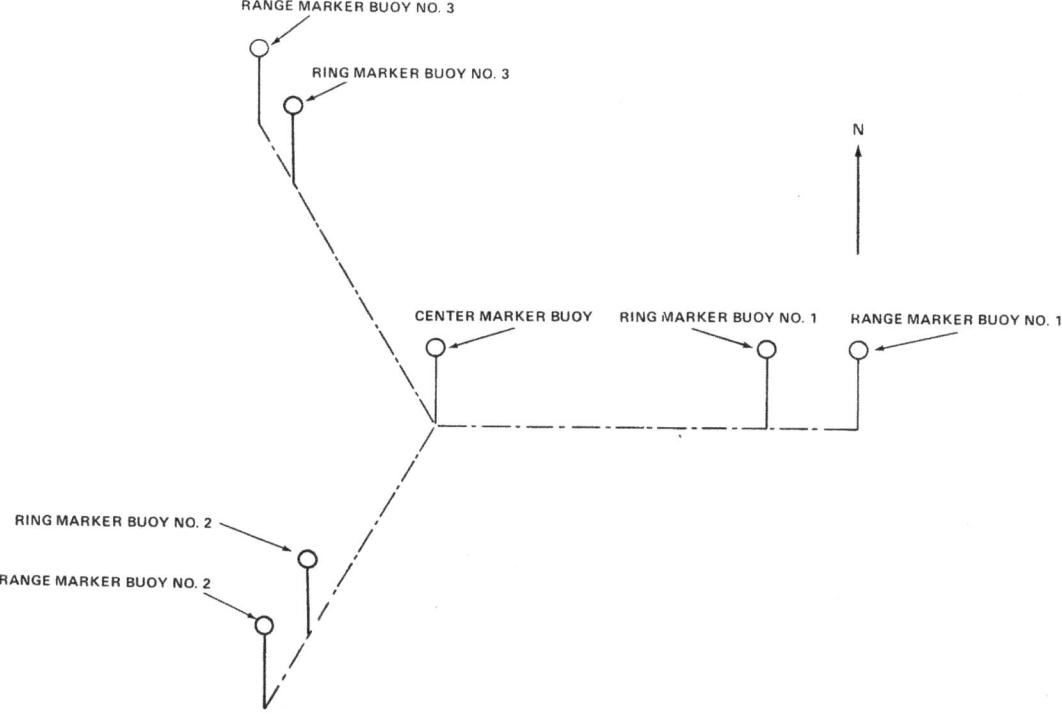

8.1.5.2. Other Positioning Equipment

Correct positioning of marker buoys is determined by the use of transits, theodolites, electronic distance measuring (EDM) devices, and other similar equipment. Marker buoy locations should be referenced to established benchmarks or fixed landmarks such as beacons, towers, smokestacks, or corners of buildings. If the mooring is to be installed close to a pier, a dolphin, or a fixed structure, the correct position of the markers can be determined by direct measurement using a measured line or similar aid. Marker buoys may be placed from a small boat or any other suitable vessel. Marker buoys should be checked before and during mooring installation to verify that they have remained in their correct positions.

8.1.6. Pre-Installation Inspection

A final inspection shall be made of all mooring material before any item leaves the storage area or prior to the material being laid out on the crane barge deck. This will include a check of all chain connections, joining links, and other fittings for proper and secure assembly.

8.1.7. Field Changes Of Design

Moorings are usually designed for a specific application at a specific site. The chain size, anchor weight, anchor fluke angle, length of a riser chain, buoy size, and other factors have been determined based on the holding power requirements and planned location of the mooring. Field changes to design specifications and planned installation procedures should not be made without approval of the cognizant design engineer because incorrect actions could adversely affect the performance and reliability of the mooring.

The choice of the type, size, and configuration of anchors is a design consideration for a particular mooring or a particular mooring location, and is discussed elsewhere in this book. The important consideration for maintenance personnel is that anchor selection is a complex design problem and that small changes to items such as fluke angle, stabilizer length, and orientation when placed on the bottom, can drastically affect the holding capacity of the mooring.

8.1.8. As-Built Drawings

Accurate as-built records (such as shown in Figure 8-3) must be maintained on all installed moorings. As-built drawings should be prepared immediately following the installation.

Figure 8-3 Typical As-Built Documentation

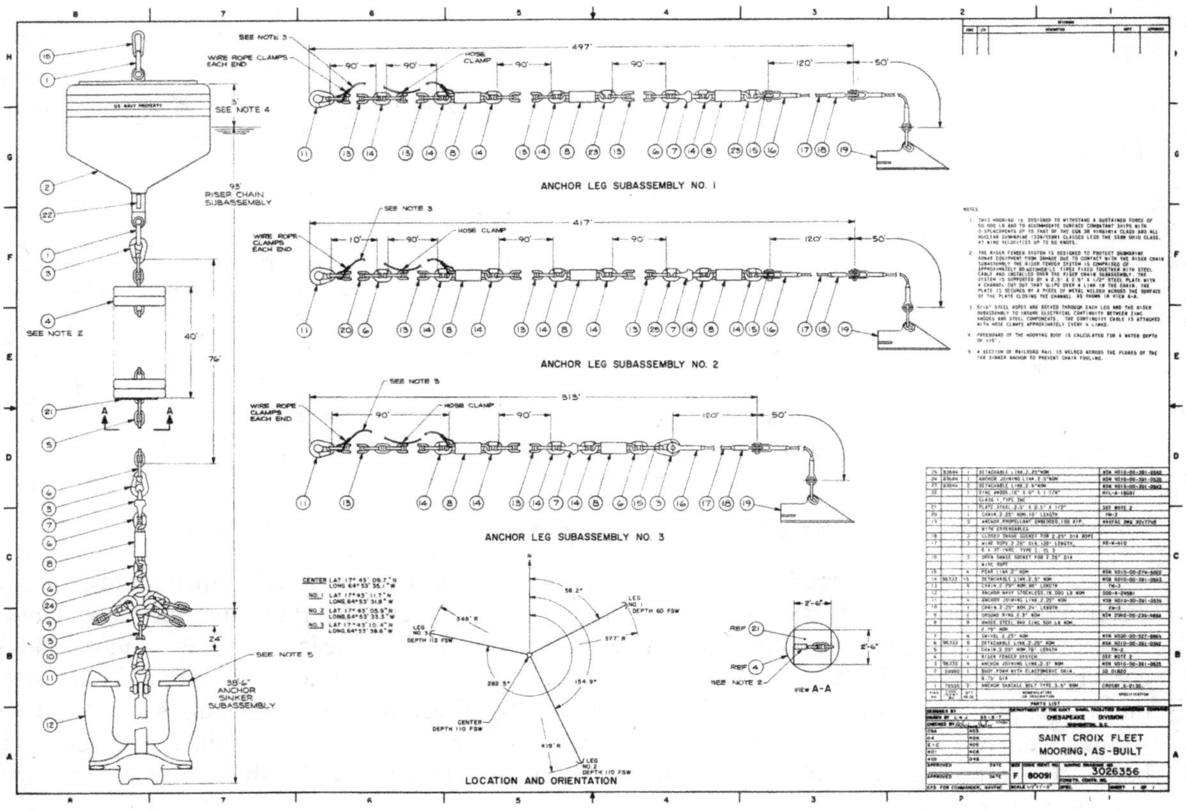

8.2. *Installation Instructions*

8.2.1. General

General installation procedure for the typical three-legged riser mooring system is presented below. The procedures are preceded by a description of the main parts of the mooring system which should be assembled before offshore operations begin.

8.2.2. Riser-Type Mooring System

Installation of this mooring system shall generally follow the procedure set forth below.

8.2.2.1. *Preinstallation Assembly*

A three-legged riser-type mooring system is normally assembled in three parts:

- First anchor chain subassembly with a sinker and anchor.
- The buoy, the riser chain subassembly from the buoy to the ground ring, and the second anchor chain subassembly with a sinker and anchor (see Figure 8-4).
- The third anchor chain (subassembly with a sinker and anchor.

Figure 8-4 Typical Riser-Type Mooring Material Pre-Installation Layout

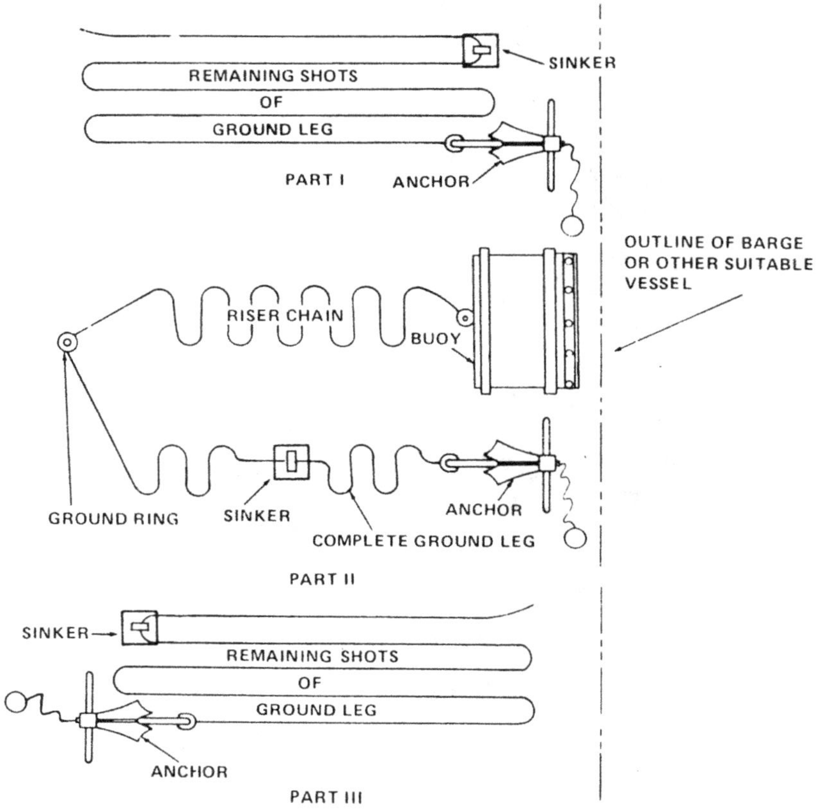

8.2.3. Installation Procedures

The marker buoys will be placed in their desired locations. The center marker buoy will be placed in the desired position of the mooring buoy. The ring marker buoys will be installed 25 feet past the point that the anchors will be lowered to the bottom and released. Predicted drag distance of the anchor is needed to determine this point. Appendix E provides tables of predicted drag distances. The desired final location of the anchors should be indicated on the design drawing provided by the mooring designer. After pulling the anchor to set it and test its capacity, the final position of the anchor must be within a 40 foot by 20-foot box with the desired location at the center of the box. The range marker buoys should be installed about 50 feet beyond the ring marker buoys on the extension of the lines from the center marker buoy to the ring marker buoys.

Before beginning installation of the system, the free end of the first anchor chain subassembly of Part I should be attached to a pickup buoy for easy recovery during the placement operation.

These general installation procedures should then be followed:

- Position the crane barge near one of the ring marker buoys (the one within 25 feet of the desired position for the anchor of the first anchor chain subassembly of Part I). During some installations, it may be necessary to weld the anchor flukes (see Figure 8-5) to a predetermined angle.[25]

---

[25] Wind and current conditions will usually dictate which subassembly is laid first.

Figure 8-5 Welding Anchor Flukes to a Required Angle

- The first anchor is slung by a bridle in a horizontal position and has attached to it a crown marker buoy and one anchor leg subassembly. The anchor is fitted with a pelican hook or a toggle bar quick release system as shown in Figure 8-6. The crane lowers the anchor and chain simultaneously over the side. When installing moorings equipment with chain study anodes, care must be exercised that the chain does not drag over sharp edges that can result in some of the anodes being stripped off.

Figure 8-6 Lifting Bridle and Release Mechanism

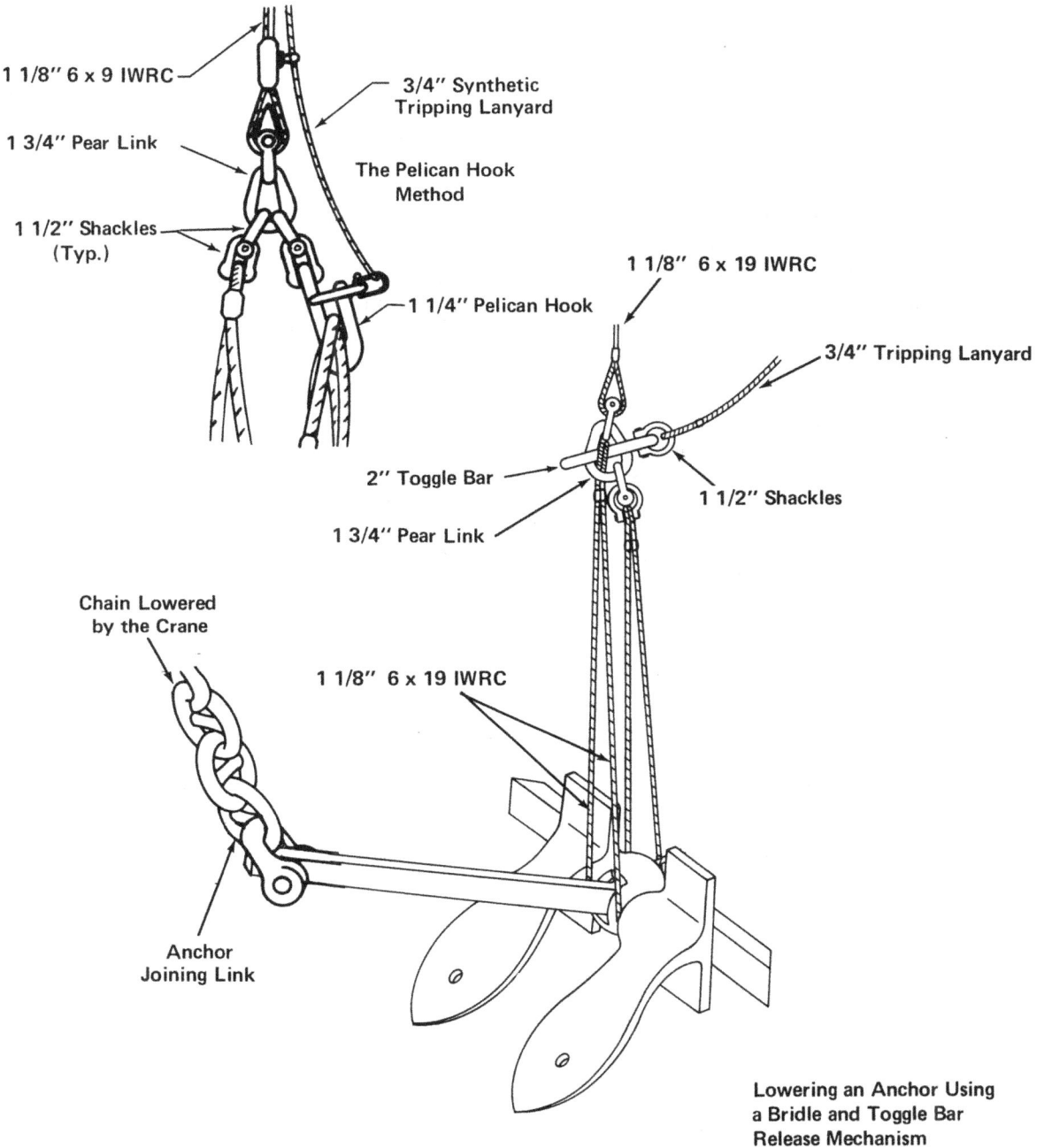

- Upon reaching the bottom, release the anchor and recover the bridle. Move the barge toward the center marker buoy while slowly lowering the chain with a flat catenary.
- Upon approaching the center marker buoy, pull the subassembly taut so that the anchor is properly set. Then, lower the bitter end of the chain (with a pickup buoy attached to it) to the bottom.
- The crane barge now proceeds to the second marker buoy and lowers the anchor (of Part II) 25 feet from the marker toward the center marker buoy. The anchor is fitted with a quick release mechanism and has a crown buoy attached to it.

- Upon reaching the bottom, release the anchor and recover the bridle. Move the barge toward the center marker buoy while slowly lowering the chain with a flat catenary.
- Upon approaching the center marker buoy, pull the subassembly taut so that the anchor is properly set. Then, lower the bitter end of the chain (with a pickup buoy attached to it) to the bottom.
- The crane barge now proceeds to the second marker buoy and lowers the anchor (of Part II) 25 feet from the marker toward the center marker buoy. The anchor is fitted with a quick release mechanism and has a crown buoy attached to it.
- Upon reaching the bottom, release the anchor and recover the bridle. Move the barge toward the center marker buoy while slowly lowering the chain with a flat catenary. Upon approaching the center marker buoy, pull the subassembly taut so that the anchor is properly set. Then, using the pickup buoy, retrieve the end of the first anchor chain subassembly and attach it to the ground ring. Attach the bitter end of the third anchor chain subassembly (Part III) to the ground ring also. Then lower the ground ring, riser, and buoy into the water alongside the center marker buoy.
- The crane barge will slowly lower the third anchor chain subassembly while proceeding toward the third ring marker. This ring marker and its range marker should be used to ensure that the chain is being installed in a straight line.
- When approaching the ring marker buoy, pull the anchor until the chain leg is taut and then lower the anchor (in a bridle with the flukes pointed downward) to the bottom and release it using the quick release mechanism (see Figure 8-7).

Figure 8-7 Typical Riser-Type Mooring Installation

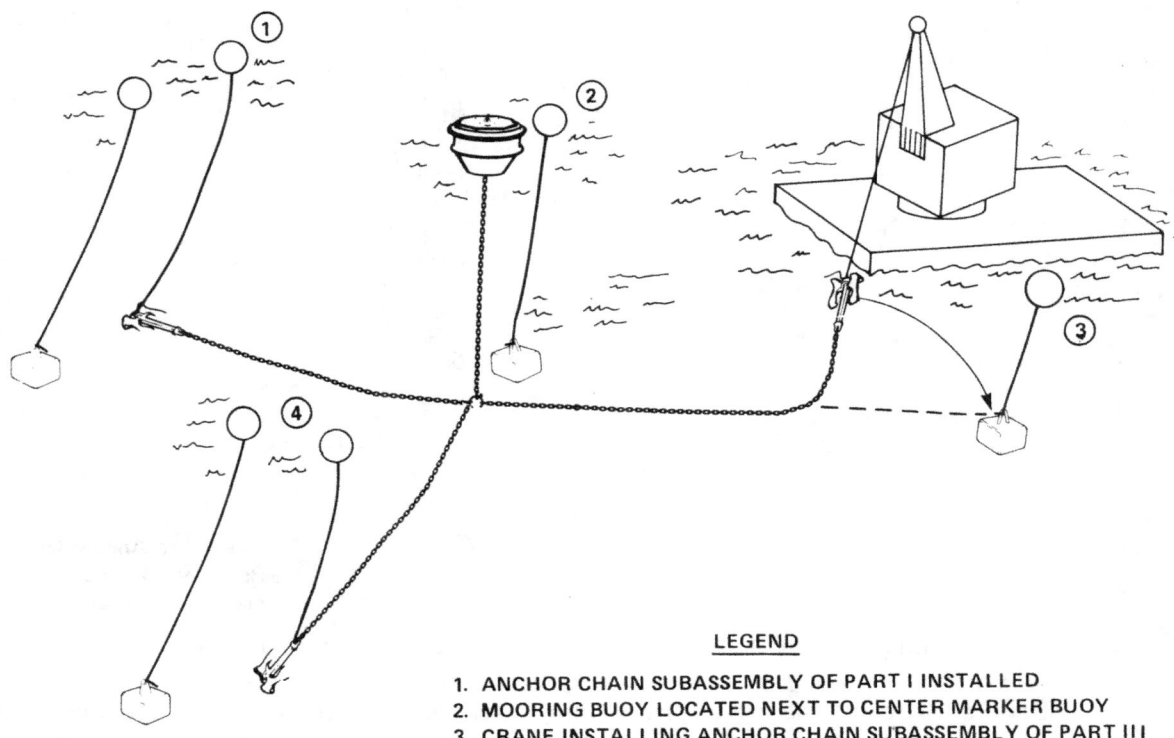

LEGEND
1. ANCHOR CHAIN SUBASSEMBLY OF PART I INSTALLED
2. MOORING BUOY LOCATED NEXT TO CENTER MARKER BUOY
3. CRANE INSTALLING ANCHOR CHAIN SUBASSEMBLY OF PART III
4. ANCHOR CHAIN SUBASSEMBLY OF PART II INSTALLED

- Conduct a final inspection of the mooring. Site the three crown marker buoys from ashore. The positions of these three markers will be the positions of the anchors. If available, have divers make an underwater inspection of the mooring installation.
- Remove all marker buoys with their cables and anchors.

8.2.3.1. Pull Testing of Anchors

Procedures for pull testing anchors are contained in 3.3.7.2. Fleet moorings will be pull tested to the holding capacity of the mooring class listed in Table 8-1.

Table 8-1 Capacities of Navy Fleet Moorings

Class	Holding Capacity (lbs)	Type of Anchor Chain Subassemblies	Chain Size (in) Riser	Chain Size (in) Anchor
AA	300,000	Twin	4	2-3/4
BB	250,000	Twin	3-1/2	2-1/2
CC	200,000	Single	3-1/2	2-1/4
DD	175,000	Single	3	3
A	150,000	Single	2-3/4	2-3/4
B	125,000	Single	2-1/2	2-1/2
C	100,000	Single	2-1/4	2-1/4
D	75,000	Single	2	2
E	50,000	Single	1-3/4	1-3/4

8.2.3.2. Installation Barge

Whenever possible, the YD or similar barge type craft to be used for mooring installation should be equipped with two stern winches to be used for pulling on kedge anchors. The installation barge should have the capability to anchor itself.[26]

8.3. Recovery Instructions

8.3.1. Riser-Type Mooring System

These systems are recovered by removing one anchor chain subassembly at a time. Proceed as follows:

- Sling the buoy from the top jewelry.
- Lift the buoy and riser until the ground ring is level with the deck of the crane barge (see Figure 8-8). In the case of a taut mooring, one anchor chain subassembly may have to be separated from the ground ring by cutting the first A-link below the ground ring with a torch.

---

[26] Divers may be used to inspect connections and to check the orientation and tautness of the anchor chains. They may also be used to jet the anchors into the bottom if included as part of the design specification.

Figure 8-8 Lifting the Ground Ring to Deck Level

- Stopper off the ground ring (see Figure 8-9).

Figure 8-9 Ground Ring Stoppered Off on Deck

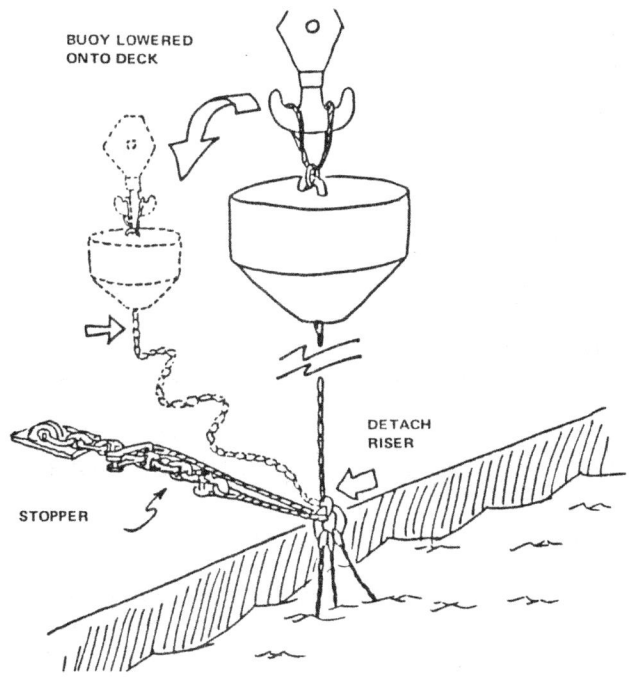

- Lower the buoy down to the deck on its side. Disconnect the riser, and either block the buoy on its side or place it on blocks to avoid damaging the tension bar.
- Disconnect the riser from the ground ring and buoy. If the joining link cannot be removed, cut the first A-link with a torch
- Sling the ground ring and lift it until the anchor chain subassemblies are accessible.
- Stopper off one subassembly (see Figure 8-10).

Figure 8-10 One-Leg Stoppered Off

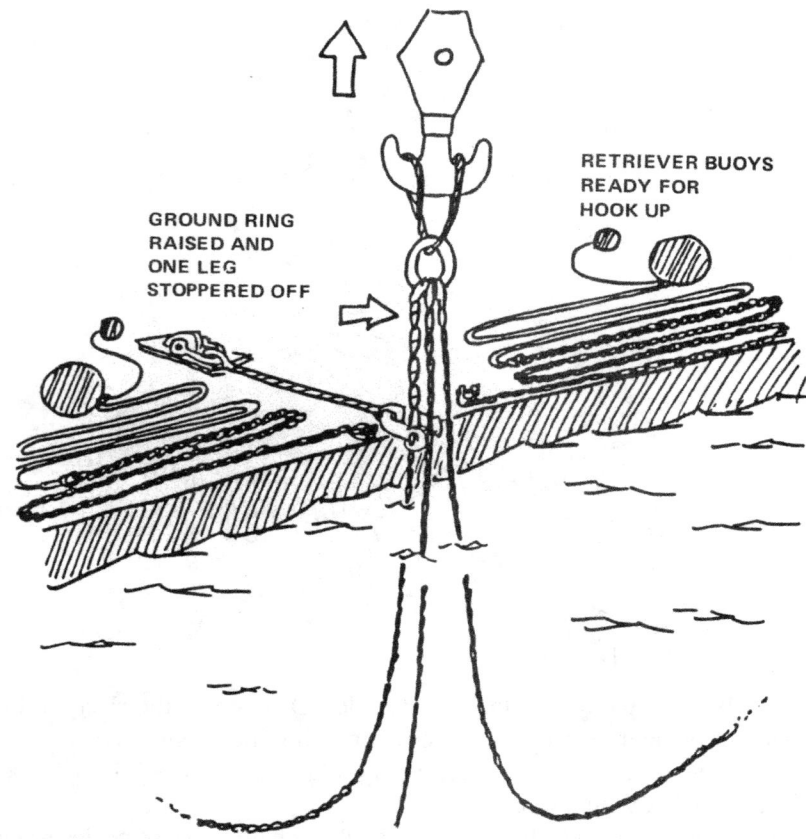

- Attach a retrieval buoy to each of the other two subassemblies. The third subassembly and the ground ring will be considered together. Fake the retriever buoy lines on deck to allow easy running.
- Cut one subassembly free from the ground ring one link below the chain joining link. Allow the chain to drop and retriever buoy to run free and over the side.
- Repeat with the other leg that has a retriever buoy attached. Lower the ground ring on deck (see Figure 8-11). Disassemble the chain joining link, if possible, and disconnect the ground ring from the subassembly.

Figure 8-11 Two Anchor Chain Subassemblies Overboarded

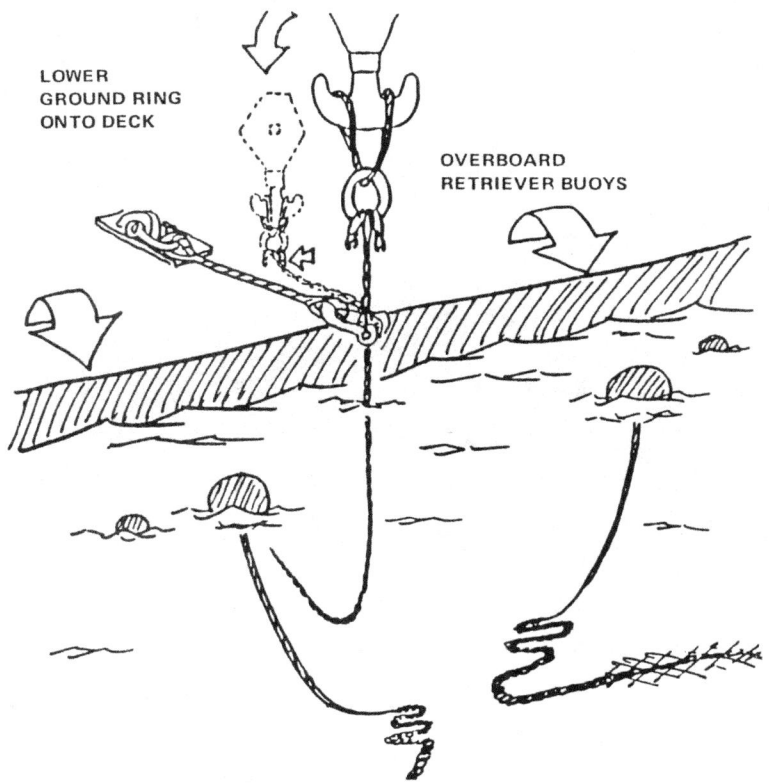

- Sling the chain to the main hoist, raise and remove the stopper.
- Continue raising until the next chain joining link is above deck.
- Stopper the chain, 3 links below the chain joining link, and lower the chain joining link to the deck for disassembly.
- When severed, move this shot of chain aside.
- Sling the chain and continue lifting and detaching chain shots as before. All components recovered should be washed down with seawater before being brought aboard the barge. Use a high-pressure hose for this purpose (see Figure 8-12).

Figure 8-12 Using High Pressure Hose for Cleaning

- When all of the chain has been recovered, bring the anchor aboard.
- Pick up a retriever buoy and, using the same procedures, recover a second subassembly and anchor. Then recover the third.

Chapter 9. Inspections

9.1. General

9.1.1. Overall Requirements

All mooring components, either in use or in storage, must be periodically inspected to determine their current material condition and their future maintenance requirements. The importance of these inspections cannot be overemphasized because the effectiveness of any maintenance program will always depend on how often and how well these checks and services are performed. Inspection plans, therefore, should take into consideration the critical elements of frequency of inspections and the thoroughness, completeness, and quality of work.

9.1.2. Inspection Classifications And Types

For the purposes of this manual, inspections are classified as either in-service or out-of-service. In-service inspections are performed on installed moorings; out-of-service inspections are performed on components stored ashore. There are four basic types of in-service inspections, as follows:

- Annual surface inspections.
- Underwater inspections.
- Lift inspections.
- Damage/failure inspections.

The above in-service types are addressed and discussed in this chapter. Out-of-service inspections are covered in later chapters.

9.1.3. Purpose

The primary purpose of in-service inspections is to determine the general physical condition of the buoy(s) and chain assemblies. The results of these inspections are used to decide if a mooring is safe for continued use. Routine inspections also provide an opportunity to detect and remedy minor material deficiencies. Future maintenance requirements are strongly dependent on the results of periodic in-service inspections.

9.1.4. Personnel

Because in-service inspections are conducted offshore, and often under less than ideal conditions, experienced personnel, as well as reliable equipment, are required to accomplish these tasks. Topside personnel or divers will be used to clean and inspect a representative portion of the buoy hull and chain assemblies. The inspection report of their findings will then be used to assess the condition of the entire mooring. Measurements taken, observations made, and data accumulated must, therefore, be highly accurate and complete in these instances.

9.2. Inspection Procedures

9.2.1. General

The following paragraphs contain descriptions of the four basic types of in-service inspections and provide guidelines for performing these inspections.

9.2.2. Annual Surface Inspections

Shore activities that operate and maintain fleet moorings must inspect the visible portion of each mooring buoy at least once each year. The purpose of this annual surface inspection is to ensure that the buoy and its topside hardware, fenders, and chafing strips are in satisfactory condition, and to verify that the mooring has not been dragged from its proper location.

9.2.2.1. Buoy Inspection

The buoy should be closely examined to determine its overall condition. The following should be documented:

- Caliper measurements of the upper jewelry whenever these appear to be excessively worn out or in marginal condition (see Figure 9-1).
- In addition, any excess top jewelry or wire rope cables attached to the buoy should be reported.
- Physical damage such as holes, dents, metal distortion, or listing (see Figure 9-2).
- Measurement of the buoy's freeboard.
- Condition of fiberglass on fiberglass-coated buoys. (Fiberglass should be inspected and any cracks, wear, peeling, or rust bleeding identified.)
- Condition of paint on painted buoys. (Paint should be checked for cracking, chipping, and/or peeling.)
- Condition of water drains and buoy surface penetrations. (Examine for broken parts, surface rust, and surface pitting.) (See Figure 9-3).
- Condition of fenders and chafing strips. (Check for physical integrity and secure connections to the buoy's surface. Fender/chafing strip brackets or studs will be inspected for corrosion and/or cracks.)

Figure 9-1 An Example of Severely Worn Top Jewelry

Figure 9-2 Severe Buoy List

Figure 9-3 Pitting/Rusting on the Side of a Buoy

9.2.2.2. Buoy Location

If it is suspected that a mooring has been dragged from its desired geographic location, the current position of its buoy will be verified by sighting from known positions ashore using a transit, a theodolite, or any instrument of comparable precision.

9.2.2.3. Documentation

The results of the surface inspection shall be fully documented (see Figure 9-4, "Sample Surface Inspection Form") and filed in the inspecting activity's maintenance files for future reference. If a buoy is found to be in relatively poor condition or in need of repair or overhaul, complete information concerning funding and the material required to correct the observed deficiencies should be reported.

Figure 9-4 Sample Surface Inspection Form

BUOY SURFACE INSPECTION FORM

ACTIVITY/STATION NAVSTA Key West   BUOY (MOORING) NO. T-4   BUOY TYPE Peg Top (12'0" X 9'6")

INSPECTION DATE 26 November 1986   CHIEF INSPECTOR J.A. Carlos   OBSERVED FREEBOARD (IN) 42

TOP JEWELRY

COMPONENT TYPE	AS-BUILT DIAMETER	MEASURED DIAMETER	COMMENTS
End Link	3 1/4"	3 1/8"	> 90%
Bow Shackle	2 3/4"	2 1/2"	> 90%
Shackle	2 1/4"	2"	Between 80% and 90%. Should be replaced.

BUOY

COMPONENT	TYPE	DAMAGE			RUST			CONDITION			COMMENTS
		MAJOR	MINOR	NONE	HEAVY	MODERATE	LIGHT	NONE	GOOD	FAIR	POOR	
HULL	STEEL			X			X		X			10° List
COATING	FIBER___ PAINT___ GLASS___ OTHER___ POLYURETHANE___		X				X		X			Light Rust Bleeding on Top. Two inches Marine Growth.
FENDER	RUBBER___ WOOD___ BRACKET___ STUD___ OTHER___									X		Missing 20% of Lower Fender. Upper Splintered.
CHAFING STRIP	RUBBER___ WOOD___ BRACKET___ STUD___ OTHER___									X		Two Strips.
OTHER	HATCH BOLTS					X						

NOTE: FOR ADDITIONAL COMMENTS, USE SEPARATE SHEET OF PAPER

9.2.3. Underwater Inspections

The purpose of mooring underwater inspections is to determine the general condition of buoys and chain assemblies and to verify or update existing as-built drawings and maintenance records. Divers will inspect each fleet mooring every 2 to 3 years. Divers assigned for each inspection will inspect only a portion of the submerged buoy hull and chain assemblies in order to compile a general description of the mooring's condition. Consistent measurements obtained during each inspection will provide a good indication of the mooring's overall condition. Obviously, underwater inspections cannot fully substitute for a complete inspection involving recovery of the mooring and the measurement and evaluation of each component.

9.2.4. Life Inspections

In the past, a buoy and riser chain were lifted as high as possible out of the water and a visual inspection of the buoy, its upper and lower jewelry, and accessible portions of the riser chain subassembly was conducted to determine wear, corrosion, or deterioration. Lift inspections have been discontinued in favor of underwater diver inspections and will no longer be conducted. To avoid disturbing the anchors, a mooring should be lifted only when the buoy and/or riser chain require repair or replacement, or when the mooring is being completely removed.

9.2.5. Mooring Damage/Failure Inspections

Fleet moorings can be damaged by collisions or dragged out of position by heavy winds or seas. They can also fail because of broken mooring components. An inspection of mooring damage, drag, or failure should be conducted as soon as possible after detection. If a collision has occurred, then the purpose of the inspection is to determine the extent of the damage and whether the buoy is in danger of sinking. If the buoy is found to have hull damage and is in danger of sinking, a marker buoy should immediately be attached to it so that the buoy's position will be marked in the event that it sinks. Arrangements should then be made to recover the buoy at the earliest opportunity so that repairs can be accomplished. If the buoy is undamaged, but the mooring has been pulled off location, arrangements should be made to recover the mooring and reinstall it in its proper position.

9.2.5.1. Inspection Procedures. Inspection will include the following:

- Visual inspection of the buoy's hull and associated fenders for damage (dents, broken fenders, scrapes, hull punctures, etc.).
- Check for buoy drag and the excursion of the entire mooring (new positions should be sighted from known benchmarks ashore and recorded).
- Check for riser chain failures that will cause buoys to float free from their intended position. Free floating buoys should either be towed to shore or temporarily attached to an anchor.

9.2.5.2. Documentation

The results of the damage inspection shall be fully documented. In all cases of damage or suspected damage to a fleet mooring, the cognizant port authority should be notified so that the mooring will not be used. A failure report must also be submitted.

Chapter 10. In-Service Maintenance And Repair

10.1. General

10.1.1. Scope.

In-service maintenance and repair will be limited to the following:

- Minor underwater repairs.
- Minor buoy and riser assembly repairs.
- Replacement of damaged buoys and/or riser assemblies.
- Minor repairs to cathodic protection systems.

It should be noted that sufficient lift capability would be needed for replacement of buoys and/or riser assemblies and, in many instances, for repairs. All of the maintenance functions listed above are covered in this chapter except for cathodic protection.

10.1.2. Equipment

The following equipment must be readily available for use as needed:

- Crane barge or floating crane.
- Tugboat, mule, or other vessel (for maneuvering and positioning the crane platform).
- High-pressure water pump and hose (for cleaning).

10.2. Procedures

10.2.1. Buoy Replacement (Riser-Type)

A buoy in a riser-type system can be replaced without removing the mooring. Proceed with the replacement as follows:

- Lift the buoy out of the water and wash it down with seawater from a high pressure hose.
- Secure the riser chain to a bitt or cleat with a wire rope sling.
- Detach the buoy from the chain by removing the anchor joining link or buoy shackle located directly under the buoy.

If a replacement joining link is not available, the removed joining link can be cleaned, recoated with a preservative grease, and reused to secure the new buoy to the riser chain.

- Secure new buoy to riser chain.
- Lower new buoy into position in the water.

When a tension bar type buoy is lifted onto a barge deck, it should be placed on railroad ties or on chocks to prevent damage to the lower portion of the tension bar.

10.2.2. Riser Replacement

Riser replacement will be accomplished as follows:

- Lift the buoy and its attached riser chain out of the water and wash down with a high-pressure hose.
- Hoist buoy and riser chain aboard the barge.
- Hold the ground ring on a deck stopper.
- Disconnect riser from the ground ring and buoy.
- Attach new riser to the buoy and ground ring.
- Reinstall buoy and riser in the water.

The above steps are standard procedure for moorings installed in shallow water. In deep water, however, the ground ring, in all probability, will not be able to be lifted on the deck of the barge without disturbing the anchors. In this situation the mooring will have to be recovered in order to replace the riser, and then reinstalled.

10.2.3. Buoy Replacement (Non-Riser-Type)

A non-riser-type buoy will be more difficult to retrieve than a riser-type. If the mooring has been properly installed, there will be a catenary section of chain suspended in the water between the buoy and the anchor. If the catenary angle is large (as in a taut, properly installed mooring), then it may not be possible to stopper off all four-anchor chain subassemblies on the barge deck simultaneously. It should also be noted that, in this type of installation, the buoy is kept in place by balanced opposing forces created by the catenaries of the anchor chain subassemblies. When one of the subassemblies is cut, the buoy will be pulled in the direction of the opposing leg. This pull will result in a potentially dangerous horizontal force on the crane boom, especially if the buoy is being held aloft when the chain is cut. Connecting the replacement buoy to the anchor chain subassemblies is also difficult under these conditions. Therefore, in the case of a taut mooring, it is recommended that the non-riser buoy system be completely recovered prior to replacing the buoy, and that the mooring be reinstalled. In many non-riser installations, significant slack exists in the anchor chain subassemblies directly below the buoy. In such cases it may be possible to lift the buoy and simultaneously stopper off the legs on the barge deck. The new buoy can then be connected to the anchor chains and replaced in the water.

10.2.4. Minor Repairs

In-service minor repairs to a buoy, such as replacing a fender, repairing an upper hull puncture, patching fiberglass or polyurethane, replacing anodes, or spot painting the buoy, can be accomplished without taking the buoy ashore. Although divers can accomplish emergency replacement of smaller sizes of mooring chain components underwater, mooring component replacement, welding, and other minor repairs should be accomplished aboard a crane barge. If it is necessary to enter a buoy, this should be done only when the buoy is ashore or aboard a barge where there is no possibility of the buoy's sinking. Special care must be taken to reseal the manhole and assure watertight integrity.

WARNING: Buoy manhole covers should not be removed while the buoy is in the water.

During the inspection of a mooring buoy, its associated top and bottom hardware should be closely inspected to determine if any components must be reconditioned or replaced. When reconditioning/replacement is required, the following should be observed:

10.2.4.1. Welding/Cutting

Welding chain appendages or cutting out retaining pins or rivets with a torch should never be done because heating will introduce internal stresses and reduce the strength of heat-treated steel components.

10.2.4.2. Shackles/Joining Links

Ensure that shackles, joining links, and other such fittings with removable parts are treated with an appropriate grease preservative and refitted. Care should be taken not to interchange matched parts of joining links. This can be avoided by tagging each part of the joining link with a unique identification number or by matching the stamped numbers on the parts as shown in Figure 10-1. Locking pins of joining links and shackles should never be welded in place due to the probable resultant loss of tensile strength of the component.

Figure 10-1 Mooring Material Markings

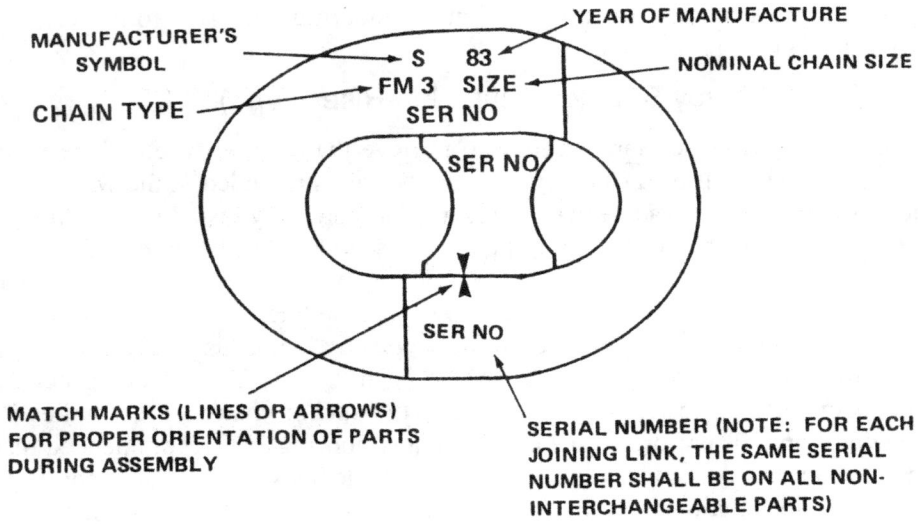

10.2.5. Buoy Coatings

Because protective coatings are frequently damaged by impact or abrasion, it may be necessary to make in-service repairs to coatings of mooring buoys. In order to repair these coatings, it is necessary to first clean the exposed steel and the area surrounding the steel or rust, salt, and loose material. This can be accomplished by wire brushing the steel (preferably power wire brushing) or by scrubbing with a bristle brush and then drying the area with an air hose connected to an oil free compressor (when required). The intact coating surrounding the damaged areas should be abraded to ensure proper bond of the repair material. There are several proprietary putty-like coatings available called splash-zone compounds that can be spread over the cleaned area, wet or dry. MIL-P-28579(YD) describes such a curing epoxy-polyamide formulation. There are also available a number of proprietary brush on coatings that can be applied to damp surfaces and will cure either above or under water. Either type of coating should completely cover the steel and extend at least on-half inch over the cleaned coating surrounding the steel.

Chapter 11. Ashore Inspection And Refurbishment Of Buoys

11.1. General

11.1.1. Scope

Ashore inspection and refurbishment of buoys will include visual inspections, repairs, tests, and replacement of damaged components. There are two types of buoy inspections: preliminary and detailed. The purpose of a preliminary inspection is to determine whether the buoy is in a condition for a further, more detailed inspection and subsequent refurbishment or whether it should be disposed of at this stage. If the results of the preliminary inspection indicate that refurbishment will be cost-effective, then a detailed inspection will be conducted. The buoy must be cleaned, completely inspected, and tested for airtight integrity to determine all repair requirements. The buoy can also be sandblasted to near white metal and, if required, a liquid dye penetrant or magnetic particle test must be conducted.

11.1.2. Preparation For Ashore Inspections

Prepare the buoy(s) for ashore inspection as follows:

- Clean off buoy with high-pressure water during recovery.
- When brought ashore, place the buoy on chocks (railroad ties, cinder blocks, etc.) to keep the tension bar clear of the ground. For ease of working, the peg-top type buoy should be placed inverted on chocks.
- Remove the top and bottom jewelry, fenders, chafing strips, and manhole covers. Mark manhole cover positions before removal for later replacement in the same location.
- Remove shackle and joining links from the buoy, as necessary.

After removal, shackles and joining links should be reassembled as complete units, including pins. Shackle pins and tapered locking pins should be used only in their original parent component.

11.2. Inspection Procedures

11.2.1. Preliminary Inspection

Before beginning the preliminary inspection, a 1'-square section of the top and bottom and four 1'-square sections (two above and two below the water line) of the side hull plates should be cleaned to bare metal. Visually inspect these four sections for pitting. In addition, perform a visual inspection as follows:

- Inspect the fenders and chafing strip fastenings for corrosion and wear.
- Check hull and deck plates for corrosion, cracks, pitting, and watertight integrity (see Figure 11-1 and Figure 11-2).
- Check the hawsepipe (if the buoy is so constructed for rust, cracks, pitting, or other material abnormalities.
- Inspect the upper and lower ends of the tension bar (if the buoy is so constructed) for wear, cracks, rust, or pitting.
- If the buoy is a non-riser (telephone) type, check the padeyes for wear or cracks.
- Check the overall condition of the paint, fiberglass, or polyurethane coating.
- Check the interior of the buoy for rust or corrosion.

Figure 11-1 Heavy Rusting of Top Deck and Jewelry

Figure 11-2 Severe Top Deck Corrosion

Based on the results of this inspection (i.e., internal structural weaknesses, hull plate cracks, severe pitting or rusting, broken tension bar, excessive corrosion, etc.), a decision will be made whether to prepare the buoy for a detailed inspection or to dispose of it.

11.2.2. Detailed Inspection

Perform detailed inspection as follows;

11.2.2.1. Ultrasonic Testing

Conduct this test as follows:

- Sandblast the buoy to near white metal in accordance with SSPC-SP-10.
- Inspect the buoy for damage, cracks, etc.
- Conduct an ultrasonic thickness test at four points on the buoy top, four points on the bottom, and eight points around the circumference of the hull. Four of the foregoing eight points will be below the waterline and four above the waterline.
- Using a pitting gauge, measure the depths of any pits observed on the hull of buoy.

11.2.2.2. Pitting Inspection

A visual inspection of the buoy hull plates will be made for pitting. The extent of pitting will determine the remaining life of the plates. ASTM G46-76, "Examination and Evaluation of Pitting Corrosion," will be the standard reference used to evaluate the damage and to formulate a quantitative expression that will indicate its significance. To obtain a quantitative expression, ASTM G46-76 recommends that the deepest pit be measured, and that metal penetration be expressed in terms of the maximum pit depth or the average of the 10 deepest pits. Metal penetration can also be expressed in terms of a pitting factor. This is a ratio of the deepest metal penetration to the average metal penetration as shown in the following relationship:

Equation 11-1: Pitting Factor = (Deepest Metal Penetration)/(Average Metal Penetration)

A pitting factor of "one" represents uniform corrosion. The larger the number, the greater the depth of penetration. Pits will be rated in terms of density, size, and depth.

11.2.2.3. Welds

Carefully check all welds, both internal and external, for cracks or corrosion. If any cracks, fissures, or other flaws are found or are suspected, then a liquid dye penetrant or magnetic particle test will be performed to determine the extent of the defects.

11.2.2.4. Air Test

Air test each chamber/compartment in the buoy. Maintain 2 psi design pressure in each compartment for 30 to 45 minutes.

11.2.2.5. Documentation

The results of all inspections (both preliminary and detailed) will be fully documented and filed for future reference. If a buoy or its components are found to be in unsatisfactory condition, the inspection results should be used in planning and estimating future repair or replacement material requirements and associated labor costs.

11.3. Buoy Repairs And Modifications

11.3.1. General

Steel buoy repairs and modifications will include manhole cover replacement, test plug and aperture maintenance, fender and chafing strip repairs, welding requirements, and air pressure testing of the buoys.

11.3.2. Procedures

Buoy repair and modification will be accomplished as described below.

11.3.2.1. Manhole Cover Replacement

Manhole cover replacement will be accomplished as follows[27]:

- Clean and lubricate the studs and use chaser nuts where required.
- Replace old gasket with a new 1/8" silicone rubber gasket held in place with RTV silicone gasket adhesive sealant (MIL-A-46106); apply sealant only to bottom surface of gasket.
- Lift manhole cover by the extension lip and position it over studs.

CAUTION: Exercise care in lowering the cover on the studs so that stud threads and gasket are not damaged.

- Lower manhole cover on studs.
- Secure cover bolts. Tighten in at least three steps using an opposite bolt tightening sequence. Apply final 45 foot-pounds of torque.

11.3.2.2. Test Plugs and Hull Apertures

Clean and check test plug threads before placement of the plugs in the hull apertures. Teflon sealant tape should be applied to the threads to ensure a watertight seal.

11.3.2.3. Fenders and Chafing Strips

Overhauled steel buoys will be provided rubber fenders and chafing strips. In most cases, this will require removal of the wooden fenders and their channeling, and the wooden chafing strips and connecting brackets. Channeling and brackets will be replaced by stainless steel studbolts, 3/4" in diameter by 2 1/2 inches long (10 threads per inch). The bolts will be positioned and welded to conform to predrilled holes in the rubber fenders and chafing strips. If the rubber fenders and chafing strips are not predrilled, they will be drilled to conform to the positions of the studbolts. Spacing of the bolts shall not exceed 16 inches on center. Figure 11-3 illustrates the recommended attachment of fendering and chafing strips.

---

[27] Each manhole cover will be replaced on the opening from which it was originally removed. Matchmarks made on cover and deck plate prior to removal will facilitate its replacement in the correct location and position.

Figure 11-3 Recommended Installation of Fenders on Fleet Mooring Buoys

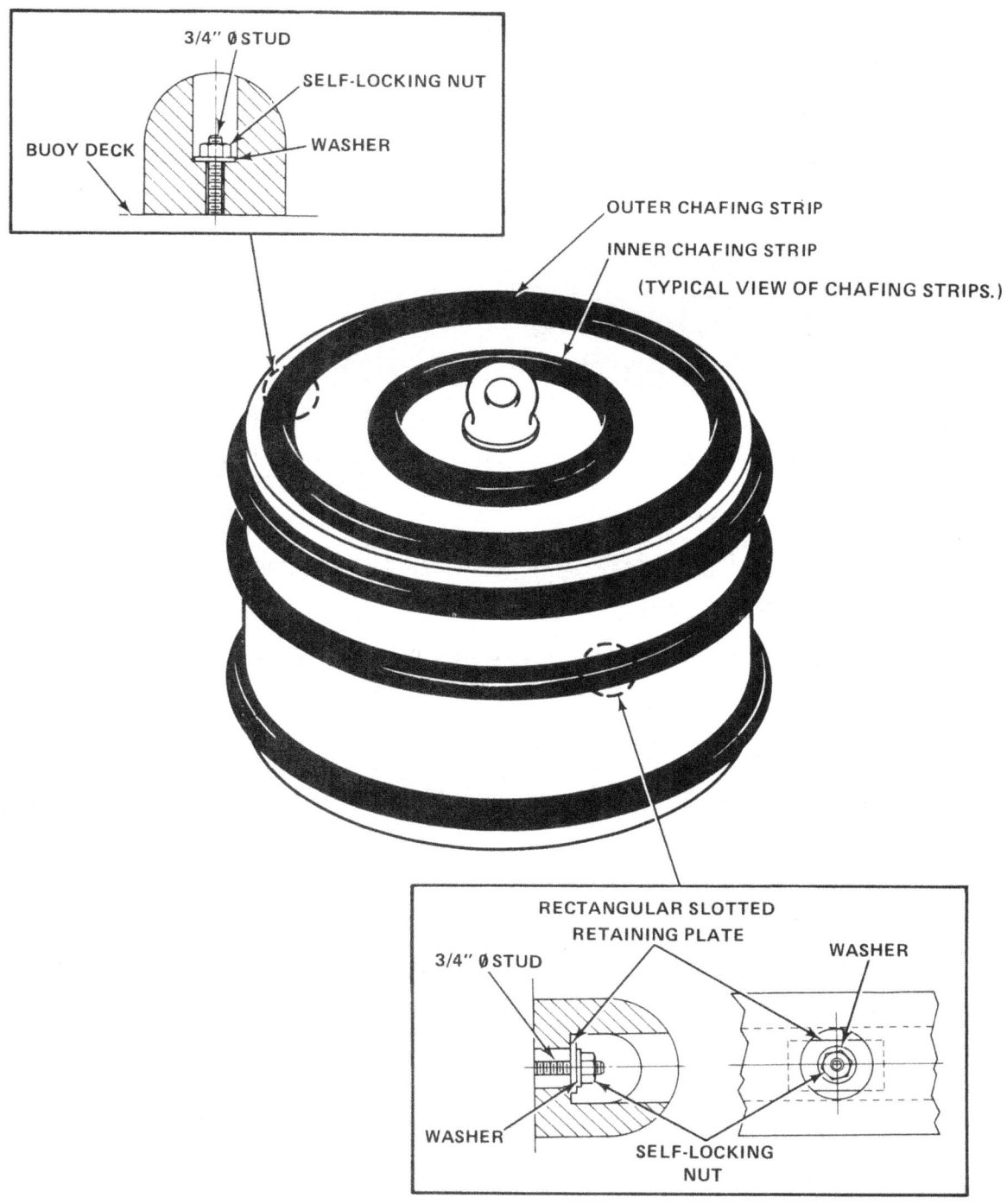

11.3.2.4. Welding

Trained and qualified personnel following accepted procedures and standards contained in the latest edition of AWS D1.1, "The Structural Welding Code", will accomplish all welding.

11.3.2.5. Buoy air-Pressure Test

Each buoy compartment will be tested for tightness at the joints by the application of air pressure. Unless the buoy is under cover, at least 2 hours of clear weather will be required for the test. The gauge used for the test must have a current calibration certification. Proceed as follows:

- Install a test plug fitted with a gauge in the top of each buoy compartment.
- Pressurize each compartment with 2 pounds per square" of air pressure for a minimum of 30 minutes after stabilization of pressure.
- Brush all joints and seams with a solution of commercial leak testing fluid. Two percent potassium dichromate may be added to the solution to inhibit the formation of rust.
- Leaks detected will be repaired and the buoy retested.
- When all tests and repairs are completed, completely remove leak-testing solution before applying surface primers.

11.4. Protective Coatings

11.4.1. Preparation For Application

Preparation of buoys for application of protective coatings will include the following:

- Remove fenders, chain links, steel plates, etc. from the mooring buoy.
- When possible, open buoy manhole and check the interior of the buoy for rust and water damage.
- Examine hull for areas that may need repair or replacement.
- Pits found that are 3/16" (4.8 mm) deep or more will be filled with clad-welding or epoxy repair compounds conforming with MIL-C-24176 and in accordance with the applicable sections of the Naval Ships Technical Manual, NAVSEA S9086-AA-STM-000.
- Buoys which are fiberglassed should have a steel reinforcing ring welded around the outside edge of the manhole opening if one is not already present. The purpose of the ring is to provide a clean, secure surface on which to seat the manhole cover gasket as well as to reinforce the buoy deck. The ring should be of 1/2" steel plate and should extend a minimum of two inches outward from the edge of the manhole opening. If a ring is not used, then the manhole opening must be welded closed using flush steel plates, which are reinforced in the underside by steel backup strips. If this is the case, the buoy will have to be cut open from subsequent inspections[28].
- Sandblast exterior surfaces of the metal hull in accordance with the latest edition of the "Steel Structures Painting Manual, Vol. II, Systems and Specifications," Specification SSPC-SP-10. All sharp and irregular edges will be ground smooth.

11.4.2. Foam Filled Elastomer Covered Buoys

In the event of small rips, tears, punctures, or gouges in the skin and underlying foam, a repair kit containing the components and procedures required to accomplish minor repairs can be obtained from commercial vendors. If the buoy should be severely damaged and major repairs are required, the manufacturer of the buoy should be contacted for advice and/or assistance.

11.4.3. Fiberglass Polyester Resin (Fpr) Coating Repairs

Fiberglass patches will be applied as instructed below:

- Immediately after sandblasting the areas to be repaired, apply one coat of pretreatment primer (MIL-P-15328, Formula 117), 0.3 to 0.5 mil thickness. The thickness of the primer shall not exceed 0.5 mil. Film thickness will be checked with a microtest thickness gauge or a comparable instrument.

---

[28] Do not apply paints or fiberglass coatings to the top surface of the reinforcing ring.

- After the pretreatment primer has dried, apply one coat of clear polyester resin (MIL-R-21607) to the surfaces that are to receive the FPR coating patches. No sandblasted surface will remain uncoated for more than 4 hours.
- Commence with the first FPR lamination, which consists of the polyester resin and chopped fiberglass mat (MIL-M-43248). Accomplish lamination as follows:
- Apply the mat to the pre-coated surface and roll or squeeze to remove all lumps and air bubbles.
- Lay on additional polyester resin until the mat is thoroughly wet.
- Roll or squeeze until smooth, adding additional resin as necessary.
- Immediately after the first lamination apply three additional laminations, as follows, to give a maximum dry film thickness of 3/16" (4.8 mm).
- One lamination of fiberglass woven roving (MIL-C-19663) will be applied in a manner similar to the initial lamination.
- Apply one lamination of fiberglass mat.
- Apply one lamination of fiberglass woven roving. Adjacent portions of mat or woven roving in all laminates shall overlap a minimum of 2 inches to a maximum of 6 inches.
- Apply additional polyester resin coatings for each successive lamination before and after the individual lamination reinforcement. The reinforcement will be rolled and squeegeed as in the initial lamination, and the polyester resin will be added and distributed, as needed, before starting the next lamination.
- Apply a generous, smooth-finished topcoat of the polyester resin mixture on the final lamination. The topcoat will be pigmented with 4 ounces of white color pigment per gallon of resin. Personnel applying the FPR may, in lieu of the laminations of fiberglass mat, use a chopped fiberglass-polyester lamination sprayed on the surface being refinished at the rate of 2 ounces per square foot.
- If breaks occur on the surface such as around the tension bar, padeyes, and bolts and studs, the FPR coating will be edge-finished carefully using polyester resin.

11.4.4. Paint Coatings

General coating operations, materials, and safety precautions are described in NAVFAC MO-110, "Paints and Coatings." The recommended coating system for mooring buoys is the Navy epoxy-polyamide system for interior and exterior ship surfaces (MIL-P-24441). Procedures for its use are thoroughly described in the Naval Ships Technical Manual (NAVSEA S9086-AA-STA-00). For optimum results, this coating should be applied to dry steel cleaned to a near white metal surface.

- The above coating should be applied in three coats, each at about 4-mil film thickness, to give an approximate 3-mil dry film thickness per coat and a minimum 8-mil total dry film thickness. There should be 16 hours of curing time between coats. The first coat should be the green primer (Formula 150); the second, haze gray (Formula 151); and the third, white (Formula 152).
- The two components of all MIL-P-24441 coatings should be mixed in equal volume by first thoroughly stirring each component separately and then stirring them together. After mixing, there should be a waiting period of about 2 hours at 50 to 60° F; 1 to 1 1/2 hours at 60 to 70° F; and 1/2 to 1 hour above 70 degrees before applying the coating to ensure complete curing later. The mixed paints do not require thinning, but the low temperature application properties can be improved by adding 10%, by volume, of a mixture of equal parts of n-butyl alcohol and AMSCO Super High Flash Naphtha, or an equivalent mixture.

Usual paint spray equipment, either conventional or airless, can be used. The pot life of the mixed coating is about 6 hours at 73° F. If more than 7 days elapse between epoxy coats, the surface should be cleaned with water and detergent, rinsed with fresh water, dried, and then a tack coat (1 to 2 mils wet film thickness) of the last coat applied before application of the next full coat.

The above coating system, when properly applied, will provide at least 3 to 5 years of protection, depending upon the severity of the environment. Experience has shown that an antifouling paint is usually unnecessary because fouling will not damage the coating, add a significant amount of weight to the buoy, or otherwise adversely affect the mooring. In addition, the effective life of antifouling paint is usually much shorter than the time between buoy overhauls, so that significant fouling will still occur before the next overhaul. Should an antifouling paint be desired for the underwater portion of the buoy, MIL-P-15931, Formula 121/63, applied in two coats, is recommended. The antifouling paint, which is red, must be applied while the topcoat of epoxy (MIL-P-24441, Formula 152) is still tacky (within 4 to 6 hours after its application). If the epoxy has hardened, a tack coat (1 to 2 mils wet film thickness) must be applied and allowed to cure for 4 hours before the first coat of antifouling paint is applied.

11.4.5. Quality Of Work

All the workmanship on the coating systems shall be in accordance with the Naval Ships Technical Manual, NAVSEA S9086-AA-STM-000. The work shall be performed by or under the immediate and direct supervision of skilled personnel who have demonstrated a continuing proficiency in the application of multilayered coatings on extensively contoured areas similar and comparable to the exterior of mooring buoys. Quality of workmanship shall meet the highest standards as set forth in the specifications and manuals noted herein.

Chapter 12. Ashore Inspection And Refurbishment Of Chain And Accessories

12.1. General

12.1.1. Scope

The purpose of inspecting chain and accessories is to make a sound qualitative decision as to the disposition of these items. There are two types of inspections: preliminary and detailed. The preliminary inspection is performed in the known wear areas and will determine whether the chain and accessories are in adequate condition for continued use, whether a detailed inspection is required, or whether the items should be disposed of. In the detailed inspections, each link of chain and each accessory is thoroughly inspected for corrosion, cracks, wear, pitting, elongation, or other abnormalities. Based on the results of this inspection, a decision is made either to refurbish and retain the material or to dispose of it.

12.2. Inspection And Refurbishment

12.2.1. Preliminary Inspection

Before commencing the inspection, the chain assembly will be disassembled into individual shots or into the shortest whole chain lengths. The ground ring, swivels, joining links, and shackles will be removed. Shackles and joining links should be kept together as complete units, including pins. Shackle pins and tapered locking pins should be used only in their original parent component. Preliminary inspection will include the following:

- Visually inspect all accessories and chain for abnormal wear, deformation, cracks, or missing studs. Studs should not be removed from chain links for any reason. Links found without studs should be cut from the length of chain and discarded.
- Take single or double link measurements in known or suspected high wear areas, such as the first two to four links on each end of a shot, the wear zone on each anchor leg assembly (where the chain links are shiny), and components used in the riser subassembly or used as buoy jewelry.
- Classify chain and accessories as good, fair, or poor. All material in good or fair condition will undergo a detailed inspection before it is recoated and returned to inventory for future use. Any material classified in poor condition (wire diameter less than 80% of original size, cracked or deformed, etc.) is unsuitable for future use and should be disposed of through normal channels.
- Isolated links in poor condition may be cut from a length of chain to salvage the usable portion. In such cases the remaining links must all be carefully examined during the subsequent detailed inspection.

12.2.2. Detailed Inspection

After the preliminary inspection, (if it is recommended that a detailed inspection be performed), the chain will be sandblasted in accordance with the latest edition of the "Steel Structures Painting Manual, Vol II, Systems and Specifications," Specification SSPC-SP-6. Detailed inspection of chain and its accessories will include the following:

Take single link caliper measurements of the wear areas at the ends of the chain links and accessories (anchor joining links, shackles, chain joining links, etc. (see

- Figure 12-1). The condition of the chain will be determined as follows:

CONDITION	AMOUNT OF DETERIORATION
Good	90% or greater of original wire diameter measured.
Fair	80-90% of original wire diameter measured.
Poor	80% or less of original wire diameter measured.

Figure 12-1 Location of Caliper Measurements of Wear Areas In Chain and Accessories

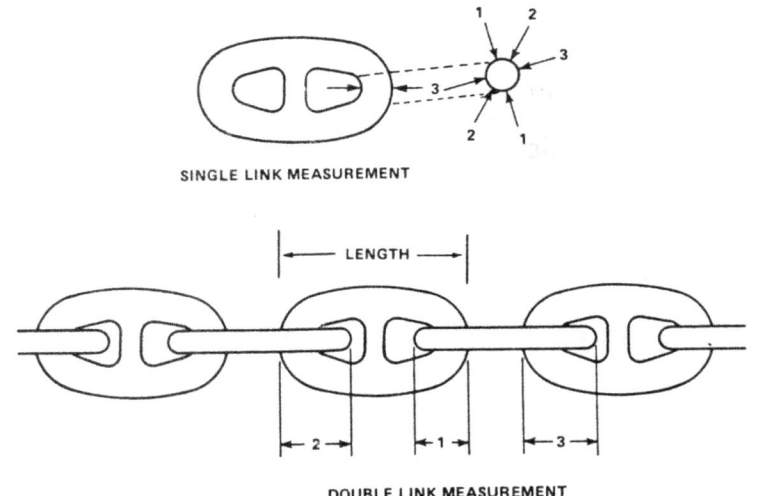

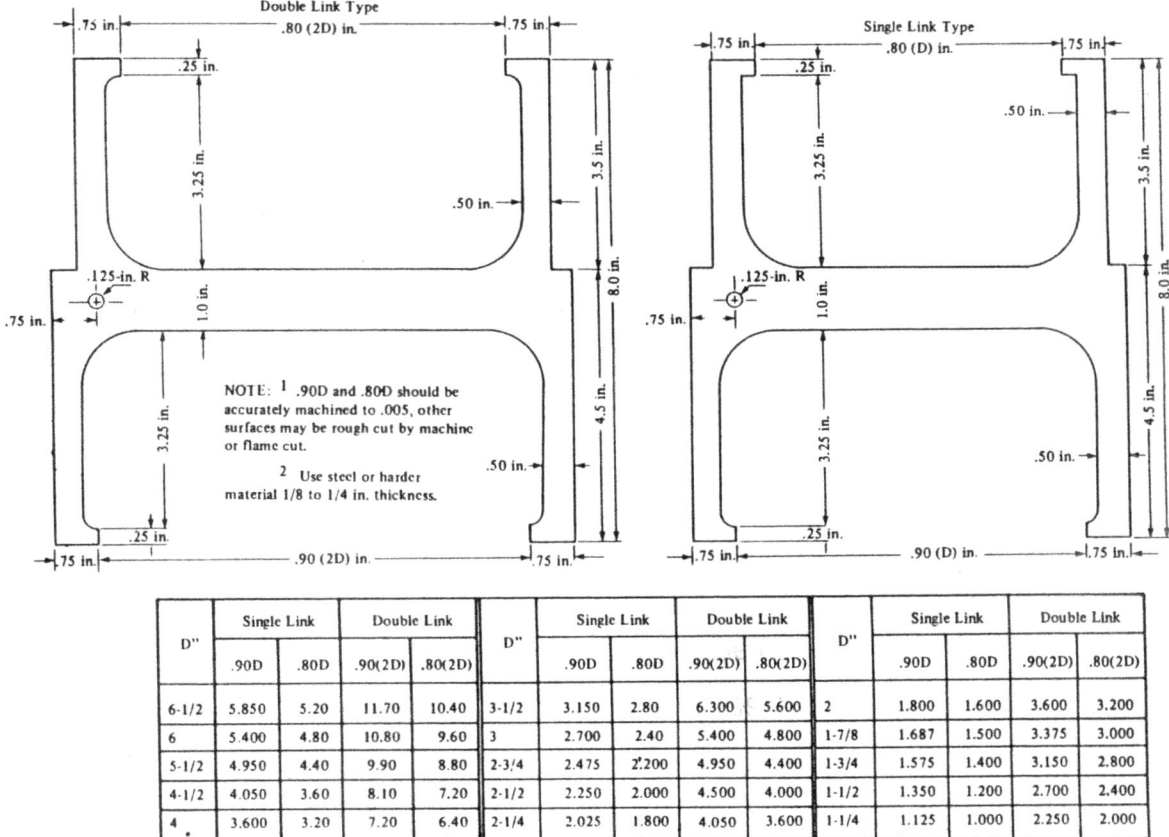

D"	Single Link		Double Link		D"	Single Link		Double Link		D"	Single Link		Double Link	
	.90D	.80D	.90(2D)	.80(2D)		.90D	.80D	.90(2D)	.80(2D)		.90D	.80D	.90(2D)	.80(2D)
6-1/2	5.850	5.20	11.70	10.40	3-1/2	3.150	2.80	6.300	5.600	2	1.800	1.600	3.600	3.200
6	5.400	4.80	10.80	9.60	3	2.700	2.40	5.400	4.800	1-7/8	1.687	1.500	3.375	3.000
5-1/2	4.950	4.40	9.90	8.80	2-3/4	2.475	2.200	4.950	4.400	1-3/4	1.575	1.400	3.150	2.800
4-1/2	4.050	3.60	8.10	7.20	2-1/2	2.250	2.000	4.500	4.000	1-1/2	1.350	1.200	2.700	2.400
4	3.600	3.20	7.20	6.40	2-1/4	2.025	1.800	4.050	3.600	1-1/4	1.125	1.000	2.250	2.000

- Carefully inspect each accessory or chain link for cracks, corrosion, elongation or other deformations, and other abnormalities. Any components containing cracks, deformations, or severe localized corrosion will be classified in "poor" condition.

- Material in good or fair condition should be recoated, tagged, and returned to the inventory for use. Material in poor condition is unsatisfactory for use in fleet moorings and should be discarded.
- Isolated links in poor condition may be cut from a length of chain to salvage the portions in good or fair condition.
- Keep shackles and joining links together as complete units (including pins).
- Coat the mating surfaces of the joining links and swivels with a molybdenum disulfide grease (MIL-G-23549) or equivalent lithium based grease. The results of all inspections (both preliminary and detailed) will be fully documented and filed for future reference. Inspection results should be used in planning and estimating future repair or replacement requirements and associated labor costs.

12.2.3. Protective Coatings

Application of protective coatings should be performed as follows:

- Spray, dip, or brush the abrasive blasted chain with an approved rust preventative (MIL-C-16173 Grade 111).
- Place the chain on a clean surface to dry. Coating operations should be avoided on cold, damp days.

Chapter 13. Ashore Inspection And Refurbishment Of Anchors

13.1. General

13.1.1. Scope

Anchors for fleet moorings generally require very little maintenance and are seldom subject to failure. Although little maintenance is expected, this chapter will cover the maintenance checks that should be performed whenever an anchor is removed from service. Maintenance will encompass inspections, tests, and repairs of the anchors.

13.2. Inspection And Refurbishment

13.2.1. General

The primary purpose of inspecting an anchor is to determine its general physical condition and its suitability for reuse. The inspection of an anchor should only be conducted when the anchor is either temporarily aboard a barge of in a storage area ashore. There are two types of anchor inspections: preliminary, which can be conducted either aboard a barge or ashore, and detailed, which is normally conducted ashore. These inspections should be accomplished as soon as possible after the anchor is removed from the water.

13.2.2. Preliminary Inspection

The purpose of a preliminary inspection is to determine whether an anchor is in satisfactory condition for continued operational use. If abnormalities are found and the condition of the anchor is suspect, then a detailed inspection of the anchor is performed. The preliminary inspection of an anchor will include the following:

- Clean the anchor with a stream of water from a high-pressure hose.
- Take at least two caliper measurements of the wire diameter of the anchor shackle. One measurement should be taken at the wear zone of the lugs of the shackle. If the wire diameter of the shackle measures greater than 90%, the shackle is accepted. If the measurement is 90% or less, the shackle must be replaced.
- Visually check the anchor for casting irregularities, cracks, or obvious mechanical damage. On STATO, NAVMOOR, and other welded anchors, all welds should be checked for cracks, fissures, corrosion, or other defects. If present, stabilizers and their attachment points should be inspected for cracks or other signs of weakness.
- Conduct a hammer test to determine if there are any invisible cracks or other abnormalities. To accomplish this test, a crane is required to lift the anchor off the ground by its anchor shackle. The palm of each fluke is then struck with a 5-pound (or larger size) hammer. A ringing tone indicates that there are no abnormalities in the structure of the fluke. A dull "thump" indicates that the fluke structure contains serious cracks or other abnormalities.

The results of the above preliminary inspection measurements, observations, and tests should be documented. If the inspected anchor is determined to be in satisfactory condition as a result of this inspection, no further effort will be expended, and the anchor will be made ready for further use or placed in a designated storage area.

13.2.3. Detailed Inspection

This inspection will be conducted only if, as a result of the preliminary inspection, abnormalities are suspected or if the condition of the anchor is questionable. The detailed inspection will include the following:

- Clean anchor by sandblasting to near white metal.
- Visually inspect anchor for cracks or casting irregularities. Pay particular attention to anchor flukes and welds.
- Perform a liquid dye penetrant or magnetic particle test on any suspected cracks or abnormalities in accordance with MIL-STD-271, "Non-Destructive Testing Requirements for Metals."

If the results of this testing indicate that cracks or other abnormalities do exist, a decision will have to be made to determine if these abnormalities can be corrected by grinding and welding or if they are too numerous or too deep for economical refurbishment of the anchor. If repairs can be made, then as soon as they are accomplished, the anchor should be protectively coated with a black-gloss solvent-type paint (MIL-P-2430). If it is determined that refurbishment of the anchor is not economically feasible, then the anchor may be retained in storage for future use as a sinker/clump or disposed of. All findings/results of the detailed inspection shall be fully documented and filed.

Chapter 14. Cathodic Protection Systems

14.1. General

14.1.1. Scope

Shore activities responsible for the operation and maintenance of fleet moorings expend considerable time and money maintaining moorings in a safe, reliable condition. The highly corrosive effect of seawater necessitates a continuous program of preventive and corrective maintenance. The frequency of maintenance actions is directly related to the rate of corrosion and wear of the mooring chain and other components. Recent analysis of the maintenance of fleet moorings, with and without cathodic protection over extended periods of time, indicates that the use of an effective cathodic protection system (CPS) can significantly reduce the corrosion rate of mooring chain, yielding an attendant savings in maintenance cost.

14.1.2. Application

Cathodic protection of steel submerged in seawater is achieved by impressing a negative electrical potential on the surfaces exposed to the seawater. Several systems have been developed to supplement protective coatings in an effort to reduce corrosion of fleet mooring chain and accessories. Typically, sacrificial zinc anodes that are gradually consumed while cathodically protecting the steel components provide the required electrical potential. Cathodic protection of the submerged portion of the buoy may also be provided by sacrificial anodes attached to the hull. Anchors normally do not require cathodic protection because bottom mud isolates the anchors from the corrosive effects of the seawater.

14.1.3. Effectiveness

The cathodic protection system and the protective coatings on the mooring components work together to keep the underwater portion of the buoy and the chain assembly from corroding. It is estimated that such a system, properly installed and maintained, can provide complete protection for 15 years in aggressive environments, and for longer periods at milder locations, before anode replacement becomes necessary. Additional anodes can then be installed without removal of the mooring. Buoy anodes are not used when the buoy is fiberglass or polyurethane coated; only the chain and the lower portions of the tension bars on these buoys must be protected.

14.2. Cathodic Protection For Buoys

14.2.1. Environmental Considerations

In designing a cathodic protection system, the extremes of favorable and unfavorable environmental conditions should be considered. Favorable local environmental conditions (low salinity, minimal tide, quiescent water, low dissolved oxygen content, low temperatures, and moderate currents) result in longer periods of time between preventive or corrective maintenance. Unfavorable local environmental conditions (high salinity, high tidal variations, turbulent water, high dissolved oxygen content, high temperature, and strong currents) increase the wear and corrosion rates and decrease the time between required maintenance actions. Cathodic protection is beneficial in all environmental conditions but especially where unfavorable local conditions exist.

14.2.2. Anodes

Anodes for use on buoys are readily available. One commonly used is a zinc casting 36 inches long with a nominal cross-sectional area of 16 square inches. The anode weighs approximately 150 pounds, and is cast on a 48"-long 3/4"-diameter pipe (see Figure 14-1). The composition of the zinc is in conformance with MIL-A-18001. Buoy anodes must be secured in a location where they will not be subjected to impact by vessels; for example, on the bottom of drum-type buoys, in a sea chest built into core portions of peg-top riser-type buoys, or within protective cages welded to the conical side of a peg-top buoy. These anodes should be located about 4 inches from buoy surfaces.

Figure 14-1 Buoy Anode

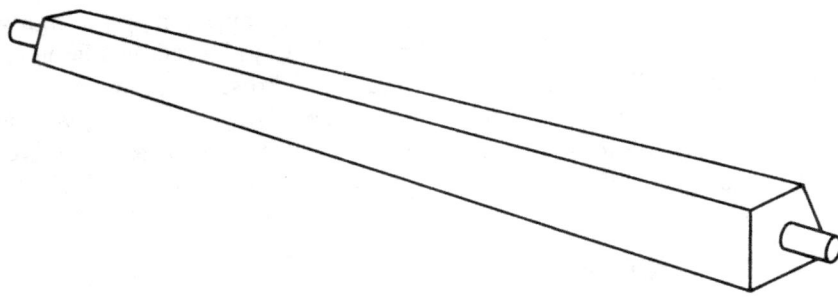

The new foam buoy hulls do not require cathodic protection and, therefore, anodes are not attached to it. The lower portion of its steel tension bar member, however, is exposed to sea water and must be protected. This is accomplished by attaching one chain stud anode on either side of the exposed portion of the tension bar.

14.2.3. Installation Of Anodes

Buoys anodes are usually installed while the buoys are being overhauled ashore. Replacement can also be accomplished at a mooring site during a regularly scheduled maintenance inspection by lifting the buoy out of the water and removing and replacing the anodes while the buoy is on deck. Whether the installation of the buoy anodes are on the bottom, in a sea chest, or within a protective cage, a pair of angle iron brackets are installed which will support each buoy anode. The brackets are welded directly to the buoy hull and generally have threaded fittings for simple anode installation and replacement.

The fittings are of clean, unpainted metal to ensure good electrical continuity between the zinc anode and the buoy hull. Where recessing of the anode is necessary, the brackets are located in a conveniently sized sea chest built to house the anode and its fittings (see Figure 14-2). Alternately, protective cages can be fabricated so as to have sufficient diameter and length to fit over the anode installation (See Figure 14-3).

Figure 14-2 Buoy Anode Recessed Within a Sea Chest

Figure 14-3 Anode Protective Cage

14.3. Cathodic Protection For Mooring Chain

14.3.1. Anodes

Zinc anodes are used for cathodic protection of mooring chain. There are basically three types of commercially available sacrificial anodes (zinc composition conforming with MIL-A-18001). These are chain stud anodes, link anodes, and clump anodes.

14.3.1.1. Chain Stud Anode

This cathodic protection system consists of an anode and a bolt assembled to the stud link as shown in Table 14-1. For each size chain, a particular size bolt and anode are required. Each anode is stamped with its type. During anode installation on each chain line, the recommended torque to be applied to the bolt is 25-35 foot-pounds. If, when attempting to replace a depleted anode, a bolt head is found to be missing, the center of the remaining bolt should be tapped with a counter-clockwise 1/8" thread so that it can be removed using a counter-clockwise threaded bolt and a hand-or electric-powered wrench.

Note that oversized anodes may be used to extend the anode life and increase the time interval required for anode replacement.

Table 14-1 Properties of FM3 Chain Anodes[29]

NOMINAL SIZE (inches)	1.75	2	2.25	2.5	2.75	3.5	4
ANODE WEIGHT (lbs)	0.80	1.10	1.38	1.70	2.04	3.58	4.41
SCREW LENGTH (inches)	1.25	1.50	1.75	1.75	2.00	2.25	2.25
ANODE WIDTH (inches)	1.50	1.62	1.75	1.94	2.06	2.38	2.69
LINK GAP (lbf)	3.74	4.24	4.74	5.24	5.74	7.48	8.48
ANODES PER FULL DRUM	1106	822	615	550	400	158	122
WEIGHT PER FULL DRUM (approx. lbf)	976	979	917	993	869	602	550

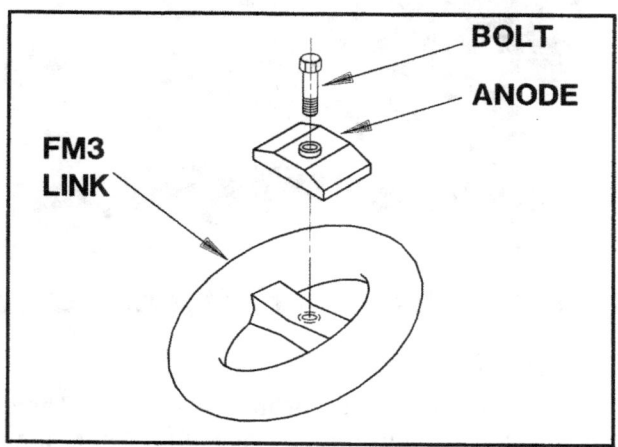

---

[29] All screws are 3/8"-16UNC-2A, grade 5, hex cap. 4" anodes fit all chain sizes, All screw heads are 9/16".

14.3.1.2. Link Anode

This design incorporates an elongated chain link that is an in-line component of the mooring. The anode (see Figure 14-4) consists of approximately 500 pounds of zinc cast onto the chain link. If the chain is in tension, the electrical connection to the mooring chain is provided by the exposed metal-to-metal surfaces in the grip area of the links. A modification of the link anode incorporates a wire rope that provides the electrical continuity throughout the chain when the links are not in tension.

Figure 14-4 Link Anode Design

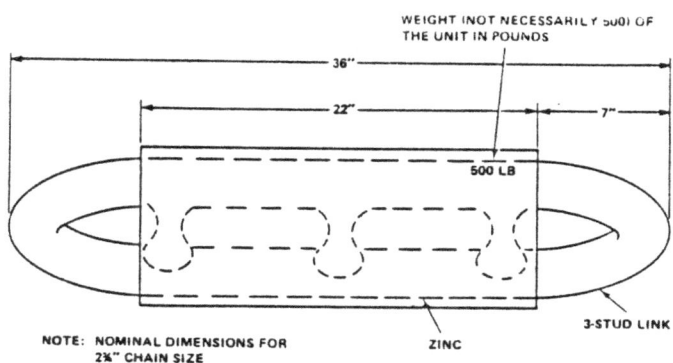

14.3.1.3. Clump Anode

This anode is similar in configuration to a small sinker weight. The required amount of zinc is cast around a steel "hairpin" and attached to the chain with a standard sinker shackle or with modified attachment hardware.

This anode also requires the use of a wire rope, which is woven through the chain to provide electrical continuity. Wire clips or hose clamps are used to connect the wire rope to the anodes and to the chain links.

14.3.2. Installation

For proper operation of the cathodic protection system, the anodes, wire ropes, and fittings should be free of paint, grease, dirt, or coatings. The anodes will cathodically protect the wire ropes and fittings. A cathodic protection system installation can be most easily and economically accomplished at the time of the regularly scheduled mooring overhaul and prior to installation of the mooring system.

14.3.2.1. Chain Stud Anode

This type anode is attached to each study in the chain link. The chain manufacturer will normally accomplish the attachment. However, divers or ashore personnel may be required to replace missing or deteriorated anodes. Installation requires bolting the proper anode to the chain link stud as previously shown in Table 14-1. In some cases new chain shots may be shipped with the studs drilled for these anodes, but with anodes not installed. In these cases, the threaded holes will be lubricated with an electrically conductive graphite lubricant (VV-G-671) to reduce the probability of corrosion, and the hole fitted with steel screw to prevent it from being filled with coating material (see Figure 14-5). The lubricant will not degrade the cathodic protection system and does not have to be cleaned out before installing the anode.

Figure 14-5 Protecting Drilled Studs

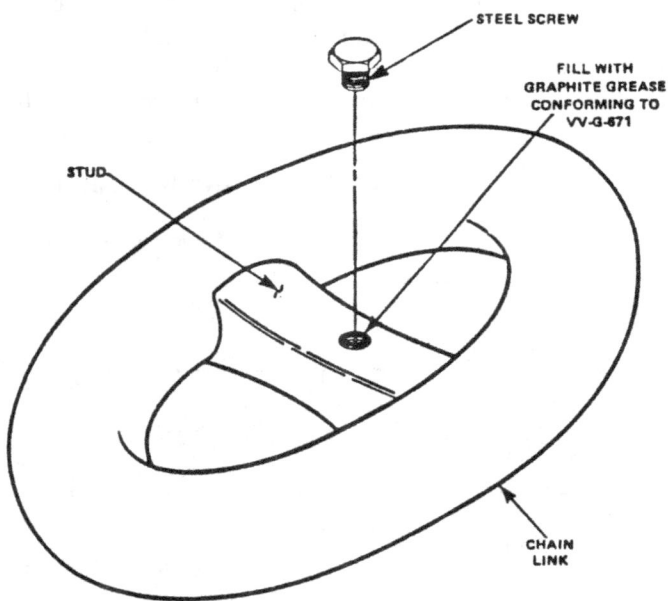

14.3.2.2. Link Anode

Link anodes are installed between lengths of chain with two chain joining links (see Figure 14-6). The installation must be completed onshore or on the crane barge before the mooring is installed. A wire rope continuity cable clamped to the link portion of the anode and attached to the chain normally provides a reliable electric path.

Figure 14-6 Link Anode Installed in a Chain Section

14.3.2.3. Clump Anode

Clump anodes are usually installed ashore during mooring overhaul. The anode is attached to the chain with wire rope (see Figure 14-7). To provide an electrical path from the anode to the chain, a wire rope continuity cable must be clipped or clamped to the anode's hairpin and attached to the chain as described in the following paragraph.

Figure 14-7 Clump Anode and Chain as Viewed Underwater

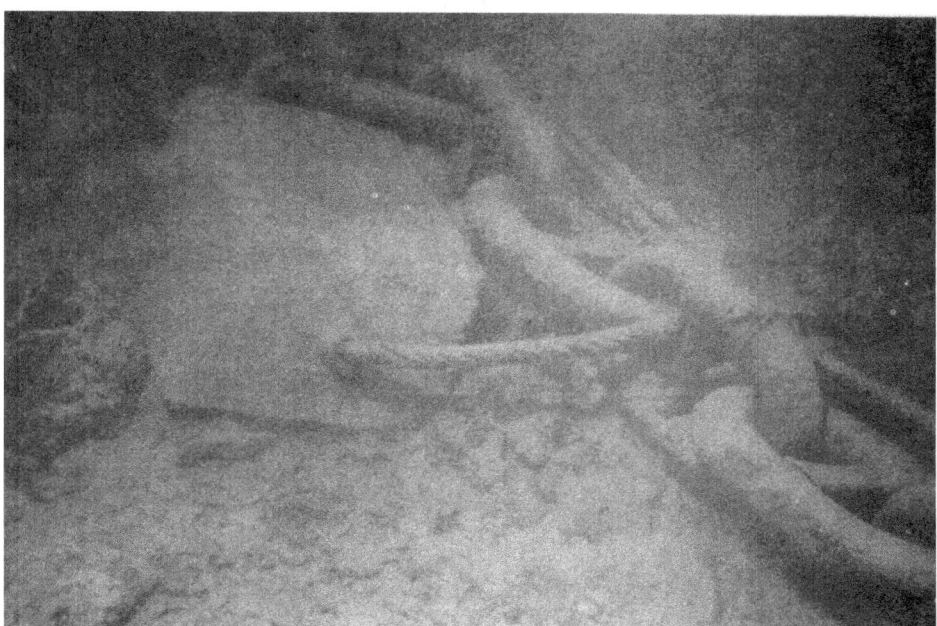

14.3.2.4. Wire Rope

The wire rope, used to provide the electrical continuity, is normally 5/8- to 3/4"-diameter galvanized steel. Wire ropes of these diameters have the strength characteristics needed for this purpose, and yet are flexible enough for interweaving through the chain links. The wire rope should be interwoven through every fourth link in a shot and attached to every eighth link with a hose clamp, wire clip, or U-bolt. Prior to attaching the continuity wire to the chain link, part of the link must be cleaned to bare metal so that the chain link and the continuity wire have good metal-to-metal contact. This is important for a sound electrical connection.

14.3.3. Anode Replacement

An anode should be replaced if less than 25% of its original weight remains. Trained personnel should accomplish replacement, and equipment and materials should be on hand, both ashore and on crane barges, to support this activity. Divers can easily handle replacement of chain stud anodes. However, replacement of in-line link anodes, clump anodes, and buoy anodes is more difficult for divers. In addition, divers will find it difficult to replace any anodes in areas where the bottom is muddy or where the anchor chain subassembly is partially buried.

Chapter 15. Storage Of Mooring Materials

15.1. General

15.2. General Requirements

Mooring components are usually stored in open areas near a coastline, thus exposing them to weather and a marine environment. To prevent deterioration while in storage, some preventive maintenance will be required in addition to the routine material handling and inventory control tasks normally performed by a storage facility. The job of handling, maintaining, and controlling stored components will be made much easier if some basic guidelines, as noted below, are followed.

- The storage area should be large enough to permit efficient movement of forklifts, cranes, and other large mobile equipment.
- Arrangement of components should allow easy access for inspections, inventory checks, and selection.
- To reduce corrosion, all components except cathodic protection materials should be coated with paint, approved rust preventatives, coal tar, or other suitable preservatives.
- All components should be tagged or labeled to ensure proper identification and accurate inventory reporting.
- Chain accessories such as joining links, swivels, ground rings, and shackles should be crated or banded together on pallets to permit easy handling. The paragraphs that follow provide additional details and other considerations for Fleet Mooring Inventory (FMI) storage.

15.2.1. Storage Area Requirements

Shore activities requiring spare mooring materials and those activities designated as stock points for the FMI will require a suitable storage area. The following should be observed:

- The FMI should normally be stored in a secure area designated by activity personnel responsible for space allocation.
- The storage area should be on solid ground or on improved surfaces, and graded for drainage.
- The area should be large enough, as well as configured, to allow easy access of equipment and personnel involved in chain handling and other maintenance operations.
- The area should be close to maintenance areas, transportation equipment, and the waterfront to reduce both maintenance and transportation costs.

15.3. Storage Procedures

15.3.1. Buoys

Store buoys as follows:

- Place all drum type buoys on chocks or dunnage, with all metal parts clear of the ground, and tilted (using additional chocks or dunnage under one side) to facilitate water runoff.
- Store peg-top buoys in an inverted position (see Figure 15-1).

Figure 15-1 Proper Storage of a Peg Top Buoy

Buoys should be periodically inspected to aid in the detection and prevention of localized corrosion areas. If corrosion is found, corrective action should be implemented. Corrosion commonly develops in the web of channel irons securing wooden fenders, in the hull behind rubber fendering, and on the nuts/studs used to secure manhole covers, fenders, and chafing strips. Ground or chock contact points are also susceptible to higher corrosion rates. Any area on the top of the buoy that could collect water is susceptible to accelerated corrosion.

15.3.2. Chain And Chain Accessories

Chain is normally stored ranged out in tiers, bailed a single shot at a time on a pallet, loaded in shipping crates, or in bundles. The ground where the chain is stored should be clear of all debris and growth, and well drained. To prevent its intermixing with older grades of chain, FM 3 chain and accessories should always be stored separately.

15.3.2.1. Tiered Chain

Store chain in tiers as follows:

- Lay chains down stretched out taut and free of turns.
- Pile tiers in multiple layers to reduce storage space (see Figure 15-2).
- Each tier should contain chain of similar construction, size, and condition for ease of access and accurate inventory control.
- The ends of each length of chain should have an identification tag attached to it that contains the chain size (in inches), type (cast/forged/etc.), length (in feet), and manufacturer.

Figure 15-2 Chain Stored in Tiers

15.3.2.2. Palletized Chain

Chain may also be stored and handled on pallets. Normally, chain is palletized in single shot lengths to reduce handling weights and to simplify inventory. Chain pallets normally consist of wooden or steel pallets onto which the chain is piled.

15.3.2.3. Crated Chain

Reusable wooden crates can be used for the shipment and storage of chain and will normally contain a single shot of chain. The crates can be stacked to reduce deck space requirements if not loaded beyond the design capacities stamped on the crates (see Figure 15-3.) Five standard-sized crates have been constructed for the various chain sizes. The design of one of these crates is shown in Figure 15-4.

Figure 15-3 Stacking Crates of Chain

Figure 15-4 Overseas Shipping Container for 3 1/2- and 4" Chain (14,000 lb. Capacity)

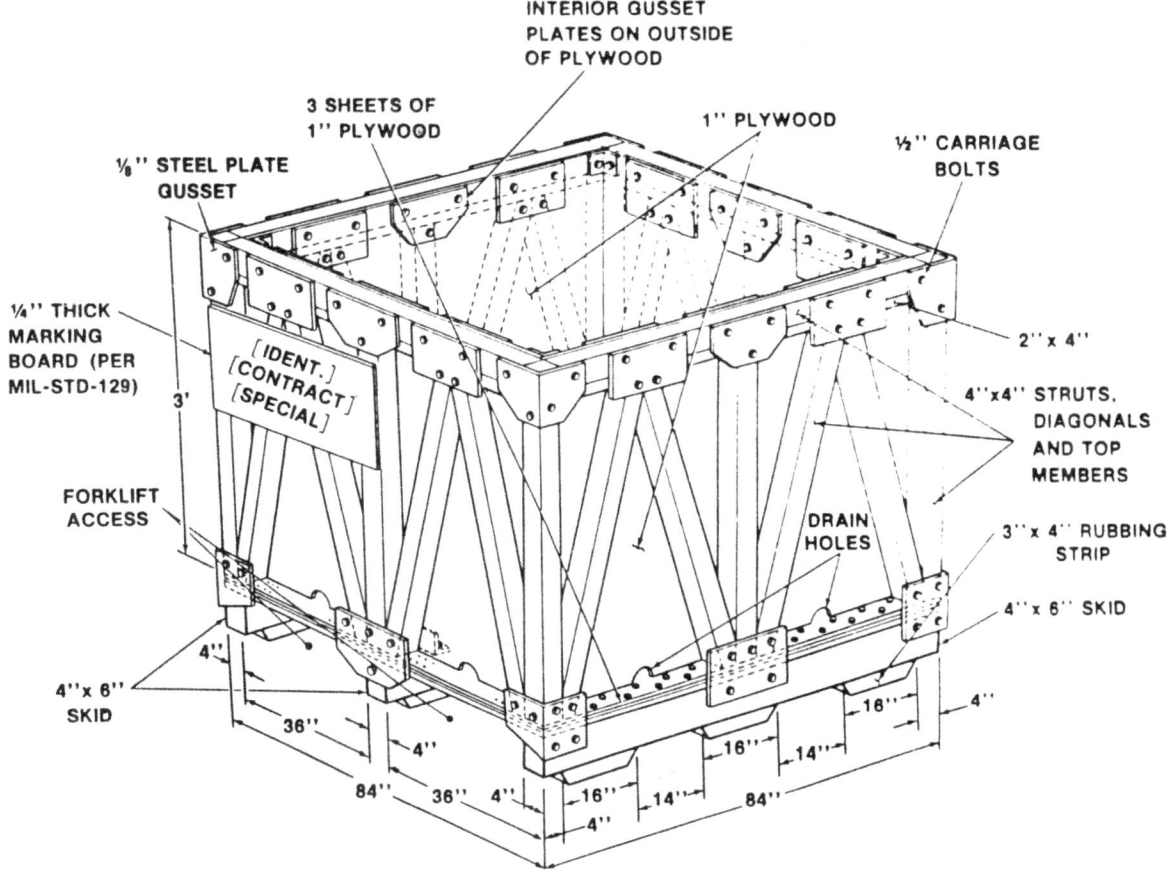

15.3.2.4. Bundled Chain

Chain will be bundled by reeving wire rope through the last link on each end of the shot and through two other links, each approximately 30 feet from the nearest end of the shot. The ends of the wire rope shall be secured to each other to form a sling that may be used to lift the entire shot. When lifted, the total length of the shot/sling combination shall not exceed 20 feet.

15.3.2.5. Accessories

Accessories should be boxed, crated, or banded to pallets. Components of the same size, type, and condition should be stored together for inventory control and ease of access. Parts for joining links and pins for shackles should not be mixed or interchanged, but stored together as matched sets to ensure proper fit in the field. All joining links and shackles should be stored clean, free of rust, well-greased, and loosely assembled. Joining links should never be disassembled and their parts stored separately.

15.3.3. Anchors

Store anchors as follows:

- Place anchors on dunnage and in a vertical position.
- Store according to type and size for ease of inventory control.

U.S. Navy anchors have identification marks cast, stamped, or cut on the anchor crown. When stored in a vertical position, these identification marks are not visible. When stored vertically, this information should be transferred to a suitable tag or stenciled to the anchor's shank.

15.3.4. Cathodic Protection Materials

Larger anodes stored outside should be sealed in plastic liners and boxed to preclude premature galvanic action. Do not store anodes in the open or near other dissimilar metals, which could result in corrosion of the anodes. Chain stud anodes should be stored in 55-gallon steel drums with removable, bolt-on type lids. Only one size of anode shall be packed in each drum. Cathodic protection material should be kept clean and should not be painted or coated with oil or grease during either storage or use.

15.3.5. Marking And Identification

Marking and identification of components will be accomplished as described below.

15.3.5.1. General

Proper marking and identification of all mooring components will assist in conducting inventories, will help prevent use of improper or substandard materials, and will speed assembly and installation times when required components must be drawn from inventory.

15.3.5.2. Color Coding

It is also good practice to color code chain to identify its condition. Recommended colors to be applied to the last link on each end of a chain length are:

- Green - for chain in new or good condition.
- Yellow - for chain in fair condition.
- Red - for chain in poor condition and ready for disposal.

15.3.5.3. Identification

Identification of chain and accessories shall be on tags of 0.031" thick aluminum alloy 1100 or 3003, attached using 0.031" diameter 300 series corrosion resisting steel wire through 0.125" diameter holes at each end of the tag. Characters shall be metal stamped using 0.25" high characters. Tag size and information content shall be as specified in Figure 15-5. On chain shots, tags shall be attached to the last link on each end of the shot. Tags will be attached snugly to each accessory in a location away from the grip area of the component, and shall be bent to conform to the contour of the component to minimize risk of snagging or damaging the tag.

Figure 15-5 Shipping Tag

```
DETACHABLE CHAIN JOINING LINK, 2-IN.

20 EACH                              05/85

WT 904 LB                        CUBE 25

N00123-85-P-1747

ACME FOUNDRY CO

BETHLEHEM, PA 12345
```

EXAMPLE TAG

9.2.5.4 Chain and Accessories. Chains and accessories are marked and identified as follows:

- The size and the name of the manufacturer are stamped or forged on each chain link or accessory during manufacture.
- New FM 3 chain will also have a unique serial number on each accessory and on the links at the end of each shot of chain (see Figure 15-6). The serial number will be used together will documentation furnished by the manufacturer (heat number, chemical composition, etc.) to monitor the performance of different heats of chain.

Figure 15-6 Chain Markings

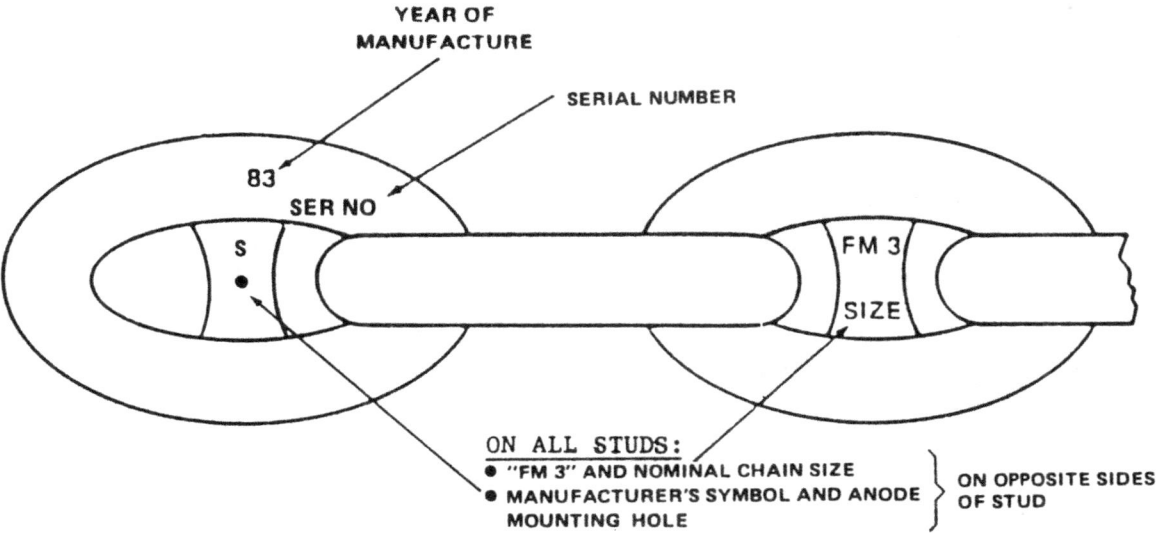

- Crates used to ship new chain and accessories from the manufacturer will be marked to show:
- Contents
- Weight
- Contract and shipping data
- Stacking limitations

- Other Pertinent Information

An artist's drawing of an overseas shipping container for chain accessories is depicted in Figure 15-7.

Figure 15-7 Overseas Shipping Container for Chain Accessories (3,000 lb. Capacity)

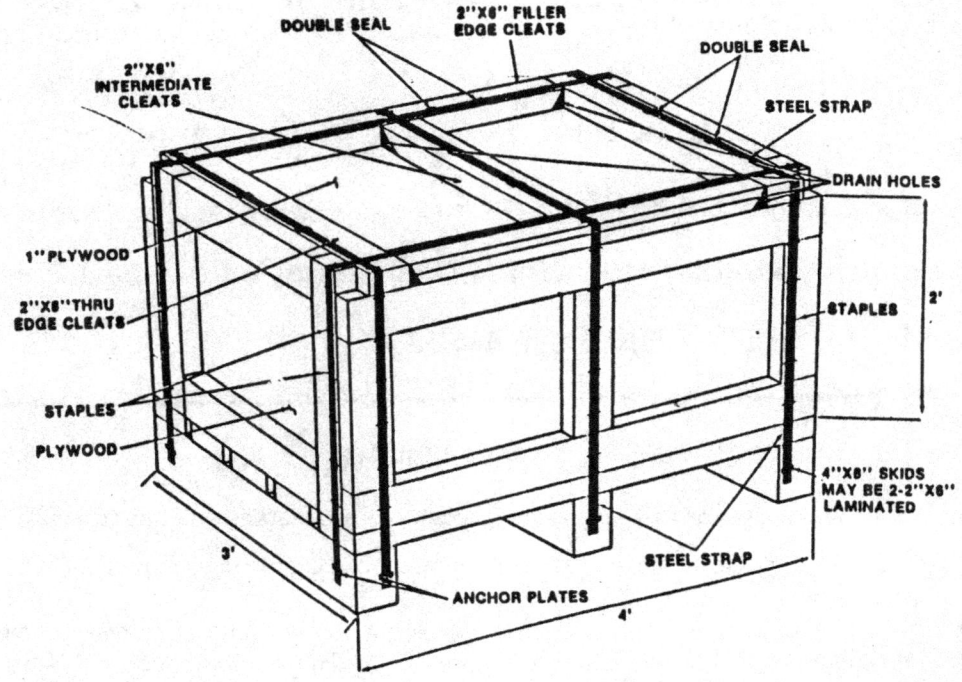

15.3.5.4. Buoys

New buoys should have an identification plate showing the following:

- Serial number
- Manufacturer,
- Date of manufacture
- Diameter
- Height
- Weight in air
- Maximum tension bar load
- NAVFAC drawing number.

The plate should be protected during sandblasting and should not be painted or coated.

15.3.5.5. Anchors

Identification information for anchors is cast on the crown, on the flukes, or on the side of the shank. Minimum information provided is as follows:

- Manufacturer
- Weight
- Serial number.

15.3.6. Pre-Issue Inspection

Items that are used from inventory should be inspected prior to shipping or movement from the storage facility. A bill of material should be reviewed to ascertain the components required and a check of all material sizes accomplished. To ensure all mooring components fit when shipped to the field, a physical fit check should be accomplished prior to shipment. The many configuration and designs of mooring hardware increase the likelihood of a misfit if a physical fit check is not accomplished.

Chapter 16. Bibliography and References

16.1. Publications Directly Used In The Preparation Of This Book

- MIL-HDBK-1026/4A, *Mooring Design*
- NAVFAC DM-26.4, *Fixed Moorings*
- NAVFAC DM-26.5, *Fleet Moorings*
- NAVFAC DM-26.6, *Mooring Design Physical and Empirical Data*
- NAVFAC MO-124, *Mooring Maintenance*

The U.S. Navy also furnished the photographs which appear in this publication.

16.2. General Publications

- American Institute of Steel Construction, Inc. *Manual of Steel Construction*, Chicago, IL (latest edition).
- American Iron and Steel Institute. *Wire Rope Users Manual*, Washington, D.C., 1981.
- American Petroleum Institute. *Analysis of Spread Mooring Systems for Floating Drilling Units*, ANSI/API RP 2P-87, Approved July 12, 1993.
- American Petroleum Institute. Recommended Practice for Design, Analysis, and Maintenance of Moorings for Floating Production Systems, ANSI/API RP 2FP1-93, Approved April 13, 1994.
- American Wood Preservers' Association, *Book of Standards* (latest revision), Washington, D.C.
- Craig, Roy R. *Structural Dynamics, An Introduction to Computer Methods*, John Wiley & Sons, Inc., New York, NY, 1981.
- Dodge, D. and Kyriss, E. Seamanship: *Fundamentals for the Deck Officer*, Naval Institute Press, Annapolis, MD, 1981.
- Gillmer, T.C., and Johnson, B. *Introduction to Naval Architecture*, Naval Institute Press, Annapolis, MD, 1982.
- Goda, Y. Random Seas and Design of Maritime Structures, Univ. of Tokyo Press, 1985.
- Karnoski, S.R. and Palo, P.A. "Validation of a Static Mooring Analysis Model with Full-Scale Data," Offshore Technology Conference Paper 5677, Houston, TX, May 1988.
- Keane, John D., "Steel Structures Painting Manuals," Volume 2, "Systems and Specifications," (latest edition) Steel Structures Painting Council, Pittsburgh, Pennsylvania
- Moffatt & Nichol, Engineers. Development of Dynamic Mooring Models for Use in DM-26.5 'Fleet Moorings' and DM-26.4 'Fixed Moorings', 21 Mar 1994.
- National Institute of Building Sciences. *Construction Criteria Base*, CD's issued quarterly (202)289-7800, Washington, DC.
- Noel, J. Knights *Modern Seamanship*, 17th Edition, Van Nostrand Reinhold Co., Inc., New York, NY, 1984.
- Palo, P.A. "Steady Wind and Current-Induced Loads on Moored Vessels," Offshore Technology Conference Paper 4530, Houston, TX, May 1983.
- Palo, P.A. "Full-Scale Vessel Current Loads Data and the Impact on Design Methodologies and Similitude," Offshore Technology Conference Paper 5205, Houston, TX, May 1986.
- Permanent International Association of Navigation Congresses. *Report of the International Commission for Improving the Design of Fender Systems*, Supplement to Bulletin No. 45, 1984.

- Saunders, H. E. *Hydrodynamics in Ship Design*, The Society of Naval Architects and Marine Engineers, New York, NY, 1957.
- Sharpe, R. CAPT. *Janes's Fighting Ships 1996-97*, 1996.
- Sorensen, R.M. *Basic Coastal Engineering*, John Wiley & Sons, New York, NY, 1978.
- U.S. Department of Transportation. *Aids to Navigation*, USCG CG-222-2, U.S. Government Printing Office, Washington, DC.
- Miscellaneous DOD Publications
- Naval Civil Engineering Laboratory. *Conventional Anchor Test Results at San Diego and Indian Island*, NCEL TN No: N-1581. Port Hueneme, CA, Jul 1980.
- Naval Civil Engineering Laboratory. *Drag Embedment Anchor Tests in Sand and Mud*, NCEL TN No: N-1635. Port Hueneme, CA, Jun 1982.
- Naval Civil Engineering Laboratory. *Stockless and Stato Anchors for Navy Fleet Moorings*, NCEL Techdata Sheet 83-09. Port Hueneme, CA, Mar 1983.
- Naval Civil Engineering Laboratory. *Current-Induced Vessel Forces and Yaw Moments from Full-Scale Measurement*, NCEL TN-1749. Port Hueneme, CA, Mar 1986.
- Naval Civil Engineering Laboratory. *The NAVMOOR Anchor*, NCEL Techdata Sheet 87-05. Port Hueneme, CA, May 1987.
- Naval Civil Engineering Laboratory. *Pile Driven Plate Anchors for Fleet Moorings*, NCEL Techdata Sheet 92-10. Port Hueneme, CA, Nov 1992.
- Naval Facilities Engineering Service Center. *Purchase Description NAVMOOR-10 Mooring Anchor*. Port Hueneme, CA, Nov 1985.
- Naval Facilities Engineering Service Center. Purchase Description, Buoy, Mooring, Foam Filled Polyurethane Cast, 11'-6" Diameter x 13'-1" High, FPO-1-88(PD 1). Port Hueneme, CA, Mar 1988.
- Naval Facilities Engineering Service Center. Purchase Description, Buoy, Mooring, Foam Filled Polyurethane, 11'-6" Diameter x 13'-1" High, FPO-1-89(PD 2). Port Hueneme, CA, Dec 1989.
- Naval Facilities Engineering Service Center. *Purchase Description for Anodes, Chain, Stud Link, FPO-1-89(PD 3) Rev E*. Port Hueneme, CA, Jun 1990.
- Naval Facilities Engineering Service Center. Purchase Description, Buoy, Mooring, Foam Filled Polyurethane, 8' Diameter x 11'-4" High, FPO-1-90(PD 1). Port Hueneme, CA, Jun 1990.
- Naval Facilities Engineering Service Center. *NAVMOOR-15 Drawing Package, NAVFAC Drawing 3050572* (8 sheets). Port Hueneme, CA, Jul 1990.
- Naval Facilities Engineering Service Center. *Assessment of Present Navy Methods for Determining Mooring Loads at Single-Point Moorings*, NFESC TR-2018-OCN. Port Hueneme, CA, May 1994.
- Naval Facilities Engineering Service Center. Guidelines for Preparation of Reports on Underwater Inspections of Waterfront Facilities. Port Hueneme, CA, Mar 1998.
- Naval Sea Systems Command. SO300-A8-HBK-010, *U.S. Navy Salvage Engineer's Handbook*, Volume 1 – Salvage Engineering. Washington, DC, 1 May 1992.
- Naval Sea Systems Command. *Metric Guide for Naval Ship Systems Design and Acquisition*. Washington, DC, Rev. Jun 1995.
- Naval Sea Systems Command. *Hitchhikers Guide to Navy Surface Ships*. Washington, DC, Feb 26, 1997.
- U.S. Naval Academy. *Viscous Drag Forces on Moored Ships in Shallow Water*, by D. Kriebel. Annapolis, MD, 1992.

16.3. *Specifications And Standards*

- Federal Specification FED SPEC RR-W-410, *Wire Rope and Strand*. U.S. Government Printing Office, Washington, DC.

- American Bureau of Shipping, New York NY.
- Rules for Building and Classing Steel Vessels (latest edition)
- Rules for Building and Classing Single Point Moorings
- *American Petroleum Institute (API)*, 1220 L Street NW, Washington, DC 20005.
- API RP 2T Recommended Practice for Planning, Designing, and Constructing Tension Leg Platforms
- API RP 2SK Recommended Practice for Design and Analysis of Station keeping Systems for Floating Structures
- *American Society of Civil Engineers (ASCE)*, 1801 Alexander Bell Drive, Reston, VA 20191-4400
- ANSI/ASCE 7-95 Minimum Design Loads for Buildings and Other Structures
- *Cordage Institute*, 350 Lincoln Street, Hingham, MA 02043
- Cordage Institute Technical Manual
- *David Taylor Research Center (DTRC)*, Annapolis, MD 21402-5067
- DTNSRDC/SPD-0936-01 User's Manual for the Standard Ship Motion Program, SMP81
- *National Bureau of Standards (NBS)*, Quince Orchard and Clopper Roads, Gaithersburg, MD 20899
- NBS Series 118 Extreme Wind Speeds at 129 Stations in the Contiguous United States
- NBS Series 124 Hurricane Wind Speeds in the United States
- Organic Coating Properties, Selection and Use; Building Science Series 7 (February 1968).
- *National Climatic Data Center (NCDC)*, 151 Patton Ave., Asheville, NC 28801-5001
- E/CC31:MJC Letter Report of 08 Dec 1987
- *Nautical Software*, Beaverton, OR
- Version 2.0 Tides and Currents for Windows
- *Naval Civil Engineering Laboratory (NCEL)* publications available from NFESC, 1100 23rd Ave., Port Hueneme, CA 93043.
- Handbook for Marine Geotechnical Engineering
- TDS 83-05 Multiple STOCKLESS Anchors for Navy Fleet Moorings
- TDS 83-08R Drag Embedment Anchors for Navy Moorings
- TN-1628 Wind-Induced Steady Loads on Ships
- TN-1634 STATMOOR – A Single Point Mooring Static Analysis Program
- TN-1774 Single and Tandem Anchor Performance of the New Navy Mooring Anchor
- Fleet Mooring Test Program – Pearl Harbor (write-up of pull test results for Pearl Harbor anchors - undated)
- *Naval Environmental Prediction Research Facility*, Monterey, CA
- TR 82-03 Hurricane Havens Handbook
- *Naval Facilities Engineering Command (NAVFAC)*, publications available from Naval Publications and Forms Center, 5801 Tabor Avenue, Philadelphia, PA 19120.
- CHESNAVFAC FPO-1-84(6) Fleet Mooring Underwater Inspection Guidelines
- CHESNAVFAC FPO-1-87(1) Failure Analysis of Hawsers on BOBO Class MSC Ships at Tinian on 7 December 1986
- MIL-HDBK-1025/1 Piers and Wharfs
- MO-104.2 Specialized Underwater Waterfront Facilities Inspections
- MO-110 Painting and Protective Coatings
- NAVFAC 1998 Ships Characteristics Database
- DM-7 *Soil Mechanics, Foundations, and Earth Structures*
- DM-26.3 *Coastal Sedimentation and Dredging*
- DM-35 *Drydocking Facilities*
- NAVFACENGCOM EPS Wharfbuilding Handbook P-715.0 (TR-Service) (November 1979)
- *Naval Facilities Engineering Service Center (NFESC)*, 1100 23rd Ave., Port Hueneme, CA 93043

- CR-6108-OCN Anchor Mooring Line Computer Program Final Report, User's Manual for Program CSAP2
- CR-6129-OCN Added Mass and Damping Characteristics for Multiple Moored Ships
- FPO-1-89(PD1) Purchase Description for Fleet Mooring Chain and Accessories
- SSR-6119-OCN D-8 Mooring Upgrade Design Report
- TR-2039-OCN Design Guide for Pile Driven Plate Anchors
- TR-6004-OCN Wind Effects on Moored Aircraft Carriers
- TR-6005-OCN (Rev B) EMOOR – A Quick and Easy Method of Evaluating Ship Mooring at Piers and Wharves
- TR-6014-OCN Mooring Design Physical and Empirical Data
- TR-6015-OCN Foam-Filled Fender Design to Prevent Hull Damage
- *Naval Research Laboratory (NRL)*, 4555 Overlook Ave., Washington, DC 20375
- NRL/PU/7543-96-0025 Typhoon Havens Handbook for the Western Pacific and Indian Oceans
- *Naval Sea Systems Command (NAVSEA)*, 2531 Jefferson Davis Highway, Arlington, VA 22242-5160
- NSTM013 Naval Ships' Technical Manual
- *Nuclear Regulatory Commission (NUREG)* publications available from Government Printing Office, P.O. Box 37082, Washington, DC 20013-7082
- NUREG/CR-2639 Historical Extreme Winds for the United States – Atlantic and Gulf of Mexico Coastlines
- NUREG/CR-4801 Climatology of Extreme Winds in Southern California
- NUREG RG 1.76 Tornado-Resistant Design of Nuclear Power Plant Structures
- *United States Army Corps of Engineers*, 215 North 17th Street, Omaha, NE 68102-4978
- Special Report No. 7   Tides and Tidal Datums in the United States
- Shore Protection Manual
- *United States Coast Guard (USCG)*, 2100 Second Street S.W., Washington, DC 20593
- CG-D-49-77   Guidelines for Deepwater Port Single Point Mooring Design
- *United States Naval Academy (USNA)*, Annapolis, MD 21402-5018
- EW-9-90 Evaluation of Viscous Damping Models for Single Point Mooring Simulation

16.4. Authored Publications

- deBoer, F.L., "Fiberglass Coatings for Fleet Moorings," Navy Civil Engineer, pp. 16-17 (March 1970)
- Flory, J.F., M.R. Parsey, and H.A. McKenna. *The Choice Between Nylon and Polyester for Large Marine Ropes*, ASME 7th Conference on Offshore Mechanics and Arctic Engineering, Houston, TX, Feb 1988.
- Flory, J.F., M.R. Parsey, and C. Leech. *A Method of Predicting Rope Life and Residual Strength*, MTS Oceans' 89, Sep 1989.
- Flory, J.F., H.A. McKenna, and M.R. Parsey. *Fiber Ropes for Ocean Engineering in the 21st Century*, ASCE, C.E. in the Oceans, Nov 1992a.
- Flory, J.F., J.W.S. Harle, R.S. Stonor, and Y. Luo. *Failure Probability Analysis Techniques for Long Mooring Lines*, 24th Offshore Technology Conference Proceedings, Offshore Technology Conference, Houston, TX, 1992b.
- Headland, J., W. Seelig, and C. Chern. *Dynamic Analysis of Moored Floating Drydocks*, ASCE Proceedings Ports 89, 1989.
- Hearle, J.W.S., M.R. Parsey, M.S. Overington, and S.J. Banfield. *Modeling the Long-Term Fatigue Performance of Fibre Ropes*, Proceedings of the 3rd International Offshore and Polar Engineering Conference, 1993.
- Hooft, J.P. *Advanced Dynamics of Marine Structures*, John Wiley & Sons, New York, NY, 1982.

- Kizakkevariath, S. Hydrodynamic Analysis and Computer Simulation Applied to Ship Interaction During Maneuvering in Shallow Channels, PhD Thesis, Virginia Polytechnic Institute and State University, May 1989.
- Myers, John J., et al. *Handbook of Ocean and Underwater Engineers*, McGraw-Hill Book Company, New York, NY, 1969.
- Occasion, L.K. The Analysis of Passing Vessel Effects on Moored Tankers, UCLA Report PTE-490x, Dec 1996.
- Weggel, J. R. and R. M. Sorensen. *Development of Ship Wave Design Information*, Proceedings of the International Conference on Coastal Engineering, ASCE, 1984, pp 3227-3243.
- Weggel, J. R. and R. M. Sorensen. *Ship Wave Prediction for Port and Channel Design*, Proceedings Ports 86, ASCE, 1986, pp 797-814.

16.5. Glossary

- *AISC*. American Institute of Steel Construction.
- *API*. American Petroleum Institute.
- *DOD*. Department of Defense.
- *DM*. Design manual.
- *MLW*. Mean low water.
- *MLLW*. Mean lower low water.
- *MSC*. Military Sealift Command.
- *NAVFACENGCOM*. Naval Facilities Engineering Command.
- *NAVSEASYSCOM*. Naval Sea Systems Command.
- *NBS*. National Bureau of Standards.
- *NCDC*. National Climatic Data Center.
- *NFESC*. Naval Facilities Engineering Services Center.
- *NUREG*. Nuclear Regulatory Commission.
- *OCIMF*. Oil Companies International Marine Forum.
- *PIANC*. Permanent International Association of Navigation Congresses.

To view and order more books from our collection please visit our website: www.pilebuck.com